Assessing Cost Recovery

Kommunalwirtschaftliche Forschung und Praxis

Herausgegeben von Wolf Gottschalk

Band 21

PETER LANG

Frankfurt am Main · Berlin · Bern · Bruxelles · New York · Oxford · Wien

Britta Ammermüller

Assessing Cost Recovery

A New Comparative Framework in Line with WFD Article 9

PETER LANG
Internationaler Verlag der Wissenschaften

Bibliographic Information published by the Deutsche Nationalbibliothek
The Deutsche Nationalbibliothek lists this publication in the Deutsche Nationalbibliografie; detailed bibliographic data is available in the internet at http://dnb.d-nb.de.

Zugl.: Leipzig, Univ., Diss., 2010

Sponsored by the „Finanz- und Wirtschaftsrat beim Verband Kommunaler Unternehmen“

15
ISSN 1435-8468
ISBN 978-3-631-61371-9

www.peterlang.de

Acknowledgements

The fascination for economic aspects of integrated water resources management was awakened during an internship with Ulrich Petschow at the Institute for Ecological Economy Research (IÖW) in Berlin. I thank Andreas R. Kraemer from Ecologic Institute for furthering this interest and introducing me to scientific work in environmental policy contexts. I greatly benefitted from my years at Ecologic and the discussions with colleagues and friends there.

The work on this thesis began when I joined the Institute for Infrastructure and Resources Management of Prof. Robert Holländer at Leipzig University. Prof. Holländer encouraged me to follow through with my idea for the thesis topic and I am very grateful for his continued scientific and personal guidance as well as for his sincere interest in my work. I also thank my second supervisor Prof. Thomas Lenk and the chair of my promotion committee Prof. Thomas Bruckner for their support and interest in this project.

I would like to express very special thanks to Prof. Antonio Massarutto, Alessandro de Carli, Vania Paccagnan and Alessandro Lodi from IEFE – Centre for Research on Energy and Environmental Economics and Policy at Bocconi University for their scientific guidance, input and encouragement during my research stay in Milan. I will always cherish the collaboration on this project and beyond. Their valuable methodological feedback and comments on my study design have improved this work. I am grateful for the support of the DAAD, which provided additional funding for my stay at IEFE.

I greatly thank the experts consulted for my illustration cases for their time and interest in this research, notably Issa Nafo, Prof. Andreas Schulz, Bernd van Bömmel and Josef Schön from the *Emschergenossenschaft* as well as Heiner Nobis-Wicherding, Andreas Schmitz, Ricarda Horlacher and Margit Höhndorf from the *Berliner Wasserbetriebe*. Over the years, the work on this thesis greatly benefitted from discussions with Sibylle Pawlowski, Hans-Peter Ewens, Jörg Rechenberg, the members of the LAWA EU Econ und the working group on economics of the Elbe river basin.

I thank my colleagues at Leipzig University, most notably Marcel Fälsch, Sabine Sorge, Sabine Lautenschläger, Enrico Thomas and Tiina Salonen for making Leipzig a wonderful place to work and live. I specially thank Stefan Geyler for his helpful comments on my work. Furthermore, I would like to thank all colleagues with whom I collaborated on WFD research projects, most notably Bernd Klauer, Herwig Unnerstall and Ingo Bräuer.

Special thanks go to my former collegues at Ecologic Institute (Thomas Dworak, Benjamin Görlach, Nicole Kranz and Eleftheria Kampa) as well as my new colleagues at Verband kommunaler Unternehmen – VKU (Nadine Steinbach, Nicole Weiß and Dirk Seifert) for stimulating discussions, feedback and moral support. I thank Thomas Abel for providing a flexible and stimulating work environment that allowed me to finish the dissertation in my new job. I also would like to express sincere thanks to Prof. Wolf Gottschalk, Thomas Papsdorf and Hermann Ühlein for their support on this publication.

I wish to thank Sara, Ana-Mari and Nadine for helping me get through the difficult times and for all the support and caring they provided. I thank my parents, my family and my grandmother for their continued encouragement throughout my doctoral studies. I am thankful to my husband Andreas, who not only became a water expert over the years and provided continued commenting on my writing, but showed incredible patience during the long time of homework this project requested. Without you this thesis would not have been possible. Finally, i thank our son Emil, who greatly furthered the completion of this project.

Table of Contents

List of Tables 11

List of Figures 13

List of Fact Sheets 15

List of Boxes 17

1 Introduction 19
1.1 European water policy: from regulation to integration 22
1.2 The European Water Framework Directive 24
1.2.1 Novelties and building blocks of the Water Framework Directive 25
1.2.2 The Common Implementation Strategy Process: harmonising procedures 29
1.2.3 The implementation schedule for the Water Framework Directive 32
1.3 The role of economics in the WFD 33
1.3.1 Explicit functions of economics in the WFD 34
1.3.2 Implicit functions of economics in the WFD 38
1.3.3 Guidance from the CIS process on implementing the economic aspects 40
1.4 Article 9 and its multiple objectives 42
1.4.1 Recovering financial and environmental and resource costs of water services 46
1.4.2 Setting incentives for efficient water resource use 48
1.4.3 Making the polluter pay and ensuring adequate contributions by sectors 50
1.4.4 Establishing transparency of financial flows in the European water sector and guaranteeing the sustainability of infrastructural systems 51
1.4.5 Putting the objectives of Article 9 into perspective 53
1.5 Research questions 54

1.6 Methodology and scope of the thesis 56
1.7 Plan of the thesis 58

2 Theoretical considerations on cost-recovery and pricing of water services 61
2.1 Water and its dimensions: implications for water services provision ... 61
2.1.1 Water as an environmental good 63
2.1.2 Water as a social good 64
2.1.3 Water as an economic good 66
2.1.4 The market structure for the provision of a multi-faceted good 69
2.2 Cost-recovery of water services: a multi-dimensional concept requiring interpretation 75
2.2.1 Possible interpretations of the concept of cost-recovery 76
2.2.2 Which costs to recover? Definition and delineation of cost categories according to Article 9 86
2.2.3 Cost factors in water services provision 104
2.2.4 Framework conditions and cost drivers 112
2.3 Water pricing as a tool for achieving cost-recovery and beyond 117
2.3.1 Potential objectives and limits of water pricing 118
2.3.2 Operationalising water pricing: an introduction to water pricing models 127
2.3.3 Specificities of user groups: pricing water services for households 135
2.3.4 Specificities of user groups: pricing water services for agriculture 135
2.4 WFD Article 9 and its implications for cost accounting and pricing of water services 140
2.4.1 The definition of water uses and water services and related implications 141
2.4.2 Environmental and resource costs: substantiating the concept . 145
2.5 Difficulties related to a European comparison of cost-recovery and water pricing 154

3 Methodology for a comparative assessment of cost-recovery in line with WFD Article 9 157
3.1 Analytical prerequisites: the ideal path to cost-recovery according to WFD Article 9 157
3.2 Cost-recovery in line with WFD Article 9: the basic assessment framework 161

3.2.1 Definition and delineation of parameters for the cost assessment 163
3.2.2 Accounting for financial costs of water service provision 165
3.2.3 Accounting for revenues in water service provision 180
3.2.4 Assessing environmental costs and their recovery 186
3.2.5 Assessing resource costs and their recovery 198
3.3 Linking the assessment components: implementation pathways and adaptation needs 204
3.3.1 Complementary information for a comparative evaluation: background information on the investigation area 205
3.3.2 Testing the relevance of Article 9 exemptions in justifying divergences from the '*ideal path*' to cost-recovery 207
3.3.3 Mapping out present approaches to cost-recovery and adaptation needs 211
3.4 Assessing the implementation of the complementary objectives of WFD Article 9 211
3.4.1 Assessing the adequacy of incentives for efficient water resource use 214
3.4.2 Assessing the implementation of the polluter-pays principle and the adequacy of contributions from other water uses 216
3.4.3 Assessing the sustainability of infrastructural systems and the transparency of financial flows 218
3.5 Plan of procedure for assessing cost-recovery and overall attainment of Article 9 objectives 220

4 Illustrative cases: testing the methodological framework 231
4.1 Aim and scope of the illustration cases 231
4.2 Background information on the selected illustration cases 232
4.2.1 The German water sector and the illustration cases of Emscher and Berlin 233
4.2.2 The Italian water sector and the illustration case of Emilia-Romagna 253
4.3 Testing the framework: illustrative examples on key steps 261
4.3.1 Testing Step 1: accounting for financial costs in water service provision 261
4.3.2 Testing Step 2: accounting for revenues in water service provision 273
4.3.3 Testing Step 4: assessing environmental costs and their recovery 286
4.3.4 Testing Step 5: assessing resource costs and their recovery 311

4.3.5 Testing Step 7: evaluating the relevance of exemptions from Article 9 317
4.3.6 Testing Step 9: analysing the complementary objectives of Article 9 321

5 Conclusions and outlook 331
5.1 The problem statements of the thesis revisited 331
5.2 Main findings of this thesis 333
5.2.1 The requirements of WFD Article 9 and the prerequisites for a methodological assessment framework 334
5.2.2 Lessons learned from the empirical illustration cases 335
5.2.3 Potential influence of the applied analytical procedure on results 337
5.3 Outlook on further research needs 338

6 Literature 341

7 Annex 371

List of Tables

Table 1:	Milestones of WFD implementation	32
Table 2:	Classifying water services under full and spare system capacity	73
Table 3:	Different approaches to cost-recovery	79
Table 4:	Possible financing approaches for water services provision	83
Table 5:	Examples of administrative costs	97
Table 6:	Operation and maintenance costs in water services provision to households	107
Table 7:	Summary overview on water pricing models	134
Table 8:	Environmental costs of water services	153
Table 9:	Indicators on the general characteristics of the investigation area	205
Table 10:	Remaining information requirements on additional Article 9 objectives	213
Table 11:	General characteristics of the German water sector	235
Table 12:	General characteristics of the Emscher river basin	242
Table 13:	General characteristics of drinking water service provision in Berlin	247
Table 14:	Basic facts on the technical system (Berlin)	247
Table 15:	Evolution of water demand in Berlin, 1992–2007	249
Table 16:	General characteristics of irrigation water service provision in Emilia-Romagna	260
Table 17:	Overview on cost categories for drinking water service provision in Berlin	268
Table 18:	Fixed tariff component (Berlin)	277
Table 19:	Connection fee (in Euro) as of 28 January 2009 (excluding 7 percent VAT) (Berlin)	278
Table 20:	Conditions for investment contributions on the replacement of connections (Berlin)	279
Table 21:	Tarification scheme for irrigation water provision from open canals (CBRO)	285
Table 22:	Status of implementation: reconstruction of the Emscher river basin	291

Table 23: Internalisation instrument 'Abwasserabgabe' 299
Table 24: Selected measures from the Programme of Measures for the Emscher river basin 305

List of Figures

Figure 1:	The economic aspects of the WFD	35
Figure 2:	The explicit and implicit objectives of Article 9	45
Figure 3:	Transparency of costs and financing mechanisms of water services provision and sustainability of infrastructural systems	52
Figure 4:	Article 9 objectives and associated requirements for the implementation process	54
Figure 5:	Plan of the thesis	58
Figure 6:	Different Dimension of Water	62
Figure 7:	Total economic value and its composition	68
Figure 8:	Classifying goods and services: excludability and rivalry	70
Figure 9:	The WFD cost categories put into perspective	87
Figure 10:	Capital versus asset maintenance	89
Figure 11:	Elasticity of water demand along the demand curve	124
Figure 12:	The requirements of Article 9 for water services and water uses	144
Figure 13:	Depicting the 'ideal path' to cost-recovery according to Article 9	158
Figure 14:	Overview on cost categories for the assessment of cost-recovery	162
Figure 15:	Environmental costs	187
Figure 16:	Step-wise approach for the assessment of environmental costs	189
Figure 17:	Step-wise approach for the assessment of resource costs	200
Figure 18:	The Emscher river basin (embedded in the area commonly known as 'Ruhr Area')	239
Figure 19:	Organisational structure of the Emschergenossenschaft	246
Figure 20:	Organisational structure of the Berliner Wasserbetriebe	252
Figure 21:	The case study area of the Consortio di Bonifica della Romagna Occidentale	257
Figure 22:	Water protection zones operated by the Berliner Wasserbetriebe	267

Figure 23: Overview on financing scheme for existing internalization measures 292
Figure 24: Analysis of the degree of internalisation of Type I EC with present measures 296
Figure 25: Financial flows in water supply service provision in Berlin 329

List of Fact Sheets

Fact Sheet 1:	Operation and maintenance costs	167
Fact Sheet 2:	Depreciation	170
Fact Sheet 3:	Opportunity cost of capital	173
Fact Sheet 4:	Administrative costs	176
Fact Sheet 5:	Other direct costs	179
Fact Sheet 6:	Revenues collected through endogenous revenue components	182
Fact Sheet 7:	Revenues collected through exogenous revenue components	185
Fact Sheet 8:	Rough analysis for detecting Type I EC	192
Fact Sheet 9:	Identifying Type II EC	196
Fact Sheet 10:	Rough analysis of resource costs stemming from ecosystem services	202
Fact Sheet 11:	Accounting for financial costs (Berlin)	269
Fact Sheet 12:	Accounting for endogenous revenue components (Berlin)	280
Fact Sheet 13:	Rough analysis for detecting Type I EC (Emscher)	303
Fact Sheet 14:	Identifying Type II EC of wastewater services provision (Emscher)	310
Fact Sheet 15:	Rough analysis of resource costs stemming from ecosystem services (Emscher)	313
Fact Sheet 16:	Rough analysis of resource costs through irrigation water service provision (Emilia-Romagna)	318

List of Boxes

Box 1: Economics in the WFD – scattered evidence 34
Box 2: Article 9 of the Water Framework Directive 44
Box 3: Approaches to cost-recovery – four polar cases 80
Box 4: Economic valuation of benefits: revealed and stated preference methods 146
Box 5: General guidelines for the cost assessment 163
Box 6: Evaluation of existing internalization instruments 191
Box 7: The role of expert judgement 203
Box 8: Earmarking of revenues (internalisation instruments) 217
Box 9: The financing scheme for the reconstruction of the Emscher river basin 293

1 Introduction

Water is not a commercial product like any other but, rather, a heritage which must be protected, defended and treated as such.

EC Water Framework Directive (2000/60/EC), Preamble (1)

Fresh water resources are essential for sustaining life on our planet, enabling economic development and maintaining environmental services. An integrated approach for ensuring the sustainability of fresh water resources use has, however, long been neglected in European water policy. Today, the sustainability of many European river basins is at stake, both in terms of quantitative availability as well as in qualitative terms (EUROPEAN COMMISSION 2007, GLEICK ET AL. 2001, EUROPEAN PARLIAMENT AND EUROPEAN COUNCIL 2000). Over-abstractions put many aquifers and wetlands in Europe at risk. This problem is strongest in Southern Europe, but extends more and more to regions in the north of Europe. In consequence, the ecological status of river basins degrades, ecosystem services can no longer be fully provided for and the survival of aquatic species is threatened. With increasing imbalances at the regional level between supply and demand, intersectoral and interregional competition for water resources increases. These water quantity problems often amplify existing problems of water quality and pollution (EUROPEAN COMMISSION 2000). While the number of heavily polluted rivers has decreased over the past twenty years, also due to concerted actions by international river basin authorities, quality improvements are mainly recorded in large rivers and diffuse pollution, particularly from agriculture, remains problematic throughout Europe. Against this background of increasing water scarcity and pollution problems, economic instruments and principles have increasingly been recognised by national and European policy makers alike as a valuable addition to the traditional water management tool box (EUROPEAN COMMISSION 2001, OECD 1999A).

Water was recognised early on as an important field of environmental policy for the European Community. But until the ratification of the EC Water Framework Directive (WFD, 2000/60/EC) in 2000[1], a large variety of independent directives on singular water policy issues hampered the move towards an integrated management of Community waters. With the WFD a number of decisive novelties were introduced to European water policy. Overall, the Directive aims to achieve "good status" for all Community water bodies by the year 2015[2], acknowledging the fact that "*common principles are needed in order to coordinate Member States' efforts to improve the protection of Community waters in terms of quantity and quality, to promote sustainable water use, to contribute to the control of transboundary water problems, to protect aquatic ecosystems, and terrestrial ecosystems and wetlands directly depending on them, and to safeguard and develop the potential uses of Community waters*" (WFD preamble (23)).

With the WFD, the integration of economics into European water policy and management gained significant momentum. It is the first environmental policy directive at the European level that draws on economic instruments, methods and principles throughout its implementation process to reach its ambitious objectives. The use of economic approaches gives credit to the understanding that the Directive's goals can only be reached within the foreseen timeframe, if economic rationales are invoked regarding water use and allocation, and implementation costs are minimised. In its preamble, the Directive acknowledges the complex and multifaceted dimensions of water, all of which translate into different and at times contradictory claims on its usage and allocation. Despite its economic take, the WFD regards water "*not* [as] *a commercial product like any other but, rather,* [as] *a heritage which must be protected, defended and treated as such*" (WFD preamble (1)).

With its economic elements, the WFD fosters a shift away from the traditionally dominant water supply management towards water demand management approaches, recognising that supply side management often does not provide sustainable solutions and may shift problems locally to other areas or over time, i.e. onto next generations (Massarutto 2004a, Roth 2001). Demand management approaches instead try to better use available resources through efficient allocation. To this end, Article 9 of the WFD demands that account be taken of the principle of cost-recovery for water services, including environmental and resource costs, and for an adequate contribution of water uses

1 Directive 2000/60/EC of the European Parliament and of the Council of 23 October 2000 establishing a framework for Community action in the field of water policy, OJ EC L 327, 22/12/2000, p.1.

2 The Directive provides the possibility of extending the deadline for reaching its objectives by a maximum of two six-year implementation cycles, i.e. to 2027 at the latest.

(disaggregated into at least industry, household and agriculture) to this cost-recovery according to the polluter-pays principle.[3] Furthermore, it calls for water pricing policies that provide adequate incentives for efficient resource use by the year 2010. While water pricing is invoked for reaching the Directive's objectives, it is not considered a one-size-fits-all solution to European water management problems, but rather as an opportunity that "*should be given due consideration, to ensure it promotes more efficient and less polluting use of our scarce water resources*" (European Commission 2000).

The overarching aim of Article 9 is to establish transparency of financial flows in the European water sector, by asking which costs are encountered in water services' provision and how these costs are recovered.[4] However, achieving the overarching aim of Article 9, namely transparency of financial flows and associated water pricing policies that aim at sustainable water services provision and efficient resource use, is severely hampered by the difficulty of comparison among Member States. The definition of cost categories to be considered in the assessment of cost-recovery as well as the accounting for subsidies and cross-subsidies varies considerably across Member States. As pointed out by the European Commission in its Communication on water pricing in 2000, "*The existing accounting rules used by Member States imply different ways of calculating costs. Also, the costs of different services can be included into water prices. As a result, comparisons between the costs of water supply and treatment services, water prices and existing levels of cost-recovery are often misleading. The adoption of common definitions for key cost variables would facilitate the comparison between costs and prices and benchmarking for different water services, uses and countries.*"

This study aims to address this need and proposes a comparative accounting framework for assessing cost-recovery of water supply and sewerage services for private households and agriculture under different institutional, geographical, economic, legal and social-cultural conditions. The assessment framework is set-up with a view to delivering a pragmatic tool that is able to expand over the years as more reliable data becomes available and is equipped to accommodate national particularities.

The following sections give an introduction to the evolution of European water policy up to the WFD (Chapter 1.1) and highlight the objectives and

---

3 Unless otherwise noted, references to articles refer to the EC Water Framework Directive (2000/60/EC).

4 Transparency would imply that comparable assessments are conducted in all European Member States, which would also help to facilitate the to-date severely hampered water price comparisons (cf. Holländer et al. 2008, Metropolitan Consulting 2006, OECD 1999, UBA 1998).

novelties introduced by the WFD in general (Chapter 1.2) and its economic elements in particular (Chapter 1.3). Next, the partly contradictory aims of Article 9 are analysed with a view to gaining a deeper understanding of its explicit as well as its implicit objectives (Chapter 1.4). The chapter sets the basis for the selection of relevant theoretical underpinnings in Chapter 2 and the methodological development for the comparative assessment of cost-recovery in Chapter 3. It presents the research questions addressed in this thesis (Chapter 1.5) and outlines the methodology employed in this study (Chapter 1.6). The chapter concludes with a brief overview of the structure of the thesis (Chapter 1.7).

1.1 European water policy: from regulation to integration

While water policy was recognised early on as an important topic for European environmental policy and coordination, it has undergone fundamental changes over the last three decades and moved from a regulation-centred policy framework to one of integration. Three distinct phases can be distinguished, when assessing the evolution of European legislation on water (EUROPEAN COMMISSION 2009, BLÖCH 1999). The first phase was mainly concerned with reducing or preventing water pollution through standard setting for Community waters and related uses.[5] The directives enacted during this phase, which entered into force between 1975 and 1980, are characterised by a regulatory approach towards water management. They specify environmental quality targets or limit values for individual parts of the aquatic ecosystems, categorised by a specific function (e.g. the Surface Water Directive 75/440/EEC 'for drinking water') or by the function attributed to them ('shellfish' and 'fish water') (HOLTMEIER 1997, MOSTERT 2003). In retrospect, the first phase of Community water protection policy is criticized for its scattered legislative framework, providing different regulatory and management requirements for different water types. This led to implementation difficulties for the responsible administrative agencies and hampered a coherent implementation process across Member States.

5 The first phase of European legislation on water was initiated by the European Commission's five-year Environmental Action Programmes (EAP) of 1973, which lay down the objectives and principles of European environmental policies (EUROPEAN COMMISSION 2002). The evolution from a regulatory to an integrative stance towards water and environmental policy in general can also be witnessed by the changing focus of the EAPs over the following years: the fourth EAP (1987–1992) widened the perspective of environmental policy to other EC policy fields, while the fifth EAP (1993–2000) recognised environmental protection as an equally integral and important element for decision-making as social and economic considerations.

By 1990, problems with surface water pollution became increasingly apparent across Europe, with rising eutrophication levels in fish and coastal waters, visible algae bloom and an overall deterioration of aquatic ecosystems. The start of the second policy phase was marked by the Ministerial Seminar on Water in Frankfurt[6] in 1988, which reviewed the existing legislation on water and identified a number of gaps and related policy options. In response, two new legal instruments were adopted in 1991. They set stricter rules on water pollution emanating from urban settlements and the agricultural sector. With the Urban Wastewater Treatment Directive (UWWTD) (91/271/EEC), new standards for wastewater treatment were established and made obligatory even for small agglomerations.[7] The Nitrates Directive (91/676/EEC) addressed water pollution from agriculture through legally binding measures, limiting the use of animal fertilizer in agriculture. It was complemented by a Directive on Pesticides (91/414/EEC) containing provisions on the authorisation and use of pesticides in agriculture. Finally, the Directive on Integrated Pollution and Prevention Control (IPPC) (96/61/EEC) focused on large industrial installations and established new rules for emissions control relevant to the water sector.[8].

Despite these regulatory interventions at the Community level, however, criticism of the lack of consistency in water protection policy continued, and pressures for a 'fundamental rethink' (BLÖCH 1999) of European water policy increased. European activities in the field of water were constrained by a lack of integration: existing Directives usually addressed a specific problem (e.g. dangerous substances 76/464/EEC), focused on specific sectors (e.g. urban settlements, the agricultural sector) or established water quality targets for individual types of waters (drinking water, fish waters, bathing waters, groundwater). This segmented approach was increasingly considered an inadequate response to the pressing problems in river basins across Europe, which seemed to require coordination across policy fields and administrative boundaries. To address these needs, the European Water Framework Directive (2000/60/EC) was set up. It entered into force on 22 December 2000, and established the third phase

---

6 The conclusions of the Community Water Policy Ministerial Seminar in Frankfurt in 1988 highlighted the need for Community legislation covering ecological water quality (WFD preamble (2)).

7 The UWWTD provided for secondary (biological) wastewater treatment in agglomerations with more than 2000 person equivalents (p.e.), and even more stringent treatment for settlements with more than 10000 p.e. in designated sensitive areas and their catchments.

8 In addition, the so-called Seveso II Directive (96/82/EEC) was adopted during this policy phase, which contains guidelines for controlling and dealing with dangers from major accidents.

of European policy on water protection.[9] It introduced substantial changes to this long-established policy field (BOSENIUS AND HOLZWARTH 2006, BLÖCH 1999) and may be regarded as the most important piece of European legislation in the field of water management in decades. The following section gives an overview of the objectives and innovations introduced by this new legislative framework.

1.2 The European Water Framework Directive

The European Water Framework Directive (WFD) establishes a comprehensive framework for the long-term protection of all water bodies in the Community, including inland surface waters, transitional waters, coastal waters and groundwater. The key objective of the WFD is to establish *'good water status'* in all Community waters by 2015. More specifically, the Directive's aims include (EUROPEAN COMMISSION 2007, BLÖCH 1999:67)[10]:

- Achieving *'good status'* of all Community waters by the year 2015 and preventing the deterioration of the status of aquatic ecosystems (Article 4)[11],
- Establishing water management in accordance with the hydrological entities of river basins (Article 3),
- Promoting sustainable water use facilitated by economic approaches and incentive pricing policies (Articles 5 and 9),

---

9 On the basis of the European Commission Communication (COM/1996/59/final) to the European Council and European Parliament in February 1996 a first draft of the Directive establishing a framework for Community action in the field of water policy or, in short, the Water Framework Directive, was already developed in 1996.

10 The relevant articles from the WFD are cited as a reference for the interested reader. In general, it is recommended to consult the English version of the WFD, the language in which the Directive was negotiated, as inaccurate translations gave rise to ample discussions (cf. e.g. UNNERSTALL 2005, SCHMUTZER 2005 and Chapter 1.3 below)

11 For groundwater, good status is measured in terms of both water quantity and chemical water quality, and for surface waters, ecological water quality and chemical water quality have to be taken into consideration in the assessment of 'good status'. These objectives are much broader than the objectives of previous water directives. In recognition of the fact that changes to morphology may make good ecological status very difficult to meet in surface waters, another category was introduced: for those water bodies, which qualify as *'Heavily Modified Water Bodies'* (HMWBs) and *'Artificial Water Bodies'* (AWBs), a *'good ecological potential'* has to be reached by 2015. In addition, the WFD also requires that no deterioration in water status takes place and that protected area objectives (established under existing Community legislation) are met (cf. CIS Guidance Documents No. 13 *"Overall Approach to the Classification of Ecological Status and Ecological Potential"* for further details, CIS 2003B).

- Establishing a *'combined approach'* that addresses both point sources through emission limit values and diffuse pollution through quality standards (Article 10),
- Progressively reducing discharges and emissions of certain priority substances and phasing-out discharges of priority hazardous substances (Article 16),
- Contributing to the mitigation of the effects of floods and droughts (Article 1),
- Getting Europe's citizens more involved with water resources management through public participation during the entire implementation process (Article 14).

In addition to the qualitative objective of improving water status across Europe and moving to an integrated management of water resources, the WFD aims to streamline legislation on water resources at the European level. After a transitional period, it replaces seven of the *'first wave'* directives[12] and harmonises the implementation of the overall set of legislative acts on water by referring to existing directives in river basin management plans, the primary tool for achieving the Directive's objective.

1.2.1 Novelties and building blocks of the Water Framework Directive

With the WFD, a number of novelties were introduced to European water policy. In general, it can be classified as a hybrid type of directive (MOSS 2004), containing attributes of traditional command-and-control approaches, together with a reliance on new tools such as cost-effectiveness analysis or processes of inter-agency negotiation and public participation. The following paragraphs review briefly the most prominent features of the WFD.

Managing Europe's waters by hydrological entities

For the entire WFD implementation process, the hydrological boundaries of river basins (i.e. the natural geographical and hydrological units) constitute the

---

12 The replacement of the old directives is foreseen in two steps: by 2007, the Surface Water Directive (75/440/EEC), the Exchange of Information Directive (77/795/EEC) and the Sampling and Analysis Directive (79/869/EEC) are repealed. By 2013, the Shellfish Water Directive (79/923/EEC), the Fish Water Directive (78/659/EEC), the Groundwater Directive (80/68/EEC) and the Dangerous Substances Directive (76/464/EEC) are revoked. Special daughter directives were enacted on groundwater (2006/118/EC) and on priority substances (2008/105/EC).

main points of reference. Theoretically, administrative and political boundaries will accordingly become less relevant to water management in the Community. While river basins already figured prominently in the management of water resources of a number of European Member States (e.g. the '*Agence de l'Eau*' in France or the river basin authorities in Spain), this was not the case for most Member States and required the creation or empowerment of appropriate institutional bodies.[13] The necessary overhaul of existing management structures and administrative procedures poses a considerable challenge to Member States. Such challenges are often described as 'problems of fit' (MOSS 2004), since ecosystem units are usually incongruent with existing political-administrative spatial units. In addition to a spatial reorganisation, the WFD requires an increased coordination with other policy-fields, including agriculture, soil protection, land management and nature protection. However, these policy fields can often only inadequately be influenced by the water sector (RUMM ET AL. 2006) leading to 'problems of interplay' (MOSS 2004) between institutions, stakeholder groups and disciplines. In addressing these problems, the WFD is expected to exert a notable influence on administrative structures for water management within Europe (KESSLER 2006).

Programme of Measures for reaching Good Status

In order to reach the objectives of the WFD, i.e. good water status in all Community waters, a large number and variety of measures will have to be undertaken in river basins across Europe (cf. Article 5 reports and draft River Basin Management Plans). According to Article 11 of the WFD, the necessary measures for bridging current gaps in water status (i.e. for bringing all water bodies up to the level of 'good status') are to be described for each river basin district (or for the part of an international river basin district within a Member State's territory) in a so-called Programme of Measures (POM) by 2009.[14] The selection of the most suitable measures for reaching good status should be undertaken with regard to the measures' cost-effectiveness[15], in order to

---

13 According to WFD Article 3 (2) and 3 (4), Member States are required to make "*appropriate administrative arrangements*" including the identification of a "*competent authority*" in order to ensure necessary national and international coordination for a successful WFD implementation.

14 The need for measures has been assessed as part of the Article 5 reports (cf. below) on the basis of a '*risk assessment*', which investigated the risk of failing to reach good water status by 2015 without additional measures extending beyond business-as-usual activities.

15 In the past, technical measures were often favoured in water management. With the WFD approach to the cost-effectiveness analysis, all types of measures are treated equally in the

ensure that the Directive's objectives are reached at minimum cost (see Section 1.3.1.2). In general, the POM contains two categories of measures, namely 'basic measures' and 'supplementary measures'. Basic measures are required by existing legislation (e.g. in order to comply with the Nitrate Directive or the UWWTD[16]) and constitute minimum requirements for the POM. They have to be undertaken in any case and are not subject to an assessment of their cost-effectiveness. Should the basic measures be insufficient for reaching the Directive's environmental objectives, further so-called supplementary measures have to be undertaken (according to Annex VI, Part B of the WFD). The WFD provides a first non-exhaustive list of potential supplementary measures, including e.g. economic and fiscal instruments, codes for good practices, voluntary agreements and research and development projects.[17] The POM is set up on the basis of a first analysis of the risk of failing the Directive's objective if no additional measures are undertaken (according to Article 5). The Programme of Measure forms part of the River Basin Management Plan.

River Basin Management Plans

According to WFD Articles 13 and 15 (and Annex VII), a River Basin Management Plan (RBMP) has to be established and updated every six years for each river basin district. It can be regarded as the main reporting mechanism on implementation progress to the Commission and to the public. According to Annex VII, the RBMP should include a general description of the river basin district's characteristics; a review of the significant pressures and impacts of human activities on the status of surface water and groundwater within the

---

assessment instead, including technical, institutional, legal, economic or social measures. However, as basic measures which are mandated by prior European directives must be implemented, this flexibility in measure selection is again partly reduced. KESSLER IN RUM ET AL (2006:48).

16 For a detailed list of existing legislation considered under '*basic measures*', cf. part A, Annex VI of the Directive. A definition and a list of basic measures are included in Article 11(3) of the Directive. In the context of this study, it should be noted that "*measures deemed appropriate for the purposes of Article 9*" are also classified as basic measures, and therefore qualify as minimum implementation requirements.

17 In practice, the distinction between basic and supplementary measures may not be trivial as some measures may qualify under both definitions (e.g. voluntary agreements are listed as supplementary measures, but may also be used as an instrument of the Nitrate Directive). This differentiation is of particular importance for justifying exemptions from the achievement of good status on the basis of '*disproportionate costs*' of measures: basic measures are required by existing legislation and may hence not be regarded as additional WFD-related costs. See Section 1.3.2 for an extended discussion.

river basin district; the identification and mapping of protected areas; a map of monitoring networks and of the results of monitoring programmes; a list of the environmental objectives established under Article 4 for surface waters, groundwater and protected areas, including information justifying any exemptions from these objectives (cf. Section 1.3.2); a summary of the economic analysis of water use (as required under Article 5 and Annex III, cf. Section 1.3.1.1); a summary of the POM adopted under Article 11, including a report on the practical steps and measures taken to apply the principle of recovery of the costs of water use in accordance with Article 9; and a summary of the public consultation and information measures undertaken, their results and the changes to the RBMP made in consequence. A draft of the RBMP had to be made available for stakeholder consultation by mid-2008.

Using economics throughout the implementation process

The need for conserving adequate supplies of a resource for which demand is constantly increasing in most parts of Europe has been a driving force for the integration of economic elements in the WFD (Blöch 1999:71). The WFD is the first environmental policy Directive at EU level that explicitly draws on economic approaches throughout its implementation process (i.e. in Articles 5, 9, 11 and Annex II and III). The Directive integrates the polluter-pays principle, aims to establish sustainable water resource use through incentive pricing, uses the cost-effectiveness analysis for singling out relevant measures for reaching the environmental objectives at minimum costs and aims at cost-recovery in water services provision. Economic considerations can also play a role in justifying exemptions from the overarching aim of the Directive, i.e. to achieve good status of all water bodies by 2015. If reaching this objective in time should prove disproportionately costly, either the 2015 deadline may be extended, or the objective may be relaxed. The economic elements of the Directive are reviewed in greater detail below (cf. Chapter 1.3).

Involving the public

The Directive's preamble already stresses that "*the success of this Directive relies upon close cooperation and coherent action at Community, Member State and local level as well as on information, consultation and involvement of the public, including users*". In Article 14, the WFD requires the responsible bodies for WFD implementation to foster transparency of the planning and implementation processes through citizen involvement. At several points during the implementation process, information and consultation of the (interested) public are made mandatory, while active participation is only to be encouraged by Member States.[18] Implementing these participation requirements is considered essential to the responsible water agencies for winning the support of policy-makers in other policy fields, increasing the acceptance of measures among the general public and accessing expert knowledge and information about local circumstances.

Applying a combined approach

Historically, there has been a dichotomy in approaches to water pollution control at the European level, with some control instruments and regulations concentrating on limiting emissions at their source, and others focussing on the carrying capacity of receiving waters and the surrounding environment through quality objectives. The WFD addresses pollution control through a so-called '*combined approach*': pollution at its source is limited through emission controls on e.g. wastewater and agricultural fertilisers, but is amended by water quality objectives to ensure that those reduced emissions fit into the local or regional environment, complying with the objective of '*good status*' (Blöch 2004). The more stringent approach is then applied in each case. Hence, POM formulate limit values to control emissions from individual point sources along with environmental quality standards addressing the cumulative impact of point emissions and diffuse sources of pollution. Furthermore, a daughter directive of the WFD on priority substances (2008/105/EC) was passed on 24 December 2008. Among others, it addresses the phasing out of priority hazardous substances.

1.2.2 The Common Implementation Strategy Process: harmonising procedures

Experience with former implementation processes of sectoral European water legislation had shown that implementation approaches and outcomes could differ significantly across Europe. Unlike prior water legislation, where no harmonised approach was necessary for achieving the objectives of a certain piece of legislation at least in parts of Europe (e.g. the implementation success of the

18 The core public participation provisions of the WFD refer to the first two pillars of the Åarhus Convention (UNECE 2001) and its three levels of participation of information, consultation and active participation. The WFD requests public consultation specifically for i) the proposed timetable, work programme and role of the public in drafting the RBMP (2006); ii) the overview of major water management issues (2007); and iii) the draft RBMPs (2008).

UWWTD in Germany is not linked to the approach chosen in the Netherlands), an incoherent implementation process would, in the case of the WFD, harbour important drawbacks. With many European river basins transcending territorial and administrative borders, concerted and coordinated action, a common understanding and a joint approach were recognised early on as prerequisites for a successful and (cost-)effective implementation of the Directive (INTERWIES ET AL. 2003). Acknowledging the need for harmonisation and coherence, the European Commission and its Member States developed an unprecedented and new structure at the European level, the so-called Common Implementation Strategy (CIS) for guiding implementation in all river basins across Europe (BOSENIUS AND HOLZWARTH 2006).

The CIS was initiated under French presidency at an informal meeting of EU Water Directors and the Norwegian Water Director in Paris 23–24 October 2000. It is directed at (CIS 2001):

- Developing a common understanding, approaches as well as methodological support and scientific underpinnings on different parts of the WFD implementation process;
- Elaborating informal technical guidance on selected implementation steps, supported by best practice examples;
- Limiting the risk of bad application by offering targeted assistance on those parts of the implementation process requiring further substantiation and interpretation;
- Sharing experiences and resources between the Commission and the Member States and avoiding a duplication of efforts;
- Furthering the involvement of the public and promoting public awareness on the key elements of the WFD; and
- Ensuring coherence between the implementation of the WFD and other sectoral and structural policies at the European level.

In order to achieve these objectives, four main activities were identified within the CIS framework (CIS 2001):

- *Activity 1 – Information sharing*: the most important vehicle created under this activity is the internet-based information exchange platform *'WFD-Circa'*, accessible to all interested parties.[19] Furthermore, an extensive num-

19 All documents published within the CIS process are made available on this platform. Furthermore, it establishes linkages to national and river basin specific WFD information and implementation websites. Draft documents from the working groups cannot, however, be accessed by the general public in order to avoid misinterpretation of interim results (EU Commission, Circa-Website).

ber of workshops and conferences have been organised, both at the European and at the national level, to support specific implementation aspects.
- *Activity 2 – Guidance on technical issues*: this task aims to develop informal, legally non-binding technical and scientific guidance documents to support the overall practical implementation process and for testing in pilot river basins.[20]
- *Activity 3 – Information and data management*: information exchange and reporting under the WFD is partly based on geographical information systems. Therefore, this activity focused on harmonising existing systems, developing guidance and testing a GIS-prototype for WFD implementation.
- *Activity 4 – Application, testing and validation*: under this task, the guidance documents developed under Activity 2 were put to a practical implementation test under real-life conditions in selected pilot river basins to foster a harmonised implementation process.[21]

All guidance developed within the CIS process is legally non-binding, recognising that the implementation of the WFD is *"a responsibility, which resides fully within the competence of the individual Member State. A Common Strategy neither could nor should challenge this fundamental principle of Community environmental law"* (CIS 2001). The legally non-binding nature of the guidance offers the advantage that flexible adaptations to regional or local circumstances

---

20 From 2003 – 2009 23 legally non-binding guidance documents and technical reports were made available on the following aspects of WFD implementation: economics and the environment – WATECO (no. 1); identification of water bodies (no. 2); analysis of pressures and impacts – IMPRESS (no. 3); identification and designation of heavily modified and artificial water bodies – HMWB (no. 4); transitional and coastal waters (no. 5); towards a guidance on establishment of the intercalibration network and the process on the intercalibration exercise (no. 6); monitoring under the Water Framework Directive (no. 7); public participation in relation to the Water Framework Directive (no. 8); implementing the Geographical Information System elements (GIS) of the Water Framework Directive (no. 9); rivers and lakes (no. 10); Planning Processes (no. 11); role of wetlands in the Water Framework Directive (no. 12); overall approach to the classification of Ecological Status and Ecological Potential (no. 13); guidance on the intercalibration process (2004–2006) (no. 14); groundwater monitoring (no. 15); groundwater in drinking water protected areas (no. 16); direct and indirect inputs in light of the 2006/118/EC Directive (no. 17); groundwater status and trend assessment (no. 18); surface water chemical monitoring (no. 19); exemptions to the environmental objectives (no. 20); guidance for reporting under the WFD (no. 21); updated WISE GIS guidance (no. 22); eutrophication assessment in the context of European water policies (no. 23). All Guidance Documents are available for download at http://ec.europa.eu/environment/water/water-framework/

21 By 2002, a pilot river basin network was established including fourteen national or international (sub-)river basins. Testing started as early as 2002 and evolved along with the restructuring of the CIS process (cf. Section 1.3.3).

are possible. On the negative side, harmonisation is not necessarily guaranteed and important leeway for interpretation remains, which may bring the entire harmonisation process into question.

1.2.3 The implementation schedule for the Water Framework Directive

For reaching the WFD's ambitious environmental objective of good water status by 2015 in all European Community waters, a roadmap has been included in the Directive setting clear deadlines for intermediate implementation milestones.

Table 1: Milestones of WFD implementation

Year	Milestone	Reference in the WFD
2000	Directive entered into force	Article 25
2003	Transposition into national legislation Identification of river basin districts and authorities	Article 23 Article 3
2004	Characterisation of river basin: pressures, impacts and economic analysis	Article 5
2006	Establishment of monitoring networks Start of public consultation (at the latest)	Article 8 Article 14
2008	Presentation of draft river basin management plans for public consultation	Article 13
2009	Finalisation of river basin management plans, including programmes of measures	Article 13 Article 11
2010	Introduction of pricing policies in line with the objectives of WFD Article 9	Article 9
2012	Cost-effective programmes of measures operational	Article 11
2015	Compliance with environmental objectives End of first implementation cycle Finalisation of second river basin management plan	Article 4
2021	End of second implementation cycle ends	Article 4 Article 13
2027	End of third implementation cycle (final deadline for meeting WFD objectives)	Article 4 Article 13

Source: amended from http://ec.europa.eu/environment/water/.

Overall, the Directive's implementation is organised as an iterative process: following its entry into force in 2000, the first implementation cycle refers to the period up to 2015. Taking account of the fact that technical difficulties, unfavourable natural conditions or disproportionate costs of measures may prevent Member States from reaching the objectives in time, this first cycle is followed up by two additional six-year implementation cycles. Table 1 briefly summarises important implementation milestones.

1.3 The role of economics in the WFD

Economic instruments were recognised early on as important to European environmental politics. The European Council highlighted already in its first Environmental Action Programme (1973) that Community-wide rules for covering and allocating the costs of environmental protection needed to be developed. Member States and the Council highlighted the polluter-pays principle as a guiding principle for the introduction of economic instruments and for furthering environmental protection. However, environmental levies or charges have barely been drawn upon in European environmental protection law. There are but a few cases in which Member states were obliged, or even allowed, to introduce environmental levies or charges for implementing environmental policy objectives (UNNERSTALL 2005:2).

The WFD is the first environmental policy directive at the European level that explicitly draws on economic instruments along its entire implementation process for achieving its ambitious objectives. While economic aspects have been placed prominently within the Directive's preamble and in two of its articles, and evidence on economics can be found scattered along the entire Directive (see Box 1), the Directive's recourse to economics should not be interpreted as an economisation of European water policy (BRACKEMANN ET AL. 2000). Instead it is evidence of the conviction that reaching the Directive's objectives will require substantial financial resources, which should be put to their economically most effective use, i.e. reaching good status by 2015 will require the selection of those measures that are cost-effective, or exemptions from reaching the objectives can be justified by 'disproportionate costs'. Furthermore, the Directive recognises the potential of economic instruments in promoting sustainable water resources use.

Box 1: Economics in the WFD – scattered evidence

Economic issues are the focus of two articles of the WFD: Article 5 (together with Annex III) asking for an *economic analysis of water use* and Article 9 on the *cost-recovery of water services* (analysed in detail below in Chapter 1.4). Nevertheless, economic aspects can also be found, or are implied, in other parts of the Directive's text and include (CIS 2003):

- An identification of areas designated for the protection of *economically significant aquatic species*;
- The designation of heavily modified water bodies based on an assessment of the significance of the *impact* (including the *economic* impact) on existing uses and *costs* of alternatives for providing the same beneficial objective;
- The provision of support to the selections of the Programme of Measures on the basis of *cost-effectiveness* criteria;
- An assessment of the potential role of *pricing* in Programmes of Measures;
- A pathway for justifying exemptions from reaching the directives environmental objectives (in terms of time or in terms of less stringent objectives) through an assessment of the proportionality of costs and benefits and of *costs* of alternatives for providing the same beneficial objective;
- An evaluation of *costs* of measures in order to come up with *cost-effective* solutions for the control of priority substances.

Source: CIS (2003a).

As the WATECO Guidance Document (CIS 2003A) points out, the Directive's economic aspects can be grouped in two categories: those economic aspects that fulfil an explicit function within the Directive and those that can rather be classified as implicit. Nonetheless, the economic elements of the Directive should be considered and tackled as an integral and interlinked part of the overall implementation process (see Figure 1). Both, the explicit and the implicit economic elements of the WFD are reviewed in greater detail in the following sections.

1.3.1 Explicit functions of economics in the WFD

Within the Directive's text, three explicit functions can be identified for economics (CIS 2003A:19). The first explicit function is the economic analysis of water use (according to Article 5 and Annex III). The second explicit function is given by the cost-effectiveness analyses, which are made a mandatory tool for selecting measures (according to Article 11 and Annex III). The third explicit function concerns Article 9, i.e. the assessment of cost-recovery (including environmental and resource costs), the reference to the polluter-pays principle and the establishment of incentive pricing by 2010. As Article 9 is at the heart of this thesis, it's multiple and partly contradictory objectives are investigated in detail and separately in Chapter 1.4.

Figure 1: The economic aspects of the WFD

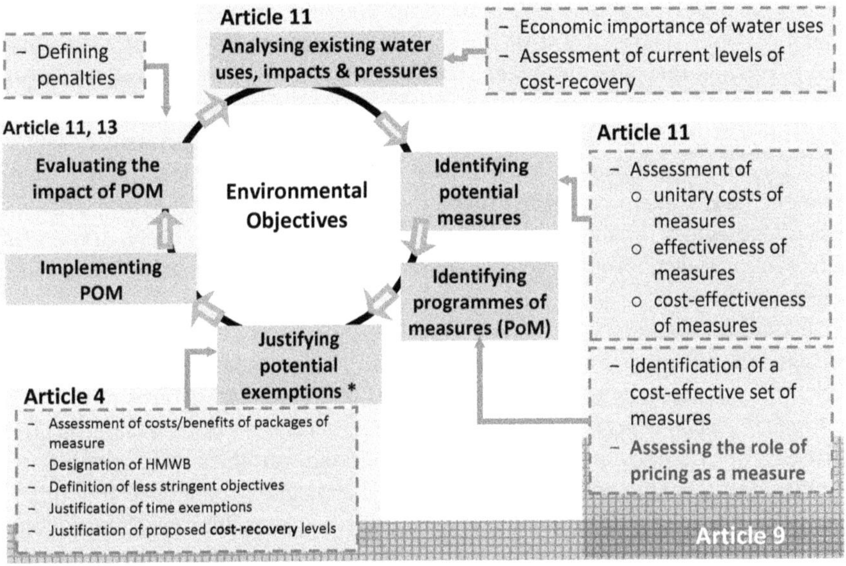

Source: adapted from CIS (2003a).
**Note: the justification of potential exemptions and the identification of programmes of measures have been switched and the reference to Articles has been added.*

1.3.1.1 The economic analysis (according to Article 5 and Annex III)

As the first guidance document from the CIS process on economics, the WATECO document (CIS 2003A) mainly addressed how to proceed with the first economic implementation milestone of the Directive, the economic analysis. The provisions for the economic analysis are given in Article 5 and Annex III but remain sufficiently vague to leave ample room for interpretation, which was concretised by means of this guidance document at the European level. According to Article 5, an economic analysis is to be prepared for each river basin district (RBD) and should "*provide enough information in sufficient detail* [...] *in order to make relevant calculations necessary to take into account under Article 9 the principle of recovery of the costs of water services* [...] *and make judgements about the most cost-effective combination of measures*". To comply with these requirements, the WATECO Guidance Document suggests including in the Article 5 economic analyses, an assessment of the following four issues (WWF AND EEB 2006, CIS 2003A).

First, an analysis of the economic importance of water uses should be conducted, in order to assess how important water is to the economy and the socio-economic development of the river basin district. The economic analysis should provide the river basin's economic profile in terms of general indicators (e.g. economic turnover, gross income, employment or number of beneficiaries of significant water uses). In a broader context, the economic analysis of water uses is intended to pave the way for the assessment of significant water management issues to be reported to the public by 2007, as well as the ensuing cost-effectiveness analysis, by initiating investigations of likely trade-offs between socio-economic development and water protection within the river basin.

Second, an assessment of future trends in economic sectors should be undertaken by means of a baseline scenario (BLS), which assesses forecasts in key economic drivers likely to influence pressures and thus water status until 2015. In the BLS, trends in water supply and water demand have to be evaluated. The focus is on changes in general socio-economic variables (e.g. population growth), in economic growth of main sectors as well as changes in the implementation of planned investments linked to existing regulation. Both hydrological as well as socio-economic drivers have to be investigated.

Third, the level of cost-recovery of water services should be assessed in accordance with the requirements of Article 9 of the WFD. Key elements to be investigated in the economic analysis by the end of 2004 include the status of water services, the institutional set-up for cost-recovery, the extent of the recovery of the costs of the water services, the contribution of key water uses to the costs of these services (i.e. for assessing the degree of application of the polluter-pays principle) as well as the incidence of subsidies.[22] Furthermore, the current role of economic instruments is to be studied, together with an analysis of present water pricing mechanisms and their role in supporting the achievement of sustainable water management and the overall environmental objectives of the WFD.

---

22 In order to complete the economic analysis, a definition and delineation of water services and water uses was required. The exact definition is of particular importance for the application of the cost-recovery principle and will hence be discussed under Chapter 1.4.

Finally, preparatory steps for the cost-effectiveness analysis are to be included in the economic analysis (see Section 1.3.1.2 below). In preparation for the cost-effectiveness analysis of possible measures and their combination, the existing gaps in cost information should be reduced and data on the unitary costs of key measures to be considered for the development of River Basin Management Plans (RBMP) should be gathered. Ranges of costs were to be estimated and collected for individual measures, along with key parameters influencing these costs. The emphasis was meant to be placed on costs that are non site-specific (i.e. financial costs of measures, indirect non-water-related environmental costs) and on basic measures.

According to the Directive, effective use should be made of the economic analysis during the following implementation steps, both for the explicit as well as the implicit functions of economics.

1.3.1.2 The cost-effectiveness analysis (according to Article 11 and Annex III)

As introduced above, programmes of measures (POM) have to be established by 2009 for reaching the WFD's environmental objectives in all river basins by 2015. Article 11 and Annex III of the Directive mandate that the selection of appropriate measures is undertaken with the help of cost-effectiveness analyses (CEA). CEA are an economic tool to identify the option that reaches a certain pre-defined target at least cost (i.e. in this case the Directive's environmental objectives).[23] To this end, a ratio that indicates the costs of achieving a per unit change in a specified physical outcome is developed and compared to competing alternatives (RPA 2004). It is not part of a CEA to investigate whether the pre-specified target that should be reached at minimum cost is justified in itself.[24] Accordingly, a CEA does not change objectives or deadlines, but only serves to identify the cheapest Programme of Measures for reaching the Directive's fixed objective (i.e. good status). For making the selection of the cost-effective combinations of measures, important input is required from technical experts. While economists deliver methodological support to the CEA, the alternatives to be tested and compared have to be provided and assessed for their technical effectiveness by technical experts. Accordingly, a high degree of transparency, coordination and exchange between economic and technical

---

23 Cf. e.g. OECD (2006) for a discussion on the advantages and disadvantages of cost-effectiveness analysis in environmental politics in relation to other economic decision-making procedures.

24 Such a justification test would entail a comparison of the social benefits derived from meeting the pre-specified target and the costs associated with doing so.

experts will be required, and parallel work streams have to be integrated, wherever possible.

With the economic analysis for 2004, first data collections and preparatory steps for assessing the cost-effectiveness of measures were to be carried out (e.g. by collecting data on costs of measures) with an emphasis on non-site specific costs and on basic measures. In a number of Member States methodological guidance has been provided on the selection of cost-effective programmes of measures[25], and pilot tests have been initiated, both within the CIS process (cf. Section 1.2.2) as well as at the national level. Drafts of the programmes of measures have to be made available for public consultation in 2008, as requested by Article 14 (cf. Section 1.2.1).

1.3.2 Implicit functions of economics in the WFD

The second group of economic aspects in the WFD is not explicitly referred to, but rather implied in other implementation requirements (cf. CIS 2003A:21 and Box 1 above). These implicit functions include most prominently the justification of exemptions from reaching the environmental objectives on the basis of disproportionate costs, the justification of heavily modified water bodies due to 'significant adverse effects' on existing water uses and the identification of economically significant aquatic species. As the justification of exemptions raises important implementation difficulties and may significantly impact the Directive's overall implementation process, it is discussed in detail in the following section.

While economic considerations do not play a role in the definition of the environmental objectives of the WFD, they constitute one argument with which exemptions from achieving good status by 2015 may be defended. If all cost-effective combinations of measures for reaching the environmental objectives prove in time to be disproportionately costly, either an extended deadline for reaching the objectives (i.e. time exemptions[26], Article 4.4) or less stringent environmental objectives (i.e. quality exemptions, Article 4.5) may be applied for. Thus, a temporal exemption may be granted if completing the necessary improvements to achieve good status before 2015 is '*disproportionately expen-*

25 For the German context cf. INTERWIES ET AL. (2004), for the Netherlands cf. VAN DER VEEREN (2005).

26 Time exemptions are limited to two six-year review cycles, i.e. up to 2027 the latest. They may also be granted, if a timely implementation is prevented by technical feasibility (cf. Article 4.4 (a)), i.e. if the scale of improvements required can only be achieved in phases exceeding the timescale.

sive'. A lowering of objectives can be justified if achieving the objective proves to be disproportionately costly irrespective of the timing. Furthermore, exemptions may be obtained for temporary deteriorations in water status due to circumstances of exceptional natural cause or '*force majeure*' (e.g. floods or prolonged droughts) (according to Article 4.6) and new modifications and new sustainable human development activities (given the conditions specified in Article 4.7 are met).

Applications for exemptions have to be justified in the RBMP for 2009 and need to be scrutinized every six years. However, the practical interpretation of the terms '*disproportionately expensive*' remains disputed (GÖRLACH AND PIELEN 2007). The two main questions pertain to which reference costs should be considered disproportionate, and what the threshold is beyond which costs are considered disproportionate.

From a welfare economic point of view, the judgement on the disproportionality of costs is indicative of a classical cost-benefit analysis: in this understanding, costs are considered to be disproportionate if they exceed the monetised benefits of achieving '*good status*' in a water body (or, possibly, if costs exceed benefits by a certain safety margin). From a theoretical perspective, monetised benefits are the most satisfactory reference. In contrast to other possible references, they are the only criterion that ensures that the chosen solution is socially desirable and efficient, in the sense that social benefits exceed social costs.

In addition to the comparison of costs and benefits in monetary terms, the distribution of costs among the affected parties, and their ability-to-pay, may also figure in the decision-making process. Accordingly, criteria that could be considered in the process are (KLAUER ET AL. 2007):

- Whether benefits accrue to those who bear the costs;
- Whether costs to firms or sectors exceed the fair proportion as demanded by the polluter-pays-principle;
- Whether the costs of the WFD lead to increases in the cost of water supply to private households, and consequently to undue social hardships;
- Whether costs to firms or sectors exceed the financial carrying capacity of firms or sectors (measured e.g. in relation to profit margins), leading to an unacceptable financial burden for firms; or
- Whether overall costs of measures for achieving good status in a river basin exceed a certain threshold (e.g. a percentage of GDP or a share of the available public budget).

From a theoretical perspective, these criteria are irrelevant to the social efficiency of the proposed course of action, since the affected parties could be compensated if they were burdened beyond their fair share in costs. However, for the practical negotiation with stakeholders, these considerations are crucial, since possible compensation mechanisms are more easily devised in theory than implemented in practice.

It can likewise be argued that all costs will eventually end up being paid by households as consumers or as taxpayers: where the WFD implementation causes costs to businesses and public utilities, the businesses and public utilities will try to pass these costs on to consumers (depending on the competitive situation). Where the WFD leads to increased public expenditure for water management, or where hardships for certain economic sectors are compensated from the public budget, these costs will be covered by taxpayers, either through increased taxes and charges, or through reduced supply of other public services. Thus, eventually, the question is what additional burden can be considered acceptable for each household. This, in turn, boils down to assessing the individual's willingness to pay, which is the measure that is also used to calculate the benefits of the WFD in monetary form (cf. Chapter 2.4).

The WFD itself does not provide any guidance on the assessment of disproportionate costs, but leaves it to the Member States to substantiate the concept. Some explanatory guidance on this issue has since been developed in the European working groups established under the WFD Common Implementation Strategy (cf. CIS 2009). In addition, several Member States have substantiated the concept to make it operational within their river basin districts. A harmonisation of approaches may only be expected with ongoing implementation. Nevertheless, the European Commission has already indicated in implementation workshops as part of the CIS process that the justification of exemptions on the basis of disproportionate costs may only be accepted, if an assessment of associated costs and benefits has been undertaken.

1.3.3 Guidance from the CIS process on implementing the economic aspects

The economic aspects of the WFD posed, from the beginning of implementation, difficulties which were widely shared by Member States. Accordingly, a Working Group on Economics (WATECO) was established within the first phase of the Common Implementation Strategy (CIS) process (under the leadership of France and the European Commission). The WATECO working group consisted of approximately forty members, including water economists from European Member States and Accession Countries as well as scientific experts and NGO delegates.[27] Its task was to clarify the general understanding of the economic aspects of the WFD and to provide guidance on their practical implementation. The work of the WATECO group focused primarily on the requirements for the economic analysis due 2004 (Article 5, Annex III). Reminiscent of the other CIS guidance documents, the WATECO document is legally non-binding. Conceptualised as a general guide to the economic aspects of the WFD, it necessitated further specification and application to the specific situation of each river basin.

The WATECO working group did not continue its activities after the adoption of the implementation recommendation. Instead, small drafting groups (DG) of experts (*'Eco 1'* and *'Eco 2'*) were formed to produce specialised information sheets on specific issues.[28] Two information sheets are of particular interest to this thesis and are hence briefly introduced:

- The *'Information sheet on assessment of the recovery of costs for water services for the 2004 river basin characterisation'* (CIS 2004A) by DG Eco 1 provides extended definitions and guidelines on how to perform a cost-recovery assessment. The guidance document is accompanied by practical examples from selected pilot investigations.
- The *'Information sheet on the assessment of environmental and resource costs in the Water Framework Directive'* (CIS 2004B) by DG Eco 2 investigates important remaining questions for the practical implementation of the environmental and resource costs (ERC) concept which could not be resolved by the WATECO guidance. The main question to be tackled by this guidance effort was how methods for assessing ERC (developed at an academic level) could be made operational in the context of the development of river basin management plans. As one main contribution to the implementation process, the drafting group provides revised operational definitions of the components of environmental and resource costs (cf. Section 1.4.1 and Chapter 2.4). However, despite illustrative cases on the nature of ERC in the context

---

27 From the start, NGOs were involved and pushed economic elements through complementary studies and evaluation reports. More recently, WWF and the European Environment Bureau (EEB) undertook a review of the Article 5 reports sent by Member States to the European Commission with a particular focus on the definition of water services (see Chapter 1.4 and Chapter 2.3), cost-recovery and incentive pricing (WWF AND EEB 2006).

28 In all, four economic information sheets have been conceptualised by the two drafting groups, namely: Information sheet on the assessment of the recovery of costs for water services for the 2004 river basin characterisation (DG Eco 1, CIS 2004A); Information sheet on the assessment of environmental and resource costs in the Water Framework Directive (DG Eco 2, CIS 2004B); Information sheet on the methodology to prepare the baseline scenario (DG Eco 1, CIS 2004C); and Information sheet on the analysis of water uses for the river basin characterisation (DG Eco 1, CIS 2004D).

of WFD implementation and an overview on potential techniques for their measurement, the guidance was perceived as too theoretical for delivering immediate implementation support.

In addition to developing guidance, pilot activities were initiated on the economic aspects of the WFD with the CIS process. They aimed to test proposed methodological procedures, which could then be used for illustrative purposes within the WATECO Guidance Document and the subsequent information sheets. Furthermore, they were instrumental in raising awareness of the economic issues of the WFD and the overall implementation process and delivered concrete examples. The pilot testing within the CIS process was undertaken as a multi-staged process: the first phase of testing in pilot river basins took place between 2002 and 2004, parallel to efforts developed in Member States for the preparation of the Article 5 report. It aimed to test and report on coherence among the different guidance documents developed within the CIS process, but did not yield clear lessons with regard to economics (CIS 2005). The second phase of testing (2005 to 2006) concentrated among others on providing guidance on the cost-effectiveness analysis (CIS 2008). During the ensuing third phase of pilot testing, the emphasis was on testing and gaining experience with proposed procedures and less on additional methodological development.

Today, economics receives less attention within the CIS process and its related activities. Since 2004, it has more and more been considered a cross-cutting issue which should not be treated in isolation but rather as a complementary aspect requiring due consideration at various stages of the implementation process. In 2009, main implementation concerns include the justification of exemptions from reaching the Directive's objectives on the basis of disproportionate costs (see Section 1.3.2 above). However, after the first implementation report of the EUROPEAN COMMISSION in 2007, adequate procedures and open issues regarding the assessment of cost-recovery for water services have also received further attention. In its first evaluation of the implementation process, the Commission emphasised that Member States' activities related to the implementation of Article 9 are considered insufficient and will require further work.

1.4 Article 9 and its multiple objectives

Article 9 of the WFD introduces a number of novelties to European environmental law and water policy. Most prominently, it establishes the principle of cost-recovery for water services, which may be considered a key element and distinct feature of the Directive (GRIMEAUD 2001:133, BOSENIUS 2001:31, HANSJÜRGENS AND MESSNER 2006:399, HÖDL 2005:93). Prior to the WFD, there had been no other European legislation on the environment with comparable prescriptions (UNNERSTALL 2005:3).[29] Due to its encompassing implications for the financial structure of water services provision within Member States, the introduction of the cost-recovery principle has been highly contested during the Directive's negotiation process: next to an economisation of environmental policy, Member States resented Community influence on the financial instruments employed at the sovereign level for reaching the Directive's environmental objectives.[30] Earlier draft versions of the Directive and the amendments made to them during the negotiation process through the EUROPEAN PARLIAMENT (1999, 2000a, 2000b, 2000c) and the European Council (1999) attest to the controversies surrounding the introduction of the principle of cost-recovery (including its extension to environmental and resource costs), the foreseen scope of application for the polluter-pays principle and the role of incentive pricing.[31] As a result of the political negotiation process, softer requirements have finally been established in Article 9 than were originally intended by some of its founding fathers (see Box 2).

Yet, the requirements of Article 9 extend far beyond earlier directives and secure a prominent role for economic considerations in the sustainable provision and financing of water services in Europe (STROSSER 2007, UNNERSTALL 2005, HANSJÜRGENS AND MESSNER 2006). In July 2000, a communication on water pricing was released by the EUROPEAN COMMISSION (2000), which can be considered an explanatory note to the Commission's intention behind Article 9. In its communication, the Commission highlights a number of building blocks that water pricing policies should incorporate in order to contribute to environmental policy objectives (EUROPEAN COMMISSION 2000). Namely, water pricing policies should aim to:

- Apply the principle of cost-recovery more rigorously;
- Extend the use of incentive pricing and increase the use of metering devices to this end;
- Take account of and internalise major environmental costs into prices;

29 Contrary to this viewpoint, REINHARDT (2006).

30 REINHARDT (2006) calls into question the interpretation that European legislation may prescribe Member States a particular financing structure for (national) water services provision, as done in Article 9, due to the subsidiarity principle. For a contradictory viewpoint, cf. UNNERSTALL (2005, 2006).

31 For a chronology and analysis of the changes made to Article 9 (respectively Article 12 of earlier drafts of the Directive), see *inter alia* UNNERSTALL (2005), KAIKA AND PAGE (2003), KAHLENBORN ET AL. (1999).

- Improve transparency of the water policy development process, including the participation of users and consumers; and
- Ensure a phased implementation of pricing policies that better integrate sound economic and environmental principles.

Box 2: Article 9 of the Water Framework Directive

1. Member States shall take account of the principle of **recovery of the costs of water services**, including **environmental and resource costs**, having regard to the economic analysis conducted according to Annex III, and in accordance in particular with the **polluter pays principle**.

Member States shall ensure by 2010

- That water-pricing policies provide **adequate incentives** for users to use water resources efficiently, and thereby contribute to the environmental objectives of this Directive,
- An **adequate contribution of the different water uses**, disaggregated into at least industry, households and agriculture, to the recovery of the costs of water services, based on the economic analysis conducted according to Annex III and taking account of the polluter pays principle.

Member States may in so doing **have regard to the social, environmental and economic effects of the recovery as well as the geographic and climatic conditions** of the region or regions affected.

2. Member States shall report in the river basin management plans on the planned steps towards implementing paragraph 1 which will contribute to achieving the environmental objectives of this Directive and on the contribution made by the various water uses to the recovery of the costs of water services.

3. Nothing in this Article shall prevent the funding of particular preventive or remedial measures in order to achieve the objectives of this Directive.

4. Member States shall not be in breach of this Directive if they decide **in accordance with established practices not to apply the provisions of paragraph 1**, second sentence, and for that purpose the relevant provisions of paragraph 2, for a given water-use activity, **where this does not compromise the purposes and the achievement of the objectives** of this Directive. Member States shall report the reasons for not fully applying paragraph 1, second sentence, in the river basin management plans.

Source: EC-Water Framework Directive, Article 9 (emphasis added).

With its communication, the European Commission reacted to the extensive room for interpretation that the formulation of Article 9 offers and presented its position on the objectives of Article 9. As the following analysis shows, Article 9 can be broken down into a number of subordinate objectives, which may not necessarily be reached effectively by the same instruments or policy responses. In the public debate on Article 9, its objectives are often narrowed down to the principle of cost-recovery, which is frequently equated with the requirement of incentive pricing, thereby mixing up two different underlying concepts. A dissection of Article 9 first delivers three explicit objectives (see also Figure 2):

1. Recovering financial and environmental and resource costs for water services;
2. Using water prices for setting incentives for efficient water resources use; and
3. Making the polluter pay and ensuring an adequate contribution of water uses.

Figure 2: The explicit and implicit objectives of Article 9

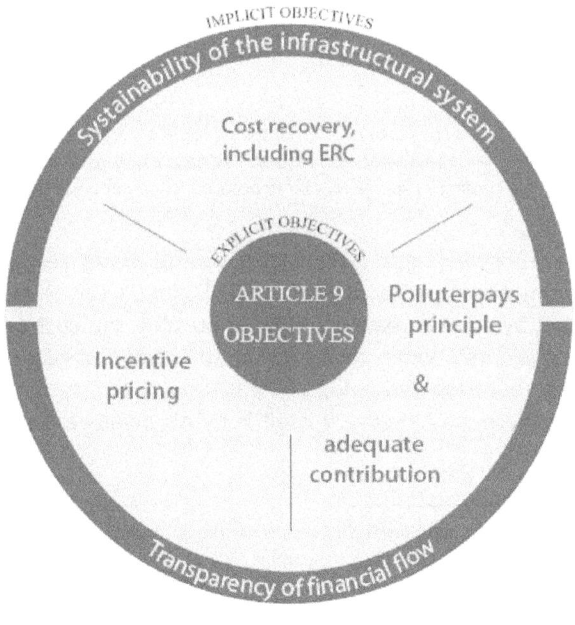

Furthermore, two additional underlying objectives can be discerned, which are not explicitly referred to in the Directive's text, but are necessary for fulfilling the role assigned to Article 9 within the overall WFD implementation process (MAESTU 2005), namely:

1. Establishing transparency of financial flows within the European water sector, and
2. Guaranteeing the sustainability of water services provision and infrastructural systems.

In the following paragraphs, the different objectives are briefly summarised in order to gain a rudimentary understanding of potential incongruence and related difficulties for the implementation process of Article 9. For a detailed assessment and discussion of the underlying theoretical concepts the reader is referred to Chapter 2.2 and Section 2.4.2.

1.4.1 Recovering financial and environmental and resource costs of water services

Establishing the recovery of financial and environmental and resource costs for water services provision constitutes the most prominent and first explicit objective of Article 9:[32]

> Article 9(1) *"Member States shall take account of the principle of* ***recovery of the costs of water services****, including* ***environmental and resource costs****, having regard to the economic analysis conducted according to Annex III, and in accordance in particular with the polluter pays principle."*

At present, cost-recovery rates for the provision of water services in Europe vary greatly among Member States (ROTH 2001, EUROPEAN COMMISSION 2000); this can be explained by various reasons. Most important, the cost categories and (sub-)services considered within such an assessment of cost-recovery vary considerably among countries, rendering comparisons of cost-recovery rates ineffective.[33] Furthermore, cost-recovery itself is by no means a clear-cut concept

32 Next to its prominent position within Article 9, the principle of cost-recovery for water services is reinforced in the Directive's preamble (38).

33 By the same logic, water price comparisons are flawed at the moment (see Chapter 2.5). An additional argument is linked to the varying interpretations of cost-recovery, as price comparisons are only meaningful if the same underlying cost-recovery interpretation is applied. Otherwise, low water prices in one region, which are enabled through a high financial contri-

(cf. Section 2.2.1). Evaluating Member States' understanding and interpretation of the concept reveals that cost-recovery is understood as a continuum of financing options, extending from a full recovery of costs from users/polluters to an entire financing of service provision through state taxes (cf. MASSARUTTO 2004A).[34] A closer analysis of this spectrum of alternatives for financing the costs of water services provision is undertaken in Chapter 2.2. For now, the assumption is made that Article 9 aims to internalise the full costs of water services provision into the decision-making process of users and polluters. Given this, cost-recovery aims to (i) integrate financial costs and opportunity costs of water services provision into water resource management and planning processes, setting the basis for economically rational decisions (i.e. costs as analytical instrument, EWRINGMANN 2006:24); (ii) establish sufficient financing for ensuring water services provision in the long-run (KRAEMER ET AL. 2003); and (iii) avoid competitive distortions within the Common Market (EUROPEAN COMMISSION 2000) through a varying application of the cost-recovery principle within the EU (see also Section 2.2.1).

In its cost-recovery requirement, Article 9 explicitly refers to three distinct cost categories that should be considered in the assessment: (1) financial costs of water services; (2) environmental costs and (3) resource costs linked to the provision of water services. However, no further guidance is provided as to what exactly should be included within these broad cost categories and how they should be accounted for, leaving this specification to the CIS process and ultimately to the Member States.[35] Most Member States have little experience with the measurement of environmental and resource costs. They were first introduced as a concept to European environmental legislation with WFD Article 9 (1) (BROUWER 2006). Accordingly, the degree of internalisation of environmental and resource costs into cost calculations varies greatly, both among Member States as well as among economic sectors. In comparison, the category of financial costs seems like a straightforward and well-established concept. However, significant differences may also arise in accounting for this cost category within a cost-recovery assessment (e.g. due to different procedures in accounting for capital costs or subsidies), as Chapter 2.2 demonstrates. Furthermore, the exact services included in the calculation of the category of

---

bution from general taxation, may be compared to higher prices in another region, in which no contributions from general fiscal budgets are present.

34 As investigated in detail in Chapter 2.1, how the concept of cost-recovery is interpreted and translated into financing strategies for water services is closely linked to a Member State's understanding and prioritisation among the different dimensions of water (i.e. it's social, economic, ecological and merit good characteristics).

35 The discussion of the definition and measurement of ERC is left to Section 2.4.2.

financial costs may also be different between Member States, rendering a comparison without a common assessment framework inherently difficult.[36]

1.4.2 Setting incentives for efficient water resource use

The second explicit objective of Article 9 highlights the incentive function of prices:

> *"Member States shall ensure by 2010 that water-pricing policies provide adequate incentives for users to use water resources efficiently, and thereby contribute to the environmental objectives of this Directive"* [...].

With this objective, water pricing is recognised as an important instrument for promoting ecologically sustainable water management by curbing water demand and water pollution. The emphasis of this objective is on avoiding the overuse of water resources and the depletion of nature capital.

There has been an ongoing debate among water policy experts, economists and policymakers alike over recent decades concerning the importance of water pricing as a suitable instrument for promoting sustainable water use (MASSARUTTO 2004A, ROGERS ET AL. 2002, DINAR 2000). With the second explicit objective of Article 9, the WFD considers prices in their incentive and information function (HANSJÜRGENS AND MESSNER 2002:67). The underlying assumption is that the price of water influences water demand (i.e. the amount of water used by different sectors, or the amount of pollution discharged) and that users will react to price increases by using water more efficiently, i.e. that water demand is to a certain degree elastic (cf. Chapter 2.3).[37] As a growing body of literature demonstrates, however, the evidence of price and income elasticities of water demand derived from empirical studies is scattered. Nevertheless, it allows a number of conclusions to be drawn about the incentive function of water prices (DALHUISEN ET AL. 2000:7, OECD 1998): first, the vast majority of studies indicates that both price and income elasticities of water demand are smaller than one (in absolute terms), but significantly different from zero.[38] Second, there is some evidence of

---

36 In principle, only those costs should be considered in the Article 9 assessment of cost-recovery, which are directly linked to the provision of a water service; costs encountered for tied-in services such as rainwater management or fire protection should therefore be excluded from the analysis (cf. Chapter 2.2 and 2.4 for a more detailed discussion).

37 The incentive effect depends on the shape of the demand curve: the more elastic the demand curve, the more effective the price signal (cf. Chapter 2.3 for an illustration).

38 DALHUISEN ET AL. (2000:7) highlight in their meta-analysis that *"most consumers are unaware of both the price of water and their water bill. Accordingly, the responsiveness to price*

water savings resulting from metering and charging by volume, or from tariff structure changes. Third, price elasticities seem to vary seasonally and with the particular use of water (e.g. in-house vs. ex-house demand, industrial vs. domestic demand, etc.).[39] Accordingly, the incentive function intended by Article 9 may play out differently for different water uses.[40] The economic rationale for water pricing and the different functions that water prices can fulfil in resource management as well as the necessary underlying assumptions for making it work are discussed in detail in Chapter 2.3.

In the public debate on Article 9, water pricing and cost-recovery are often used interchangeably, as if both concepts affect the same objectives. For a number of reasons, however, implementing a pricing policy that sets incentives for reducing the use of water resources can be in severe conflict with cost-recovery objectives. First, cost-recovery requires a constant flow of funds, while successful incentive pricing will lead to reduced water consumption and thus erode the revenue basis for the service provider. In consequence, the service provider has to increase prices further in order to refinance investments and meet cost-recovery objectives, leading to a vicious circle.[41] Second, with about eighty percent of costs being fixed in the water sector, a true representation of costs in prices is largely incompatible with an incentivising price structure. Third, incentive pricing requires that actual consumption can be monitored, i.e. individual metering is a prerequisite. This can, however, in itself be expensive and should therefore be balanced against the expected efficiency gains. Accordingly, objectives one (Section 1.4.1) and two of Article 9 may be severely at odds.

---

changes is puzzling, unless consumer reactions are enhanced by a growing awareness associated with, for example, information campaigns surrounding price reforms." Chapter 2.3 expands on the issue of water demand elasticities.

39 Dalhuisen et al. (2000) outline that estimating and comparing elasticities for water demand is linked to important difficulties. First, there is no general agreement on how the estimation should be done; while it seems well-established that estimation by ordinary least squares yields biased estimates, no generally agreed upon alternative is yet available. Secondly, studies differ in the detail in which various categories of demand have been distinguished (e.g. indoor and outdoor demand, the latter being more likely to be elastic than the former). Thirdly, studies may differ with respect to other aspects included in the analysis, e.g. location, time span, and variables that have been controlled for (e.g. temperature).

40 Besides low elasticities, the effectiveness of water pricing is limited by transaction costs, the public good components of water and potential perverse effects of pricing policies.

41 The situation can be further complicated by certain surrounding factors e.g. demographic change and depopulation leading to further reductions in consumption, stranded assets, etc. See Section 2.2.4 for a detailed discussion.

1.4.3 Making the polluter pay and ensuring adequate contributions by sectors

The third explicit objective of Article 9 gives a prominent role to the polluter-pays principle in the Directive and asks for an '*adequate contribution*' of water uses to the cost-recovery of water services (aka objective 1 above). According to Article 9 (1), Member States shall ensure by 2010:

> *"an **adequate contribution** of the different water uses, disaggregated into at least industry, households and agriculture, to the recovery of the costs of water services, based on the economic analysis conducted according to Annex III and taking account of the **polluter pays principle**."*

The polluter-pays principle constitutes a well-established principle of European environmental law.[42] However, with WFD Article 9 an unprecedented instrumental focus has been assigned to it for reaching the Directive's environmental objectives.[43] Whether the polluter-pays principle can actually develop into a potent instrument within the implementation process depends strongly upon the definition of water services and water uses. While the provision of water services should, according to objective 1 above, adhere to the principle of cost-recovery, the broader group of water uses only has to contribute adequately to this cost-recovery in line with the polluter-pays principle. Accordingly, environmental costs, which are not encountered due to a defined water service, are

---

42 The polluter-pays principle was first mentioned at the international level in a recommendation by the OECD Council in 1972 concerning international economic aspects of environmental policies. The recommendation stated that "*the principle to be used for allocating costs of pollution prevention and control measures to encourage rational use of scarce environmental resources and to avoid distortions in international trade and investment is the so-called Polluter-Pays Principle*" and amended that "*this principle means that the polluter should bear the expenses of carrying out the above-mentioned measures decided by public authorities to ensure that the environment is in an acceptable state.*" The polluter-pays principle was further included in Principle 16 of the Rio Declaration of 1992, which stated that "*national authorities should endeavor to promote the internalization of environmental costs and the use of economic instruments, taking into account the approach that the polluter should, in principle, bear the cost of pollution [...]*".

43 In addition to Article 9, the WFD Preamble refers twice to the polluter-pays principle, stating that "[...] *environmental damage should, as a priority, be rectified at source and **that the polluter should pay***" (WFD Preamble 11) Likewise, WFD Preamble (38) makes reference to the polluter-pays principle: "*The use of economic instruments by Member States may be appropriate as part of a Programme of Measures. The principle of recovery of the costs of water services, including environmental and resource costs associated with damage or negative impact on the aquatic environment should be taken into account in accordance with, in particular, the **polluter-pays principle**. An economic analysis of water services based on long-term forecasts of supply and demand for water in the river basin district will be necessary for this purpose.*"

not necessarily to be borne by their polluters; at least, this falls out of the scope of Article 9.[44] Hence, the exact definition of and differentiation between water services and water uses determines the relative weight assigned to the two objectives and outlines the scope of application of the polluter-pays principle (see Chapter 2.4 for a detailed investigation also of the potential *'opting out clauses'* from the polluter-pays principle through the qualifications included in Article 9 (3) and 9 (4). In general, the requirement of an *'adequate contribution'* is formulated vaguely within the Directive, necessitating further substantiation on what exactly may determine the adequacy of the different sectors and which indicators could be employed towards its measurement. This investigation is undertaken in Chapters 3.3 and 3.4).

Again, the objective of making the polluter pay may contradict other Article 9 objectives. The polluter-pays principle as described in Article 9 demands a water tariffication structure that reflects the true cost structure of provision (UNNERSTALL 2005, SCHMUTZER 2005). Accordingly, it contradicts the objective of establishing incentive pricing as outlined above, since the typical incentive tariff structure of a small fixed price component and a high variable part would contradict the (strict) implementation of the polluter-pays principle and equate to a cross-subsidisation between users. Furthermore, it is not a given, that meeting the first objective of recovering financial and environmental and resource costs coincides with this objective, as it may be easier to recover (environmental and resource) costs without strict adherence to the polluter-pays principle.

1.4.4 Establishing transparency of financial flows in the European water sector and guaranteeing the sustainability of infrastructural systems

Next to the three explicit objectives of Article 9 discussed above, two implicit overarching objectives can be discerned, which act like a harmonising parenthesis around the partly divergent explicit aims.

When examining the role of Article 9 within the WFD and the three associated objectives discussed above, it becomes apparent that one overarching aim of Article 9 is to achieve transparency of financial flows within the European water sector. At first instance, Article 9 is about making costs transparent – which costs are encountered in water services provision and which of these costs are so far left uncovered (i.e. environmental and resource costs or insufficient investment levels eroding the infrastructural system)? At present, there are significant differences in the definition and hence in the composition of cost

44 Member States are, however, free to employ the polluter-pays principle in a wider definition.

categories for water services provision across Member States (as illustrated in Figure 3 below).

Figure 3: Transparency of costs and financing mechanisms of water services provision and sustainability of infrastructural systems

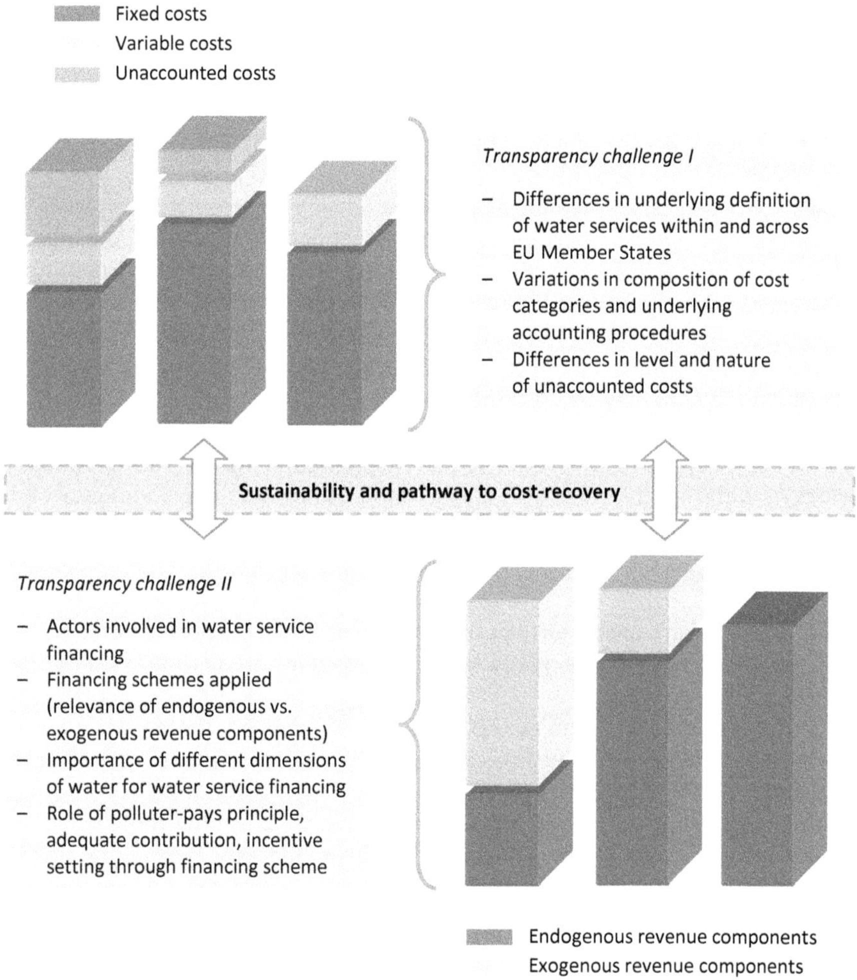

At second instance, it is about transparency in refinancing these costs, i.e. to which degree is cost-recovery achieved and by which pathway, how are costs allocated to different user groups, how does this compare to the actual contribution to the costs and in consequence, how far is the payment scheme in line with incentive pricing and the polluter-pays principle (see Figure 3)? On the financing side, different actors and institutional levels are involved in different Member States, rendering a coherent assessment of financial flows within the water sectors of different countries inherently difficult. Generally, (at least) the following actor groups can be distinguished in the water sector, namely the river basin (district) level, the owner of the resource, the state, the service provider and the water service user (cf. MASSARUTTO 2004A).[45] The degree to which these entities are involved in the financing of services may vary greatly.

The overarching objective of establishing transparency of financial flows in the water sector is a direct prerequisite for investigating the first objective of cost-recovery of financial and environmental and resource costs, as well as for the third objective of ensuring an adequate contribution of water uses and making the polluter pay. Without transparency of costs and financial flows, they cannot be addressed and implemented successfully.

Another overarching objective of Article 9 can be described as achieving sustainability of infrastructural systems. It is implied by the requirement of making the polluter pay and the approach of recovering all costs of water services provision. If one applies the definition of sustainability in terms of equity, efficiency and environment (BARRAQUÉ 1999) to the Article 9 definition of water services, three questions have to be answered (BARRAQUÉ 2003:204): first, how is capital maintained in the long-run, which includes an investigation of old subsidies, current depreciation schedules and financing schemes? Second, to which extent are new investments needed to improve the environmental and sanitary performance of existing systems in order to contribute to the environmental objectives of the Directive? And third, is it politically acceptable to pass full sustainability costs on to consumers, and can they afford them? A detailed discussion is deferred to Chapter 2.1 and Chapter 3.4.

1.4.5 Putting the objectives of Article 9 into perspective

The brief review of the multiple Article 9 objectives has already shown that some of these objectives may not necessarily be reached in parallel or with the same set of policy instruments. The in-depth investigation of the individual objectives undertaken in Chapter 2 presents further evidence on their dissonances. The following Figure 4 highlights the implementation requirements

45 The river basin district level is not explicitly included in MASSARUTTO (2004A).

emanating from Article 9 and its individual objectives as reference for the following discussions.

Figure 4: Article 9 objectives and associated requirements for the implementation process

Article 9 Objectives				
Transparency of financial flows	Sustainability of infrastructure	Cost-recovery, including ERC	Incentives for efficient resource use	Polluter pays & adequate sector contribution
Implicit overall objectives		**Explicit objectives**		
Reference in Article 9 (or CIS documents)				
CIS	CIS	Article 9 (1)	Article 9(1)	Article 9(1)
Implementation requirements				
Definition & analysis of cost categories and related revenues → *'ideal CR path'* and analysis of divergences therefrom Analysis of financial flows between actor levels	Set of indicators and guiding analytical questions for assessing the sustainability of the present framework for water service provision	Clearly defined cost and revenue categories, knowledge of historic financing approaches, information on and transparency regarding ERC	Information on the incentive properties of existing pricing systems, set of indicators and guiding analytical questions	Information on the existence and recovery of ERC (who pollutes and who pays), sectoral analysis of the contribution to the cost of water services provision
Actors involved				
State, authorities, providers, self-service producers, users, polluters				
Relevant scale				
European level (Community wide water sector)	Member State level (river basin (district) level)	Member State level and river basin (district)	Member State level and river basin (district)	Sectoral level and river basin (district) level

1.5 Research questions

The analysis above called to attention that the objectives of Article 9 are manifold and partly controversial. Accordingly, there will not be one clear-cut implementation procedure for achieving all of them simultaneously with a prescribed set of instruments, approaches and principles. Rather, national differences in water management practices, societal preferences and localised experiences with economics in water resources management will influence the relative emphasis given to the different objectives within the implementation process of Article 9 in European Member States. Comparing implementation procedures and ultimately compliance with the Directive will thus necessitate an assessment framework that allows for mapping out the objectives' respective degrees of implementation.

This thesis therefore proposes a comparative framework for assessing cost-recovery of water supply and wastewater services for private households and agriculture under different institutional, geographical, economic, legal and social-cultural conditions. Optimal data reporting rules are developed together with a set of indicators and an interrogation scheme necessary for a comprehensive and meaningful comparison of cost-recovery and ultimately prices of water services within different national settings. The assessment framework is set up with a view to delivering a practicable assessment tool that is able to expand over the years as more reliable data become available.

The approach followed is both theoretical and empirical, although the emphasis lies on the development of the theoretical assessment framework. The study addresses the following **questions**: How can the partly contradictory objectives of WFD Article 9 be translated into operational objectives? Which framework conditions are decisive for a comparative assessment of cost-recovery in different Member States and through which indicators can they be effectively measured? Which cost factors should be covered through water prices and sewerage charges (or other sources) in order to ensure long-term sustainability of service provision? How are water services financed in different countries, and in how far are different financing schemes in line with the objectives and requirements of Article 9? Will improved system sustainability be reached through improved transparency and rationalised decision-making? And finally, the question is answered of how the implementation of Article 9 and the success of economic approaches to water resources management can be compared across European Member States.

In addition to facilitating transparency of financial flows within the European water sector, the comparative assessment framework may serve multiple purposes. First, it provides a basis for evaluating the compliance of Member States with Article 9 of the Directive. Second, it offers customers of water services a tool for assessing how their services are being financed, what role is assumed by the polluter-pays principle, whether the present burden-sharing is in line with societal preferences and whether water services are affordable to all. Third, it may be used by responsible authorities for evaluating whether the provision of water services is sustainable in the long-run.

The analysis conducted in the course of this thesis strives to present new insights on the analysis of financing schemes for water services in different European Member States, thereby delivering a tool for increasing transparency of financial flows in the European water sector. It aims to provide theoretical and illustrative empirical evidence on cost-recovery of water services provision and contributes to setting the basis for meaningful comparisons of water prices across Europe.

1.6 Methodology and scope of the thesis

In order to answer the research questions, first an overview on theoretical considerations of cost-recovery and pricing of household and agricultural water services is provided. Next, the theoretical framework is developed together with a complementary set of indicators for the integrative assessment of cost-recovery, which is then applied to three illustrative examples for demonstration and validation purposes.

In order to establish a comparison of cost-recovery across Europe, two different scales are of interest for the analysis: first the national scale needs to be investigated, as the Member State has ratified the Directive and the institution and management of economic instruments is under each Member State's own responsibility. Second, the river basin or river basin district scale has to be studied, as the hydrological entity constitutes the reference frame for the Directive's overall implementation process. Accordingly, the empirical illustrations address both the national level and the river basin district level. A number of criteria were considered in selecting the illustration cases. For testing the framework's applicability and practicality, the cases were intended to prominently feature those issues which constituted the motivation for including Article 9 in the Directive, namely low rates of cost-recovery in agriculture and uncovered environmental and resource costs in water services provision in general. Although the Directive explicitly refers to three sectors in Article 9, namely, households, industry and agriculture, this study focuses on water services provided to households and agriculture to allow for an in-depth investigation. In principle, however, the methodology is conceptualised with a view to transferring its application to the industrial sector as well.

The first set of stepwise illustrative examples investigates the applicability of the assessment framework to the household sector and addresses water supply and sewerage services provision in Germany. Germany has been selected for this illustration because the principle of cost-recovery has a long-standing tradition in German water legislation. Furthermore, as water management falls in the responsibility of the German *Länder*, the problems encountered due to the Member State's federal structure may serve as an approximation of some of the harmonisation problems encountered in Europe-wide assessments (i.e. linked to problems of different accounting systems, different financing schemes, different (historical) subsidisation procedures etc.). In addition, the German case allows for addressing particularities of both Eastern and Western European historical developments and financing schemes for water services provision. For a regional illustration of households, the Emscher river basin has been selected. In the Emscher river basin, municipal discharges still constitute one of the main anthropogenic sources of pressures on water status. As the Emscher river was long used as a canal for open discharge of effluent due to mining activities within the region with severe consequences for the aquatic environment, it provides a particularly interesting case for validating the framework for wastewater services provision to households and for studying the role of environmental and resource costs. Most important, the Emscher case study also covers a number of other typical problems for an assessment of cost-recovery, namely multiple polluters which cannot be readily identified, old as well as new subsidisation schemes and most evidently environmental and resource costs. Furthermore, the provision of water supply services in the capital Berlin is drawn upon for illustrative purposes.

The second set of illustrative stepwise examples focuses on the agricultural water sector in Italy and more specifically on the region Emilia-Romagna, which forms part of the Po river basin in the north of Italy. In contrast to the German case, water scarcity plays an important role in water management in Italy, and agricultural water use is significant both in terms of quantity and quality. In addition to divergent natural conditions, the historic approaches to financing water services have also differed considerably between the two countries, with water prices playing a less important role in Italy, notably in its agricultural sector. Accordingly, different adaptation paths will be necessary for complying with the requirements of Article 9. Applying the assessment framework to the Italian agricultural sector may serve as an exemplification of problems encountered with the implementation of Article 9 in other southern European Member States as well, from data difficulties to different surrounding socio-cultural, institutional, legal and economic conditions.

In order to complete the knowledge base, the study draws next to official statistics and publications on interviews with representatives from ministries at national, state and river basin levels as well as on unpublished grey literature compiled in the course of the WFD implementation process.

While the empirical illustrations form an integral part of this study, it should be noted that the study aims mainly at delivering new insights by developing an appropriate methodological framework for assessing cost-recovery comparatively across Member States. It is intended to deliver an appropriate assessment tool for facilitating the WFD implementation process across Member States.

1.7 Plan of the thesis

This study is organised in five chapters (see Figure 5). *Chapter 2* first gives an overview of how the perception of water in its different good dimensions may influence the organisation of the provision and financing structure of water services. It investigates different interpretations of the concept of cost-recovery, both from a theoretical perspective as well as with a view to present practices in EU Member States, and evaluates these in light of the requirements of Article 9. Relevant cost categories for water services provision to households and agriculture are examined and possible cost drivers and framework conditions, which may influence the magnitude of these costs both across regions as well as over time, are discussed. Finally, water pricing schemes are contrasted with other potential financing mechanisms and evaluated in view of the requirements and objectives of Article 9. Finally, present difficulties for pan-European comparisons of cost-recovery and water pricing are discussed.

Figure 5: Plan of the thesis

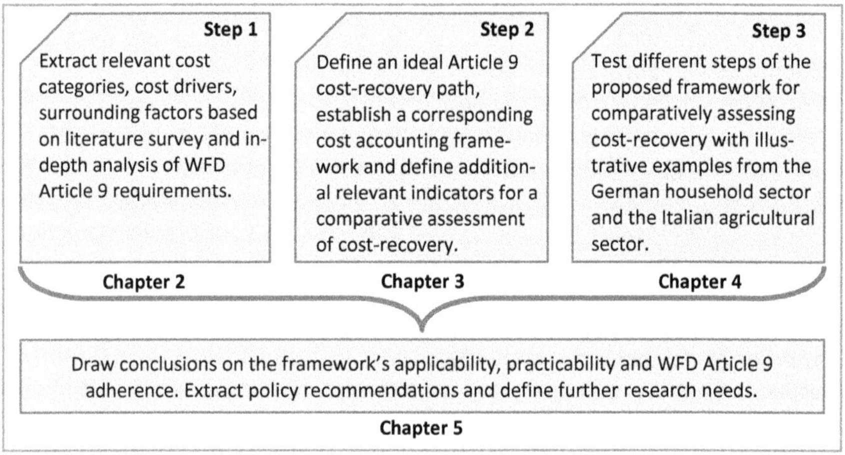

Chapter 3 develops the comparative framework for assessing cost-recovery in keeping with the requirements of Article 9. To this end, definitions of relevant cost and revenue categories, potential drivers and influencing framework conditions are provided along with a set of complementary indicators and an interrogation scheme required for a comparative assessment of the interplay of the different Article 9 objectives. Together with a water services accounting model, an ideal Article 9 cost-recovery path can then be mapped out, rendering transparent divergences from this path which would necessitate justification for noncompliance with Article 9.

Chapter 4 applies different steps of the assessment framework in illustrative examples, based on the household sector in Germany and the agricultural sector in Italy and draws conclusions regarding the framework's adaptability to national particularities and its practicability. Building on the results of the empirical illustration cases, *Chapter 5* concludes the study with policy recommendations and an outlook on potential ongoing research.

2 Theoretical considerations on cost-recovery and pricing of water services

The virtue of economics is not that it brings some better answers but some more useful questions.
Green (2000)

The second chapter first provides an overview on how the perception of water in its different good dimensions may influence the organisation of the provision and the financing structure of water services. It investigates different interpretations of the concept of cost-recovery, both from a theoretical perspective as well as with a view to present practices in EU Member States, and evaluates these in light of the requirements of Article 9. Relevant cost categories for water services provision to households and agriculture are examined, and possible cost drivers and framework conditions, which may influence the magnitude of these costs both across regions as well as over time, are discussed. Subsequently, the chapter investigates different water pricing schemes as potential financing mechanisms and discusses these in view of the requirements and objectives of Article 9. Finally, it outlines present difficulties linked to European comparisons of cost-recovery and water pricing.

2.1 Water and its dimensions: implications for water services provision

Water, considered here to be a renewable resource[46], has a number of distinct features that make it fundamentally different from other resources and prevent a simplistic approach towards its management. It has long been acknowledged in the literature that water has different dimensions (ROGERS ET AL. 2002, 1998, SAVENIJE 2000, SMETS 2000, HOEKSTRA 1998, PERRY ET AL. 1997). First, water is recognised as an *environmental good*, i.e. a localised, fugitive physical resource, providing important ecosystem services and functions (OPSCHOOR 2006, SAVENIJE 2002). In its second dimension, water is classified as a *social good* that is non-

46 In contrast to surface water, groundwater may have characteristics of a non-renewable resource. As DALY AND FARLEY (2004) point out, water in fossil aquifers qualifies a non-renewable resource similar to fossil fuels with fewer externalities.

substitutable, essential and often scarce (SAVENIJE 2002). In this function, water plays an important role in public health and in combating epidemics (SMETS 2000). The resulting classification of water as a basic human need raises important social aspects in terms of its availability and provision. Fresh water has historically been considered *"a gift to man coming from the sky"* (SAVENIJE 2002), raising issues of fairness and affordability (SMETS 2000). Furthermore, water also has a cultural dimension: it is one of the four elements of antiquity and an important symbol in religions worldwide. In its third dimension, water is recognised as an *economic good*, which has *"an economic value in all its competing uses"* (UNECE 1992).[47]

Figure 6: Different Dimension of Water

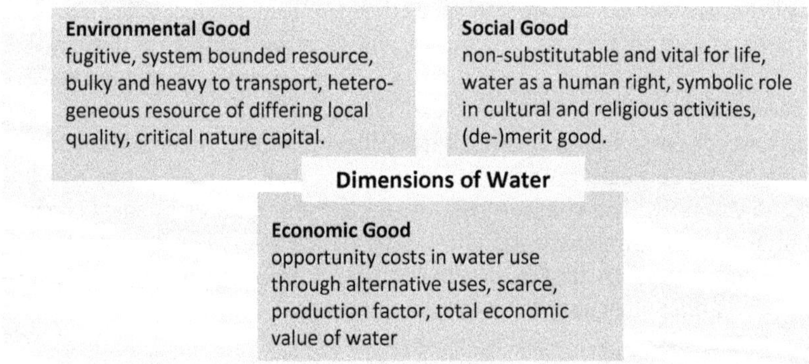

While a separate discussion of the dimensions (cf. Figure 6) is suited for a first introduction, their interdependences and contradictions need to be studied and acknowledged for the purpose of this study. Depending on the relative importance assigned to the three dimensions of water, a different instrument mix may best support their respective objectives. Ultimately, it is a political decision which dimension features most strongly in a given water policy arena and thus how much emphasis is given to economic instruments or affordability considerations in the provision of water services. As the following analysis shows, the perception and resulting integration of water's different dimensions

47 More recently, water has been classified as a political good (SCHOUTEN AND SCHWARTZ 2006) due to the political influence on its provision. However, as OPSCHOOR (2006:426) points out *"we do not really need new concepts such as 'political goods' to present the predicaments of the water sector or investments in it."* In line with this viewpoint, this classification of water is not studied further in this thesis.

has important implications for the objectives assigned to water pricing and the financing of water services.

2.1.1 Water as an environmental good

Water is a natural resource of heterogeneous quality determined by its natural surroundings and regarded as critical natural capital.[48] Although water is moving, it is restricted by its hydrological boundaries and its location of origin and thus qualifies as a localised resource (SAVENIJE 2002). Through the natural water cycle from rainfall to runoff, water is framed as a complex system consisting of several interconnected and interdependent processes (from infiltration, surface runoff, recharge, seepage, re-infiltration to moisture recycling) (SAVENIJE 2002). The natural availability of water varies both over the short and the long term as a result of climatic factors and changes in land use (GREEN 2000). In comparison to other natural resources such as gas, water is bulky and heavy, making it difficult and costly to transport over long distances (GIBBONS 1986, EWERS ET AL. 2001).[49]

In terms of water services provision, the characterisation of water as an environmental good implies:

- Exogenously determined water availability and supply conditions (GREEN 2000) and thus a need for storage capacities to compensate for fluctuations in the natural water supply (SAUER 2005);
- High costs for transporting or conveying water, which has itself a low unit value (GIBBONS 1986);
- Physical and technical problems regarding the joint transportation of water of differing quality (SAUER 2005, BRACKEMANN ET AL. 2000);

48 Critical natural capital is generally understood to be a part of the natural environment that performs important and irreplaceable functions (CHIESURA AND DE GROOT 2003). In the literature, critical natural capital is described in terms of essential environmental functions for human wellbeing (DE GROOT 1992), in terms of ecosystem services or source and sink functions (PEARCE AND TURNER 1990). For assessing the sustainability of water use, the local scale constitutes the relevant reference frame, where all relevant ecosystem services and environmental functions should be preserved (EKINS 2000, FAUCHEAUX AND O'CONNOR 1998, TURNER AND DUBOURG 1993).

49 There is a long history of moving water from regions with abundant water endowments to regions with water scarcity. DE BODISCO (2007:16) outlines examples ranging from Roman aqueducts in southern Europe; the Incas in Peru, who were moving water from the mountains to the valleys; to Californian water transfers from the Colorado River to the city of Los Angeles.

- Potential for multiple complementary or conflicting productive and consumptive uses (GRIMBLE 1999, YOUNG AND HAVEMAN 1985, SAUER 2005);
- High water pollution potential through water uses, necessitating costly treatment (GRIMBLE 1999).

The perception of water as an environmental good, which falls from the sky, leads to the second dimension of water.

2.1.2 Water as a social good

Historically, water has been considered a social good, whose availability enabled the development of human settlements and economic development. Notably, from the Roman period onwards, water supply networks were erected to ensure a safe water supply to villages without direct access to communal wells or rivers (SMETS 2000). Safe access to drinking water and sanitation was considered key to economic development and came to be regarded as a human right.[50] Today, it is generally acknowledged and embedded in a large number of international agreements on sustainable development that safe and sufficient access to drinking water has to be ensured for all segments of society.[51] Most prominently perhaps, the Dublin Statement, adopted at the International Conference on Water and the Environment in 1992 (hereafter: Dublin Conference), made explicit reference to water as a social good by stating that *"it is vital to recognise first the basic right of all human beings to have access to clean water and sanitation at an affordable price"*.[52] Generally, the definition of water as a human right is restricted to domestic water supply and sanitation services and excludes agricultural water services (CORNISH ET AL. 2004A).[53]

In terms of the provision of water services, the social dimension of water raises the issues of affordability, equity and fairness. In this understanding, socially relevant water services should be affordable to every member of

50 For a chronological discussion of the perception of water as a human right in international water politics, cf. SMETS (2000).

51 Cf. SMETS (2000:134) for an overview of international agreements and treaties that consider water a basic human right.

52 Similarly the UN Conference on Environment and Development in Rio de Janeiro in 1992 acknowledged water's social dimension. The *'Agenda 21'* proceedings of the conference make several references to the social dimension of water e.g. in § 6.12 "*to ensure universal access to safe drinking water*".

53 It can, however, be questioned whether water use for subsistence farming should fall under this classification.

society, regardless of income levels and social conditions.[54] A number of policy instruments can be drawn upon for promoting equitable access to water services, such as the socialisation of the costs of providing access to remote rural areas (i.e. through geographical cross-subsidies), socially adapted water tariffication schemes (e.g. with special tariffs for vulnerable groups or supplying a small quantity of water for free), or social assistance payments as well as legal measures safeguarding access to water (e.g. by prohibiting water supply cuts without an enforceable judgement) (SMETS 2000). These policies may, however, contradict other policy objectives on the use of water (see below). Most notably, "*keeping tariffs artificially low for all customers may result in a vicious circle of underfunded service providers, insufficient investment, collapsing infrastructure and deteriorating services that further reduce the benefits that users receive from them and therefore their willingness to pay*" (OECD 2009).

In addition to the perception of water as a human right, the social dimension of water also comprises *symbolic, cultural and religious* connotations of water use, which are culturally determined (GREEN 1998A IN GREEN 2000:207). Today, it is widely acknowledged that these social and cultural influences surrounding the use and perception of water have an important impact on the instrument mix chosen for water management. For example, using the price mechanism for allocating water to different uses can contradict "*prevailing social norms that govern the entitlements and obligations understood to exist between members of society*" and may therefore be confronted with strong disapproval, or necessitate a radical revision of these underlying norms (GREEN 2000:207).[55]

Another facet of the social dimension is water's important role in public hygiene. In this context, water can be considered a (de-)merit good[56]: consumers would not necessarily invest sufficiently in water supply and sanitation services, due to informational and risk preference considerations as well as intertemporal and intergenerational aspects (JAMES ET AL. 1978). Then the public benefit would not be taken adequately into consideration in the individual's decision-making process, resulting in sub-optimal provision or consumption of water services.

---

54 OECD (2003) suggests that a critical limit in terms of affordability of water prices is reached when households have to spend more than three percent of their disposable income on their water bill.

55 On the religious element of water BOULDING (1980:302) noted that "*the sacredness of water as a symbol of ritual purity exempts it in some degree from the dirty rationality of the market*".

56 (De-)merit goods are goods (public or private) that are to be consumed at (lower or) higher levels than consumer sovereignty would entail (EDELMAN AND OPSCHOOR 1970 in OPSCHOOR 2006:425).

While the social dimension of water still figures strongly in the political debate, the idea that the provision of drinking water is a public service that is virtually free or at any rate highly subsidised is tending to give way to the idea that water is an economic good.

2.1.3 Water as an economic good

The Dublin Conference (1992) ascertained the idea of water as an economic good by stating in its fourth principle on sustainable water management that *"water has a value in all its competing uses and should be recognised as an economic good"*. This principle can be considered a landmark shift towards emphasising the economic dimensions of water use in general and in agriculture in particular (Molle and Berkhoff 2007). The economic value of natural resources further found legitimacy in the Rio Declaration on Environment and Development of the United Nations in 1992 and its Agenda 21 (United Nations 1992), which supported the "*implementation of allocation decisions through demand management, pricing mechanisms and regulatory measures*". Earlier, the OECD had defined water as an *'economic good'* on the grounds that (scarce) economic resources (human, capital, knowledge) have to be applied to the natural resource in order to make water available in the form, quality, location and time demanded and needed by users (OECD 1987:18,129). In this understanding, water qualifies as an economic good, as it is scarce and serves as a production factor.[57] While this fact is generally acknowledged, there is a debate on its proper bearing for water policy formulation. While many economists claim that understanding water as an economic good implies that it should not be treated differently from any other economic good (e.g. Briscoe 1996), water professionals generally maintain that although water qualifies as an economic good, its distinct character still warrants special attention in the formulation of adequate policy responses to water management problems (Savenije 2002, McNeill 1998, Perry et al. 1997).

Generally, two schools of thought can be distinguished in the debate on water as an economic good (van der Zaag and Savinije 2000:51). The first view, held e.g. by the World Bank[58] (see Briscoe 1996, 1995 as an example), maintains

57 However, it must be noted that water is not always scarce, e.g. in the majority of regions in Germany or northern Europe. Accordingly, Ward and Michelsen (2002) claim that water has an economic value only when its supply is scarce relative to its demand.

58 In recent years, international organisations such as the World Bank have moved away from their earlier strong belief that *'full cost-recovery'* could bring about equitable and sustainable

that water should be priced at its economic value in order to allow the market to allocate water to those uses in which it yields the highest total economic return.[59] The second view interprets '*water as an economic good*' as the process of integrated decision-making on the allocation of scarce resources, without necessarily involving financial transactions (McNEILL 1998).[60] Rather, the concept of water as an economic good is interpreted as necessitating decisions on the allocation and use of water to be based on a multi-sectoral, multi-interest and multi-objective analysis in a broad societal context, involving social, economic and ethical considerations (HELLEGERS AND PERRY 2004, VAN DER ZAAG AND SAVINIJE 2000). Thus, while this view acknowledges water as an economic good, it recognises the relevance of its other good dimensions, too, and asks for their integration into the decision-making process. GREEN (2000) articulated this view by explaining that economics is about the application of reason to choice, i.e. about making the right choices about the allocation and use of water resources on the basis of an integrated analysis of all the advantages and disadvantages of alternative options.

For both viewpoints, it is important to note that the concept of economic value entails many complex dimensions. While no agreement has been reached on a taxonomy for the total economic value[61] (TEV) of natural resources, it is generally regarded to be of a composite nature (see Figure 7) consisting of two broad categories of value, namely use and non-use values (MARKANDYA ET AL. 2002).

---

service delivery and an optimal distribution of water among competing demands in all cases (CORNISH ET AL. 2004A).

59 This first view is termed the '*Panglossian approach*' by GREEN (2000:205). Under this approach, "water as an economic good" means managing water as a purely private good through markets and economic instruments.

60 This second view is termed the '*Pragmatists approach*' by GREEN (2000:205): Pragmatists interpret the statement "water as an economic good" to mean that because water is scarce it must be recognised that any one use means that an alternative potential use must be sacrificed (cf. the discussion of resource costs below). But they argue that economic efficiency is not the only objective to be satisfied, and markets, together with economic instruments, are only one option among the greater toolkit of policy instruments available.

61 As MARKANDYA ET AL. (2002) highlights "*the use of the term total economic value is unfortunate, because within it uses the word total in a different way from that commonly used in economics. What is being sought here is the full value of an environmental asset or service. This may be for a change in the amount of the asset or for the whole asset. It does not imply that the valuation is for the asset in its totality*" (emphasis added).

Figure 7: *Total economic value and its composition*

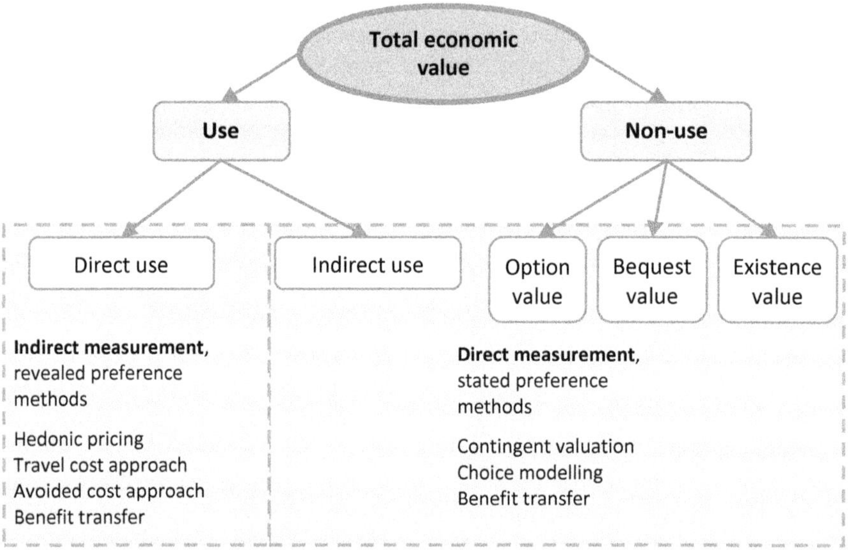

Source: based on Markandya et al. (2002) (cf. also Box 3 below on measurement).

Use values are defined as values derived by individuals from the actual use of the environment (BATEMAN AND TURNER 1992). They can again be divided into two categories, namely, direct use values (e.g. water used for irrigation purposes) and indirect use values (e.g. erosion control). While the concept of use values is relatively well-accepted among researchers in the field, the definition, and in particular the measurement of non-use values, had been subject to intensive debate, most notably in the early nineties (cf. BATE 1994, DESVOUSGES ET AL. 1993B, DIAMOND AND HAUSMAN 1993, QUIGGIN 1993, RANDALL 1993, SMITH 1993, EDWARDS 1992, KAHNEMAN AND KNETSCH 1992, KOPP 1992, LARSON 1992, LAZO ET AL. 1992, ROSENTHAL AND NELSON 1992). Non-use values (sometimes referred to as '*passive-use*' values) are defined as values that do not depend on any direct or indirect interaction with the resource or good in question[62] (TURNER 1999, PORTNEY 1994). They are generally classified into three main categories, namely, option value, bequest value and existence value. Box A2.1 in the Annex gives a brief overview on the different value categories.

62 LOOMIS (1996) points out that non-use values have a public good character due to the fact that their benefits can be enjoyed by millions of people simultaneously over an unlimited geographical area.

The sum of the economic values that an individual is ready to attribute to water is summarised by the concept of willingness-to-pay. That is the maximum amount of alternative welfare opportunities that a person would theoretically accept to pay – or the maximum amount of alternative welfare opportunities that he would be theoretically ready to sacrifice in order to have water available while remaining better off. Willingness-to-pay analyses are, however, linked to important analytical difficulties (HOROWITZ AND MCCONNELL 2002, SHOGREN ET AL. 1994, BOYCE ET AL. 1992, HANEMANN 1991, BROOKSHIRE AND COURSEY 1987, KNETSCH AND SINDEN 1984, RANDALL AND STOLL 1980, ROWE ET AL. 1980). A vast array of studies has shown that the willingness-to-pay for household water services may diverge significantly from the value of those services (GREEN 2000:207, cf. Chapter 2.4).

2.1.4 The market structure for the provision of a multi-faceted good

As the discussions above have shown, water is a multi-faceted good and is characterised by different dimensions. The interplay of these dimensions leads to or amplifies a number of market failures in water services provision. The following paragraphs review the extent to which water resources and water services (1) can be regarded as market goods, which are excludable and rival; (2) qualify as natural monopolies; and (3) are linked to externalities in their provision and are subject to information asymmetries.

The limits on excludability and rivalry of water resources and water service

Two criteria, which characterise the nature of the demand for goods and services, are generally applied in economics for determining the appropriate market structure for providing a certain good or service, namely whether it is excludable and whether it is rival in consumption:

- An *excludable good* is one for which exclusive ownership is possible, i.e. for which the owner is able to establish his *property rights* and prevent others from using it, if so desired (DALY AND FARLEY 2004). If this criterion is not met and a good or service is not owned exclusively by someone, it will not be efficiently allocated or produced by market forces.
- A *non-rival good* is one for which the use by one individual does not reduce the amount available or the benefit obtained from making use of the good to other individuals (GREEN 2003). Rivalry can occur in qualitative, quantitative or spatial terms (DALY AND FARLEY 2004)

When discussing the degree of excludability or rivalry of water (cf. Figure 8), it is important to clearly define the good or service under investigation: different conclusions will be drawn for finite fossil groundwater resources (non-renewable stock resources), ecosystem services provided by water resources (e.g. the climate-regulating function or the self-cleaning capacity of water), freshwater resources from small aquifers or the sea. For water-related services such as water supply, wastewater management and irrigation, the classification depends on the type of water resources used and the local use conditions (e.g. scarcity or abundance, seasonal and quality variations), as well as on the nature of supply (individual / small scale supply or collective network supply[63]).

Figure 8: Classifying goods and services: excludability and rivalry

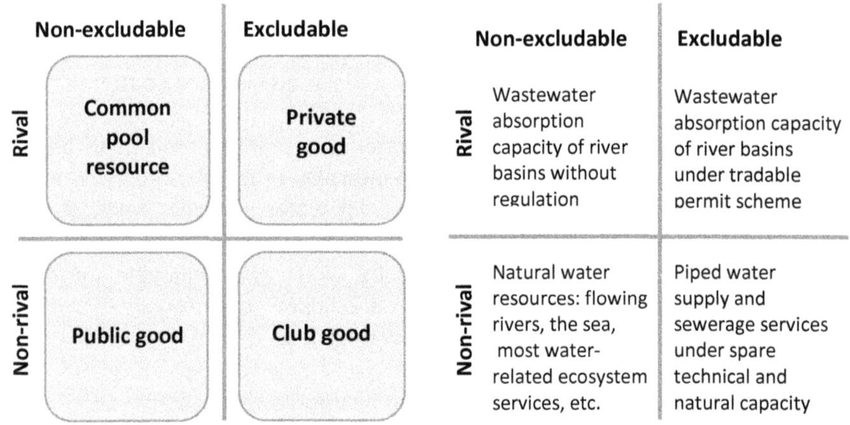

Source: amended based on Daly and Farley (2004), Green (2003), Liu et al. (2003).

In general, natural water resources display a low degree of *excludability*, as they have no regard to administrative or private boundaries and therefore cannot be owned privately (DALHUISEN ET AL. 2000). For the provision of water services, however, the non-excludability of natural water resources can often be circumvented through appropriate institutional arrangements (DALY AND FARLEY 2004). For example, in the case of a collective water supply, water suppliers are

63 Cf. Annex, Figure A2.1: GREEN (2003) studied the impact of the supply structure on the classification of water related goods and services along the two criteria of excludability and rivalness.

often assigned the exclusive right to use groundwater bodies[64] within water protection zones, and access to the water supply or sanitation networks is governed through contracts between water service providers and customers.[65] In the case of wastewater disposal, (tradable) pollution permits can turn the naturally non-excludable but rival pollution absorption capacity of a river basin from a common-pool resource into a private good (cf. Figure 8). In other cases, it is impossible to design institutions which would make the use of water resources exclusive to an individual, for example in the case of water-related ecosystem services.[66] Excludability is thus not a property of water per se, but rather of the regime that is established for managing the access to and the use of water resources. However, in order to successfully establish excludability, the institutional arrangements must be enforceable and regulation must be effective.[67]

As there is a finite amount of water resources available in a certain period of time that must be shared in common over a variety of uses and geographical areas, water resources and water services are subject to *rivalry* for consumptive uses (DALHUISEN ET AL. 2000).[68] Similarly, the waste absorption capacity of receiving waters constitutes a rival good, as e.g. the disposal of sewage in a wetland of one service provider leaves less capacity subsequently for that wetland to process another service provider's waste. Non-consumptive uses of water

---

64 Other uses are generally prohibited and controlled for in close proximity to the spring source. However, even in this case, the assignment of property rights is often only of a temporal nature and linked to protection obligations.

65 Due to the non-substitutability and the high societal dependence on water, granting exclusive property rights to natural water resources is by many regarded as against the public interest (cf. social dimension of water) (this circumstance is termed *'socially determined non-excludability'* by HARRIS ET AL. (2002) in reference to the water sector). The provision and consumption of e.g. drinking water are often made legally mandatory. However, as service disconnection is theoretically possible, service providers could act upon payment arrears of customers by cutting services and thus exclude individuals from service provision. Whether a land owner possesses the right to water resources on his land is also a matter of legal / institutional arrangements. In most European Member States, restrictions are imposed on private water abstraction or sewage disposal and licences or permits have to be obtained for uses extending beyond a minimum threshold. Often, these allowances can also be withdrawn if required by the public interest, e.g. in times of water stress.

66 E.g. granting exclusive rights on the micro-climatic regulating function of water is not a viable option.

67 E.g. licensing or permit systems must be accepted by users of the resource. If all users accept that a permit is needed for abstracting water (putting a boat on a river), water abstraction rates (access of boats to the river) can be controlled, and overabstraction (congestion, negative impacts on flora and fauna) can be avoided.

68 As long as rates of consumption/withdrawal remain below the natural rate of recharge or regeneration, water resources remain non-rival between generations (DALY AND FARLEY 2004).

resources are in quantitative terms non-rival, as more than one consumer can use the same water (DE BODISCO 2007). Recreational and environmental uses of in-stream flows are frequently cited examples (SALIBA AND BUSH 1987). However, these water uses may be subject to rivalry in qualitative terms, as the benefits received from using the resource may be decreasing: rivers and lakes congest, if too many uses are taking place simultaneously (e.g. too many boats on a lake or anglers along a river). Fixing a hierarchy of consumptive water uses in water-stressed times[69] and handing out licenses or permits for different (consumptive and non-consumptive) water uses represent institutional mechanisms for addressing rivalry among uses.

For water services under collective provision, network congestion may occur under full system capacity, leading to rivalry in access and consumption. Under spare (natural and technical) system capacity, water services are virtually non-rival, as access and consumption by other consumers or water uses is not affected by an additional customer (cf. Section 2.1.4.2). Under individual provision, e.g. of drinking water, the degree of rivalry depends on the point of analysis: if an individual consumes a litre of water, it cannot simultaneously be enjoyed by someone else. If, however, one individual abstracts water from an aquifer for his private water supply without significantly impacting the aquifer's quantitative and qualitative status, this water use is non-rival to other uses of the same aquifer. As more neighbours use the same aquifer for drinking water or irrigation self-supply, the joined uses may, however, lead to negative quantitative or qualitative impacts and the individual uses may become rivals to one another; not everyone can have as much as he would like (GRIMBLE 1999).

The following Table 2 summarises the discussion on the classification of water supply, sanitation and irrigation services under collective or individual provision and full or spare system capacity.

Water services as natural monopolies

Network based water services, such as piped water supply and sanitation or canal irrigation services, exhibit characteristics of natural monopolies (DIAKITE ET AL. 2009, BALANCE AND TAYLOR 2005, DALY AND FARLEY 2004, EWERS ET AL. 2001, SCHEELE 2000, KIM AND LEE 1998, BHATTACHARYYA ET AL. 1995, YOUNG AND HAVEMANN

69 Generally, water supply is assigned highest priority, followed by power plant cooling, industrial uses or agriculture, depending on the sectors' relative political and economic importance. Recreational uses of water resources generally have the lowest priority, except for regions where water-related tourism plays an important role for the regional economy.

Table 2: Classifying water services under full and spare system capacity

	Individual supply – private demand WS_1: Self-supply of water through private wells / small-scale decentralised systems WS_2: Sanitation through cesspits / small-scale decentralised systems WS_3: Private access to a well / pond / river for small-scale irrigation	**Collective supply – public demand** WS_1: Piped water supply network WS_2: Piped sanitation system WS_3: Canal irrigation system
Full system capacity used	– Rivalry[1] in demand with other local water uses and other self-suppliers / private irrigators[3] – If exclusive property rights can be established: excludability given (e.g. fenced-in well on own land) → private good – If no exclusive property rights: non-excludable → common pool resource	– Rivalry[2] in access and consumption – If excludability granted through institutional arrangements → private (but *'network'*) good – If excludability cannot be effectively established through institutions or is not aimed at: non-excludable → common pool resource
Spare system capacity	– Non-rival, access and consumption of other consumers/ water uses not affected – If exclusive property rights can be established: excludability given (e.g. fenced-in well/ cesspit on own land) → club good – If no exclusive property rights: non-excludable → public good	– Non-rival, access and consumption of other consumers/ water uses not affected – Excludability granted through institutional arrangements → club good – If excludability not established through adequate institutions or not aimed at (e.g. sanitation, min. drinking water provision): non-excludable → public good

[1] *Assumption: capacity constraint relates to the local availability of water resources for abstraction or the potential of receiving water resources to manage pollution inflow and not to the technical capacity (e.g. pumping capacity).*
[2] *Assumption: capacity constraint relates to a system's technical capacity in terms of customers served.*
[3] *Rivalry in consumption due to water scarcity could be avoided via institutional mechanisms, e.g. abstraction licences (combined with adequate control mechanisms).*

1985).[70] A natural monopoly occurs where there are continuing economies of scale in production. That is, the unit cost of production continues to decrease as production volumes increase (BALANCE AND TAYLOR 2005), which is also referred to as sub-additivity of costs. For the natural monopolist, marginal costs are thus always below average costs (cf. Section 2.3.2.1). Consequently, a single supplier will have lower costs than two competing suppliers, making the introduction of competition very difficult (GREEN 2003).[71]

Furthermore, natural monopolies are characterised through high capital intensity and irreversibility of investments. The presence of these so-called '*sunk costs*' hampers free entry into and departure from the market.[72] In addition to economies of scale and high capital intensity, NEWBERRY (2000) identifies the following characteristics of natural monopolies: non-storability with fluctuating demand, locational specificity and direct connections to customers. For collective water conveyance and distribution networks (i.e. piped water supply and sanitation systems and canal irrigation), all of these characteristics are met, although to varying degrees.[73] Network supply systems are generally constructed with significant excess capacity due to the irreversibility of the investment and the high adjustment costs associated with changes in the level of capacity. As a result, low marginal costs of transporting additional water and high sunk costs of network investment form an effective barrier to new entry (BALANCE AND TAYLOR 2005).

If natural monopolies are left unregulated or uncontrolled through public institutions, the incumbent supplier has the capacity to exert his market power to the detriment of the consumers and society in general. Allocative inefficiency can result, as the incumbent supplier has an incentive to charge prices significantly above the continuing costs of supply, leading to a lower supply of goods and services to customers than would be demanded in a competitive market (CHAVEZ AND QUIROGA 2002, BALANCE AND TAYLOR 2005).

---

70 Most economists attribute considerable economies of scale to water storage, conveyance and distribution (for a numerical discussion, see e.g. SAUER 2005).

71 Cf. e.g. FINGER ET AL. (2007), WACKERBAUER (2007), BALANCE AND TAYLOR (2005), MARR AND GORDON-WALKER (2002), EWERS ET AL. (2001) for a discussion of the potential, the limitations and the risks associated with the introduction of competition for the market (up or down the distribution network service), or *in* the market of water services provision.

72 The relevant market is then not a contestable market in the sense of the theory of '*contestable markets*' (WACKERBAUER 2007).

73 In this context it is important to distinguish between short-run and long-run economies of scale. In the short-run context, providing increased volumes of water supply or extending service provision to more customers is progressively less expensive due to economies of density within a network or a service area of a fixed or given size. In the long-run context, network size is not treated as fixed. (BALANCE AND TAYLOR 2005)

Additional aspects

Two additional points that need to be considered in the analysis, relate to externalities in water services provision and information asymmetries:

- The provision of water services is generally linked to both positive and negative externalities on an economic, environmental or inter-generational level.[74] Positive externalities materialise e.g. in terms of health benefits, public hygiene or landscape conservation. Negative externalities may result e.g. from overabstraction, soil salinity, diffuse pollution or erosion. The WFD aims to internalise negative environmental externalities linked to water services (cf. Chapter 2.2 and 2.4).
- Information asymmetries or deficiencies may be present, i.e. consumers may not be adequately equipped to judge water quality in contrast to service providers (BALANCE AND TAYLOR 2005). These information asymmetries have to be accommodated through a system of regulation or control.

The good characteristics of water services and the related market failures in water services provision need to be kept in mind when discussing appropriate water financing approaches and cost-recovery objectives, as their perception influences the appropriate instrument mix chosen at the Member State level.

2.2 Cost-recovery of water services: a multi-dimensional concept requiring interpretation

As the discussion of the objectives of Article 9 in Chapter 1 has shown, the interpretation of the concept of cost-recovery is not necessarily straightforward. A number of studies has pointed at the marked differences in the understanding and hence implementation of the concept of cost-recovery across European Member States (METROPOLITAN CONSULTING GROUP 2006, ROTH 2001, EUROPEAN COMMISSION 2000, OECD 1999A, OECD 1999B, OECD 1999C). Therefore, the following Section 2.2.1 begins with an overview on possible interpretations of the cost-recovery principle and by exploring the WFD's understanding of cost-recovery. Section 2.2.2 then investigates which costs should be recovered in order to achieve a sustainable provision of water services. By providing defi-

74 The classical example of a negative externality is that of pollution, where the price paid for a product by customers does not reflect the cost to the environment of its production, and as a result, more of the product is consumed than would be the case if the price reflected the full production costs (BALANCE AND TAYLOR 2005).

nitions and delineations of the different cost categories to be considered in assessments of cost-recovery, this section delivers an important input to the development of the assessment framework in Chapter 3. Once the different cost categories have been broadly defined, Section 2.2.3 outlines specific cost factors for households and for the agricultural sector within these cost categories. Section 2.2.4 extends the analytical basis to framework conditions which may have an impact on the magnitude and relevance, as well as on the development of cost categories, in water services provision over time.

2.2.1 Possible interpretations of the concept of cost-recovery

The concept of cost-recovery is interpreted very differently in the literature and even more so in the political arena. Generally, it is understood as ensuring sufficient financial means for sustaining the long-run viability of the infrastructural system for water services' provision. Which cost categories should be included in this assessment and how costs should be accounted for, are, however, matters of debate (MASSARUTTO 2004A, EUROPEAN COMMISSION 2000). Similarly, the understanding of how and by whom the necessary financial means for recovering these costs should be generated differs.

A comparative assessment of cost-recovery thus necessitates transparency regarding the underlying cost-recovery mechanisms, i.e. which costs have been considered as well as the accompanying payment or financing scheme. The following section therefore first highlights alternative views on cost-recovery, before analysing the WFD's understanding of cost-recovery.

2.2.1.1 Views on cost-recovery in the provision of water services

According to standard economic theory the principle of cost-recovery is met if the total costs and total revenues associated with the provision of a good or service (i.e. direct and indirect, social and private costs) are in correspondence, and revenues are generated through prices and charges from the consumers or beneficiary of the good or service.[75] A review of empirical investigations on cost-recovery of water services provision and an evaluation of the (historical)

75 As outlined by MASSARUTTO (2005A), the concept of cost-recovery it is not necessarily intended in neoclassical economic terms as the need to ensure an optimal allocation of resources through marginal-cost pricing (cf. Section 2.3.2.1), but may also be regarded as a means for assessing the long-term capacity of the community to raise the required financial means for guaranteeing a viable service provision in the long-run, whether through prices or taxes.

practical implementation of the concept in selected European Member States reveals, however, that cost-recovery is rather understood as a continuum of financing options (cf. MASSARUTTO 2004A).[76] While some analysts and Member States alike consider the principle of cost-recovery to be met only if costs are fully internalised and borne by the users of a service, others interpret the principle more leniently by requesting simply that no costs are left unconsidered and transferred onto the next generation or externalised e.g. on the environment, thus allowing ultimately for a complete financing of water services through public budgets.

These differences in interpretation may be related to the fact that the three dimensions of water (cf. Chapter 2.1) have traditionally entered water policy-making to a differing degree across Europe, with some countries emphasising the need for considering water an economic good and others highlighting more particularly its non-substitutability and social dimension. How the different good dimensions are considered and weighted against each other may thus influence the financial structure chosen for financing water services provision, i.e. the degree to which fiscal or endogenous sources of finance are exploited. Generally, the merit good characteristics of water services along with the presence of market failures, foremost in terms of positive externalities generated by water services provision and accruing to society as a whole (e.g. landscape conservation, provision of fire fighting capacities and water conservation measures, public health and hygienic benefits), as well as the equity and affordability requirements on water services, may justify a divergence from a strict endogenous cost-recovery and allow for public budgetary involvement.[77]

Not only the overall conception of the concept of cost-recovery differs, but also its exact composition, i.e. the underlying cost components to be included in the assessment. While most cost-recovery definitions include operation and maintenance costs, differences relate to the treatment of initial investment costs or replacement and rehabilitation costs (CARDONE AND FONSECA 2003). Also, the role of (cross-)subsidisation varies along the spectrum of cost-recovery interpretations (KRAEMER 2003), ranging from intra-generational subsidies between user groups (e.g. households and agricultural water users) or between regions (e.g. rural and urban areas) to direct subsidies (e.g. contribution to

76 MASSARUTTO (2004A) presents possible financing options for covering costs of water services along a continuum. Alternative ways of financing the services are listed according to their distances from the pure long-run marginal costs, representing the theoretically optimal, through often practically unfeasible solution (cf. Section 2.3).

77 A clear-cut definition of the water service to be considered in the assessment is vital for the analysis, cf. Chapter 2.4. Public involvement in water service provision is not necessarily related to the core service itself, but may also relate to tied-in side-services.

investments) or indirect subsidies (e.g. low-interest grants, soft loans) paid by public authorities. The incidence of subsidies shows the importance of studying financing schemes of water services with care in cost-recovery assessments. As MASSARUTTO (2004A) points out, *"the existence of subsidies does not mean that the cost is left 'uncovered', what is in fact impossible, but rather that individuals pay as taxpayers instead than as water users."*[78] Only in those cases, in which costs are passed onto the next generation (e.g. through public debt or depreciated natural or artificial capital), may costs be considered *'uncovered'*. However, subsidies financed through general taxation may cause many potential distortions in the allocation of resources through distorted incentives, which should not be neglected in the analysis. These may, however, be softened by promoting equitable access to water services or other policy objectives (cf. Chapter 2.3).

Notable differences are also present among as well as within Member States with regard to the requirements made on cost-recovery assessments for different water services. Cost-recovery approaches for water services provided to households and agriculture often differ markedly. While it is generally acknowledged for both sectors that beneficiaries should pay the full ongoing costs of system operation, maintenance, replacement and upgrading of facilities (even if implementation does not proceed strictly to plan), major differences relate to the extent and form of capital cost-recovery for original investments (CORNISH ET AL. 2004A). In general, the recovery of costs from users or polluters is applied more rigorously to the household sector. For example, community irrigation schemes have traditionally been erected through public funds, with irrigation often seen as a regional development expenditure for backward areas, benefiting not only poor farmers but also society as a whole through lower food prices and food security (CORNISH ET AL. 2004A).[79] Where these costs are not recovered, governments pay the difference, thus subsidizing the agriculture sector, which is often regarded as a politically sensitive sector, or else the infrastructure is left to deteriorate, thus leading to a sustainability gap in service provision (cf. Section 2.4).[80]

---

78 In the case of water services which have a 100 percent rate of coverage, it may also be debated whether efficiency losses due to a recovery of costs from tax payers instead of from water service users are significant from a welfare point of view (cf. Section 2.3).

79 As MOLLE AND BERKHOFF (2007) point out, World Bank policy regarding the application of the cost-recovery principle in the agricultural sector can be seen as an illustration of how the concept's perception has changed over time.

80 The sustainability objective for agriculture may be difficult to determine, as it is not necessarily equivalent to a long term sustainment at full capacity (see discussion below).

The following Table 3 gives an overview on the spectrum of cost-recovery interpretations and the ways in which water services can be financed. Options range from a '*strong internal cost-recovery*' (case 1), i.e. a full endogenous recovery of financial and environmental and resource costs of service provision, to '*no internal cost-recovery*' (case 4), i.e. financing of water services entirely from public budgets. Box 3 provides a brief description of the four polar cases.

Table 3: Different approaches to cost-recovery

Degree of internal financial CR / Degree of internal ERC recovery	Strong internal financial cost-recovery	Medium internal financial cost-recovery	Weak internal financial cost-recovery	No internal financial cost-recovery
Strong internal ERC recovery	**Full recovery of costs from users/polluters including ERC.** *(case 1)*	O&M costs and costs of small investments recovered from users/polluters, full ERC recovery.	Operation and maintenance costs and full ERC recovery from users/polluters.	**ERC recovered entirely from users/polluters.** *(case 3)*
Medium internal ERC recovery	Full financial cost-recovery, ERC largely recovered from users/polluters.	...	...	Service related ERC largely recovered from users/polluters.
Weak internal ERC recovery	Full financial cost-recovery, some attempts/ instruments for recovering ERC from users/polluters.	...	...	No financial costs recovered from users, some attempts/ instruments for recovering ERC from users/polluters.
No internal ERC recovery	**Financial costs fully recovered from users/ polluters, no ERC recovered from users.** *(case 2)*	O&M costs and costs of small investments recovered from users, no ERC recovered.	Only O&M costs recovered from users.	**Full financing of water services through public budgets.** *(case 4)*

(O&M = operation and maintenance; ERC = environmental and resource costs)

Box 3: Approaches to cost-recovery – four polar cases

Table 3 maps out possible approaches to cost-recovery along two dimensions, namely the extent to which *financial costs* are included in the cost-recovery assessment and the extent to which *environmental and resource costs* are taken into account. In the following, only the four polar cases will be described.[81]

Case 1: Full recovery of costs from users/polluters including environmental and resource costs

The first case is characterised by a full endogenous recovery of costs from users including environmental and resource costs. Hence, a strong internal cost-recovery is given from an economic as well as from an environmental perspective. The extent to which this approach to cost-recovery corresponds to incentive-setting for efficient resource use or raises affordability issues depends on the specification of the instrument mix through which the required revenues are generated (cf. Chapter 2.3) and the level at which cost-recovery is to be achieved (community level or individual level), i.e. to what extent cross-subsidies between user groups are allowed for. If cost-recovery is to be implemented at individual level, universal water metering is a prerequisite to the allocation of costs according to use (cf. Section 2.3.2). Direct cost attribution at the individual level is, however, in many cases problematic (cf. Section 2.2.2) and may be linked to important transaction costs. Hence, achieving cost-recovery not at the individual, but rather on an aggregated community level, may be more in line with efficiency considerations.

Case 2: Financial costs fully recovered from users/polluters, no ERC recovered from users/polluters

Under the second case, there is a strong economic internal cost-recovery on financial costs, but environmental and resource costs remain externalised to society as a whole. Cost-recovery is understood in this case as a way of ensuring long term financial sustainability in service provision, without taking account of environmental and resource costs associated with the service's provision. Whether this scheme is nonetheless capable of setting incentives for (economically and environmentally) efficient resource use depends on the specification of the tariffication scheme (cf. Section 2.3.1.2). Affordability may again be an issue under this scheme if cost-recovery is to be achieved at the

81 In between these polar cases are those cost-recovery interpretations in which large infrastructural investments are financed through the public budget, and only some (effective) instruments are operated to recover environmental and resource costs from users/polluters.

individual level, in particular if the system is still in the expansionary life-cycle phase, with large capital investments.

Case 3: ERC recovered entirely from users/polluters

In the third case, only the environmental and resource costs associated with the provision of water services are recovered from their users (i.e. a strong internal environmental cost-recovery), while the financial costs of the services are financed entirely through the public budget (i.e. fully externalised on society as a whole). The provision of water services is thus fully financed through general taxation, while environmental taxes and charges are levied on users/polluters according to their relative use. This approach to cost-recovery necessitates an extensive monitoring system for determining the environmental and resource costs which must be borne by each sector or individual household/ farmer. This scheme may, however, provide incentives for users to shift to more environmentally sound practices, again depending on the exact conception of the payment scheme, e.g. if charges can be influenced by a user's behaviour (cf. Section 2.3.1.2).[82]

Case 4: Full financing of water services through public budgets

In the fourth case, the costs of water services provision are fully covered through public budgets (full external financing of the service). This approach is associated with the understanding of water as a human right, i.e. a public service free to all citizens. The long-term sustainability of water service provision is thus dependent upon the ability to secure sufficient and continuous financing from the public budget. Furthermore, if water is a scarce local resource, the incentivising function of water prices for curbing water demand is rendered ineffective, and sustainability of water use is not directly promoted through this financing scheme (cf. Section 2.1.4).

In light of these four polar cases of cost-recovery approaches, it becomes apparent that '*cost-recovery*' is always achieved in one way or another; what has to be investigated is therefore the differentiated financial structure underlying the provision of water services along with the related set of incentives that emanates from the financial structure for water service users and other actors along the process of water service provision (MASSARUTTO 2004A).

82 This does not have to be the case. One could also recover costs in lump-sums.

2.2.1.2 Cost-recovery understanding of the WFD

As the analysis of Article 9 objectives in Chapter 1.4 has demonstrated, a number of motivations are connected to the cost-recovery provisions made in the Directive. Despite the open formulations of Article 9 (e.g. *"shall take account of the principle of cost-recovery"*, *"shall ensure an adequate contribution"*), the explicit reference to the polluter-pays principle for achieving cost-recovery and the integration of environmental and resource costs as relevant cost components for the assessment propose an economic cost approach and a strong understanding of internal cost-recovery. This interpretation is confirmed by the economic guidance documents from the CIS process, which highlight that *"cost-recovery means income received from the direct sale of services or in any other way that is directly associated with the operation of a service or the degree to which revenues from users cover the costs of producing a good or service"* (CIS 2004B, emphasis added). Cost-recovery is seen as a vital step towards achieving water sustainability, both of the natural resources as well as of the infrastructural system, by assessing deficits and bringing unsustainable practices to light. This again shows that the *costs to society as a whole* are considered the relevant reference frame for decision making, and thus that a wide cost definition is to be applied in the assessment of cost-recovery.

In its evaluation of reporting on cost-recovery in the Article 5 reports of 2004, the EUROPEAN COMMISSION emphasises in internal documents of the CIS Process that cost-recovery should be *"achieved through the prices that the water service consumers have to pay to the provider directly or through any tax, charge or levy that is imposed on the service and is borne by the consumers directly or indirectly"* (emphasis added).[83] Nevertheless, justifications for a (partial-) financing of infrastructural systems for water supply, sewerage and irrigation services through general taxation may be derived from the public good components of these services[84], as well as from equity and fairness considerations (cf. Table 4). A clear definition of the water service under investigation for cost-recovery is thus needed for the assessment (cf. Section 2.4.1), along with a differentiation of potential incorporated side services.[85]

---

83 In particular for the wider water services interpretation, extending to navigation and flood protection a "general interest"/ public good nature can be claimed (although benefits are distributed unevenly on a geographical scale, e.g. in the case of flood protection, with some benefiting more from these water services than others), which could also serve as a justification for the recovery of costs through general taxation (cf. Section 2.1.4).

84 I.e. public good components cannot be clearly separated for the analysis.

85 To this end, public good components of water services and potential side-services that do not fall under the cost-recovery requirement (e.g. provisions for fire-fighting, rainwater

There is a general understanding that in terms of Article 9, the cost-recovery requirement will be more strictly applied as implementation progresses. According to UNNERSTALL (2005:12), cost-recovery rates well below 100% are in line with Article 9 up to the year 2010, due to its open formulation of "*tak(ing) account of the principle of cost-recovery*" (emphasis added), After 2010, however, cost-recovery rates diverging significantly from 100 percent will be in breach of Article 9.[86]

Table 4: Possible financing approaches for water services provision

	Possible financing approaches **Full internal financing from water services users / polluters**	**Financing through general taxation**
Pro	Incentive for efficient resource use (depending on specification). Independent source of financing; water management systems can be financially self-sufficient. Increased motivation for efficient behaviour at level of service provider through consumer dependence.	Huge investments encountered at initial infrastructure start-up are manageable. Public good components inherent in water services provision. (Equity considerations.)
Contra	Important prerequisites have to be met for costs attribution. Could lead to price levels exceeding the threshold of affordability for certain groups of society.	Dependence on public budget constraints. Risk for opportunistic behaviour and for entering into a vicious circle of public spending. Difficult to set incentives for efficient service provision. No incentive setting linked to water service use possible.

On what *scale* cost-recovery has to be achieved can, however, not be univocally extracted from Article 9 (UNNERSTALL 2005). The provisions on cost-recovery

management, drainage of public places, etc.) have to be separated in the assessment. However, as Chapter 2.2.3 and Chapter 3.1 demonstrate, a clear separation of costs is often difficult to achieve (cf. also EWRINGMANN 2006:27).

86 UNNERSTALL (2005:12) establishes a lower limit for compliance with the cost-recovery requirement at 70 percent up to 2010. Rates below this threshold could only be defended based on Article 9 (3) or (4) of the WFD (cf. below). UNNERSTALL also notes that the qualification of an '*adequate contribution*' refers only to the distribution of costs among water users and not to the level of cost-recovery itself. Accordingly, a cost-recovery rate which significantly diverges from 100 percent after 2010 will be in breach of Article 9.

seem not to address the scale of the individual household or farmer (KAHLENBORN ET AL. 1999:18)[87], but the scale of service provision, i.e. the total financial and environmental and resource costs encountered by an undertaking in delivering the service.[88] Accordingly, tariff structures may diverge from cost structures, e.g. in the form of variable tariff components, which exceed the portion of variable costs (KAHLENBORN ET AL 1999, UNNERSTALL 2005), thus allowing for incentive pricing (cf. Chapter 1.4 and 2.3). In contrast, the provisions of Article 9 for the polluter-pays principle rather demand an individualistic perspective and ask for a water tariffication structure that reflects the cost structure of provision.[89] The *adequacy* of the contribution of water uses to the cost-recovery of water services is determined by the polluter-pays principle, i.e. by the relative contribution of water uses to pollution or the costs of service provision in general (SCHMUTZER 2006:232, UNNERSTALL 2005). A cost attribution mechanism which significantly diverges from the relative contribution of user groups to the total costs then qualifies as a cross-subsidy (UNNERSTALL 2005:21). In this understanding, an incentivising price structure, which does not mirror the high fixed component in costs, would contradict the application of the polluter-pays principle at the household level (cf. Chapter 1.4).[90] These aspects again underline the controversies inherent in addressing the different Article 9 objectives simultaneously.

In addition to its specifications on cost-recovery and the polluter-pays principle, Article 9 contains additional provisions, which add further complexity to the debate:

> *"Member States may <u>in so doing</u> have regard to the **social, environmental and economic effects** of the recovery as well as the **geographic and climatic conditions** of the region or regions affected"* (emphasis added)

While this paragraph implies that divergences from a strict internal cost-recovery approach are acceptable in terms of Article 9, the addition of the

87 A critical view of this stance is presented by GAWEL (1995).

88 The issue of scale is important when dealing, for example, with the role of geographical subsidies. While the water sector in England and Wales is characterised by large districts for water services provision with only ten water supply companies, the water sector in France and Germany has a highly decentralised structure with about 10,000 and 6,400 water suppliers respectively. On a larger scale, geographical (cross-)subsidies could thus level out more easily.

89 According to REINHARDT (2006) the implementation of Article 9 does not (necessarily) imply modifications to the level of water prices and charges, but rather demands a polluter-pays oriented modification of water price calculations.

90 However, as GAWEL (1995) points out, under an economic cost perspective, the variable costs may theoretically exceed fixed costs (under high environmental and resource costs).

phrase *"in so doing"* points out that the language cannot be interpreted as allowing the cost-recovery principle to be neglected in full (UNNERSTALL 2005). Divergences from this principle warrant a clear justification and a demonstration of how the objective will be approached in the continued implementation process. However, it allows Member States to take considerations of fairness into account (e.g. by introducing cost-recovering water prices in a staged process or by introducing financial aid for vulnerable groups). In any case, this paragraph cannot be interpreted as freeing Member States from the general obligation of achieving cost-recovery in accordance with the polluter-pays principle.

In interpreting the cost-recovery understanding of the Directive, Article 9(4) is also of interest:

> *"**Member States shall not be in breach** of this Directive, if they decide **in accordance with established practices not to apply the provisions** [...] for a given water-use activity, where this does not compromise the purposes and the achievement of the objectives of this Directive."* (Article 9(4), emphasis added)

This provision, the so-called *'opting-out clause'*, has been added in response to opposition from Ireland, where water services were, at the time of the passage of the Directive, provided free of charge to households as part of the state's social politics (BROCKMANN 2003:152). It allows Member States to continue such established practices, provided that the overall objectives of the WFD are not compromised.[91] This clause should, however, be interpreted very narrowly.[92]

Overall, the Directive thus seems to promote a strict internal cost-recovery approach, although an iterative approximation of this objective will be accepted. However, the European Commission has made clear that reports on cost-recovery within the RBMP for 2009 should also contain (1) an assessment of financial cost-recovery; and (2) an investigation of the recovery of ERC and the degree to which they are taken account of in payment schemes; (3) a description of the institutional mechanisms for cost-recovery; and finally (4) an analysis of the adequacy of the contribution from key water uses to the cost-recovery of water services. Hence, *transparency of the ways in which costs are currently recovered* and a *full cost accounting approach* (i.e. which costs are encountered and by whom) are at the centre of attention.

In the first implementation phase, it could thus also be acceptable to adopt the concept of *'sustainability cost'* or *'quasi-full cost-recovery'* promoted by

91 According to SCHMUTZER (2006:232) this paragraph does not constitute an exemption from raising cost recovering prices from polluters, but is more to be understood as a minimum standard, which Member States may outstrip, but not undercut.

92 Cf. SCHMUTZER (2006) for a detailed legal discussion.

some authors (cf. TARDIEU AND PRÉFOL 2002), meaning that contributions from users should not necessarily cover the total costs, but rather be able to finance the long-term maintenance of the capital infrastructure, including natural capital. In other words, there might well be public contribution to the initial investment, provided that later on, users will be able to continue financing operation and depreciation of the system in order to prevent its value from declining (and therefore preventing the need for new public contributions in the future). In the longer term, however, these intergenerational subsidies (i.e. if the present generation does not put aside enough money for reinvestments) are not accepted under the cost-recovery understanding of Article 9. Depreciation of the infrastructural assets should then recover the initial investment costs, in order to accumulate sufficient financial resources for reinvesting after the infrastructures' technical lifespan has expired. This aspect is taken up in greater detail in Section 2.2.2.1.

2.2.2 Which costs to recover? Definition and delineation of cost categories according to Article 9

According to Article 9, three broad cost categories should be considered in the assessment of cost-recovery of water services, namely:

- Financial costs,
- Environmental costs, and
- Resource costs.[93]

By extending the cost categories to be included in the assessment beyond financial costs, the Directive takes a full economic costs approach and establishes the *costs to society as a whole* as the relevant reference frame for decision making (EWRINGMANN 2006, HANSJÜRGENS AND MESSNER 2002).[94] The WFD itself does not contain any further specifications regarding the exact definition

93 In the beginning of the CIS process, there was an intensive debate, as to whether environmental and resource costs are, in effect, two different concepts and hence two different cost components. The WORKING GROUP ECO II (CIS 2004B) investigated the definition and delineation of environmental and resource costs and provided further guidance on their measurement (cf. Section 2.4).

94 The economic cost perspective is furthermore underlined by Annex III of the WFD, which outlines the requirements for the economic analysis (cf. Section 2.2.2.1).

or content of these three cost categories.[95] With the WATECO Guidance Document (CIS 2003A) and the Information Sheet on Cost-recovery (CIS 2004A), the CIS process provided further delineations, definitions and examples, which nevertheless leave considerable room for interpretation. The following Figure 9 puts the WFD cost categories into perspective and gives an overview on the composition of full economic costs.

Figure 9: The WFD cost categories put into perspective

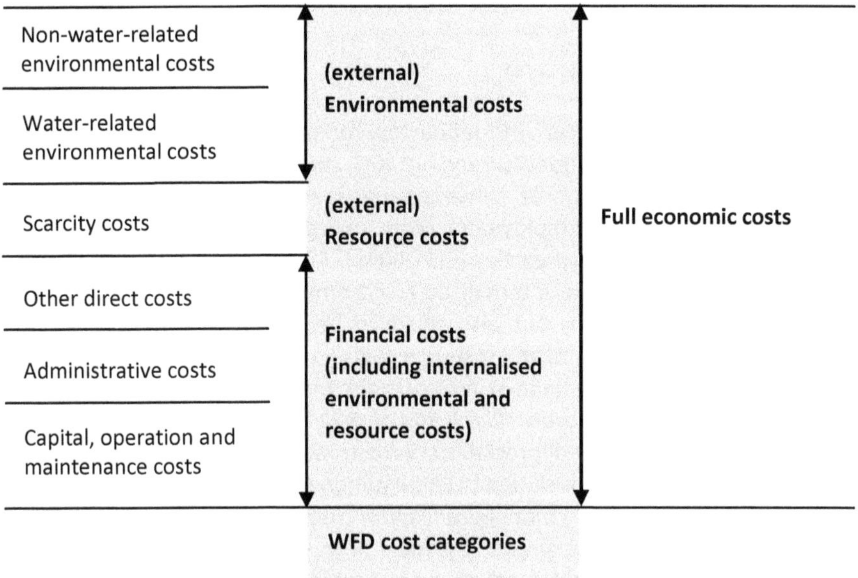

Source: adapted from Rogers et al. (1997), in: CIS (2003a).

The following sections first provide an overview on the three cost categories, taking up the debate within the CIS process. Section 2.2.3 then discusses the specificities of these cost categories when applied to the provision of water services to households and the agricultural sector.

95 The first draft of the WFD still contained a further substantiation of the cost category *'financial costs'* (in Article 2(33)), which was further expanded in the EUROPEAN COMMISSION's Communication on water pricing (EUROPEAN COMMISSION 2000:10), cf. Chapter 1.4.

2.2.2.1 *Financial costs*

The WATECO guidance document (CIS 2003A) identifies five subcomponents of financial costs, which should be explicitly addressed in the assessment of cost-recovery, namely operating costs, maintenance costs, capital costs, administrative costs and other direct costs (cf. Figure 9). An additional subcomponent, '*taxes and subsidies*', has been added to this list through the CIS Information Sheet on Cost-recovery (CIS 2004A).[96] The different subcomponents are briefly defined for the WFD context in the following sections.

Operating and maintenance costs

The WATECO guidance document defines *operating costs* as '*all costs incurred to keep an environmental facility running*' (CIS 2003A). These costs include, in the case of public water supply, sewerage management and agricultural water services amongst others, employment costs, energy costs, chemical costs and the costs of employing third parties (CIS 2004A). Authority rates, e.g. for water abstraction, which qualify as internalised environmental and resource costs, are included in this definition, but also need to be taken into account in the assessment of environmental and resource costs (cf. Section 2.2.2.2).

Maintenance costs are defined as '*costs for maintaining existing (or new) assets in good functioning order till the end of their useful life*' (CIS 2003A). Costs relating to assets, such as depreciation, infrastructure renewals or costs of financing capital, which are included in some national definitions of maintenance costs, form part of the cost component '*capital costs*'.

While not mentioned explicitly by the CIS guidance documents, atypical costs, which do not constitute part of the continuing operation of the business, such as provisions for restructuring or adaptation costs related to extreme climatic events (e.g. dry weather costs, flooding costs), should also be included under this cost category.

Capital costs

Due to the long lifespan of capital assets in the water sector, capital costs generally make up for a substantial part of total costs (cf. Sections 2.1.4 and 2.2.3). For analytical purposes, the cost category 'capital costs' is again subdivided into

96 The WATECO Guidance Document (CIS 2003A) refers to taxes and subsidies and cautions that they should be separated in the assessment of economic costs, but did not integrate them as a separate category.

two sub-categories, namely (1) depreciation and (2) opportunity cost of capital (CIS 2003A).[97] In particular, the treatment of depreciation is a complex issue and varies both across water services within Member States as well as among Member States.

1. Depreciation

Through this cost component, the annualised cost of replacing existing assets in the future is integrated into decision-making. Depreciation takes account of the reduction in value of the infrastructural system as assets wear and tear over their lifetime and lose value due to technological and economic developments (BÖTTCHER 1998). In terms of cost-recovery, depreciation is intended to ensure the sustainability of service provision by guaranteeing the financial viability and substance (depending on the approach to depreciation chosen) of the service undertaking in the long-run.

Figure 10: Capital versus asset maintenance

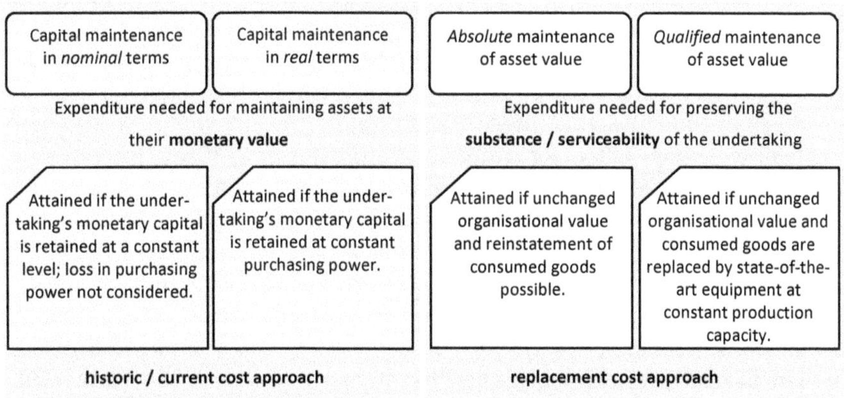

In the literature, a distinction is generally made between *maintenance of capital* and *maintenance of asset value* (i.e. the preservation of substance/ serviceability) (BÖTTCHER 1998, FRIEDL 1996, see Figure 10). The capital maintenance approach focuses on measuring the expenditure needed for maintaining assets at their monetary value. It can again be subdivided into nominal capital maintenance and real capital maintenance. The maintenance of asset value

97 The WATECO Guidance Document (CIS 2003A) identifies capital costs for new investments as a third category under capital costs and proposes to spread the costs of new capital investments over their expected lifetime, by means of the Annual Equivalent Cost (AEC) Method.

approach aims to preserve the service provider's substance and to ensure a long-term sustainability of service provision. The approach is again composed of two methodological strands, namely absolute and qualified maintenance of asset value.

Nominal capital maintenance approach

Nominal capital maintenance is achieved if an organisation's equity, measured as the difference between the initial value of the assets and the accounts payable, is the same at the beginning and at the end of each accounting period. This approach aims to retain the monetary capital at a constant level, without taking account of its loss in purchasing power. In essence, the approach aims to refinance the infrastructural assets needed for the provision of a water service through the current generation of (actual) users (FRIEDL 1996).

The nominal capital maintenance approach draws on the historic cost approach, i.e. assets are valued at the prices at which they were originally purchased. This approach has the advantage that the required information for calculating depreciation is readily available and undisputed. However, as changes in markets and inflation are not taken into account, the provision of adequate financial means for replacing assets with those of comparable capacity and quality cannot be guaranteed. Furthermore, the nominal cost approach may lead to wrong signals for water service providers, as it sets incentives to postpone new investments due to the low costs associated with a continued use of existing assets (BÖTTCHER 1998:84, BALS 1973:211).[98] The financing gap in the case of reinvestment necessitates compensating contributions from public budgets, direct contributions from users, or external financing through credits, which may lead to changes in the undertaking's capital structure. The aim of sustainable service provision through cost recovering charges may thus not be reached through the nominal capital maintenance approach (BÖTTCHER 1998, GAWEL 1994).[99]

---

98 As capital costs form a large part of total costs, prices may lose their signalling function (cf. Section 2.3) to consumers if they are based on historic costs, as they do not provide information on the scarcity of all production factors at the point of the consumption decision (BALS AND NÖLKE 1990).

99 FRIEDL (1996) illustrates with reference to BUDÄUS (1978) that the nominal capital maintenance approach bears the tendency to increase public debt of municipalities (if these are responsible for service provision), as additional funds have to be raised for replacing assets. These may then also impact other municipal duties. Furthermore, it should be noted that water prices may have to rise sharply if capital assets have to be replaced at higher reinstatement values than capital has been accumulated.

Real capital maintenance approach

The real capital maintenance approach accounts for the inflationary financing gap and aims to maintain the organisation's equity at a constant purchasing power. It is based on the current value approach for depreciation, i.e. the historical value multiplied by an inflation index. This approach is effective in maintaining the asset value over time, under the assumption that the initial investment was sufficient. However, this approach to depreciation is also marked by a number of drawbacks (CIS 2003A). First, estimating the inflation index may not necessarily be straightforward and is open to interpretation (e.g. regarding the indexation of prices, which may be based on general inflation, construction or consumer price index). Second, similar to the nominal capital maintenance approach, this method does not take account of technical progress.[100] Finally, similar to the case under nominal capital maintenance, if asset replacement turns out to be more expensive than the financial means accumulated, substantive changes in prices and charges may result, with adverse effects both from a political and from an economic perspective (BALS AND NÖLKE 1990). Accordingly, while this method is relatively easy to apply and may be more appropriate than the nominal capital maintenance approach, it is not entirely in line with the aim of Article 9 of ensuring a sustainable service provision.

Absolute and qualified maintenance of asset value

When the value of assets is maintained, the goal is a preservation of the organisation's substance, i.e. of the original production capacity. An *absolute* maintenance of asset value is attained if the organisation has at the end of the accounting period, under nominally unchanged liabilities, the same value as at the beginning of the period, or – as consumed goods are generally not replaced by identical substitutes – would be able to reinstate the goods consumed during a period (GAWEL 1994). A *qualified* maintenance of asset value is obtained, if technological developments and changes in demand are taken into consideration in the assessment. In this case, the substance of an organisation is maintained if consumed goods are replaced by goods which are in accordance with the state-of-the-art and in line with economic and technological developments, while simultaneously displaying an equivalent production capacity to that of the

100 For example, a water treatment plant that was constructed for a given amount ten years ago might cost half the original amount today due to technical progress.

originally replaced assets (HAX 1957:19).[101] Under the maintenance of asset value approach, the replacement valuation method is used for determining depreciation. It estimates the present value of an asset from the current cost of replacing it with an asset achieving an identical service level. The advantage of this method is that it allows for taking account of technical progress as well as of the costs of wear and tear of assets at the time of use. Thus a sustainment of the capacity and productivity of the undertaking is provided for (BALS AND NÖLKE 1990:215).[102] However, it might be costly and time-consuming to apply this approach to the entire capital stock. Furthermore, it necessitates a new assessment for each accounting period[103] and draws on indices and current market values, which may be subject to interpretation and selection difficulties (BÖTTCHER 1998). Some authors also claim that the replacement valuation method leads to a double burden for the first user generation (cf. discussion below). Yet, through the replacement valuation method sufficient financial means will be accumulated for reinvesting after the end of the assets' lifetime.

Discussion and comparison of the different approaches to depreciation

Under the capital maintenance approach, each generation pays for its own infrastructural assets, and there is no pre-financing of future and potentially more expensive assets. With every generation paying for its own use, the polluter-/user-pays principle is met at a meta-level (FRIEDL 1996).[104] However, as a result only insufficient financial means will be accumulated for reinvesting after the infrastructure has reached the end of its technical or economic lifetime. Hence, the sustainability of the infrastructural system, and thus the provision of water services, is not guaranteed. On the contrary, if depreciation takes account of the additional costs required for a renewal of the infrastructure due to infla-

---

101 In applying the concept of real maintenance of asset value, the capital structure of the undertaking remains unchanged i.e. the ratio of equity to foreign capital is fixed over time (GAWEL 1994). In contrast to the nominal maintenance of asset value approach, the financing aspect is taken into consideration and a constant substitution of equity with foreign capital is prevented.

102 BALS AND NÖLKE (1990:215) demonstrate that this is only valid under certain stylized theoretical assumptions, e.g. complete divisibility of the capital stock, a homogeneous age structure of assets, as well as constant and immediate reinvestment of the value of depreciation.

103 In order to minimize administrative costs BÖTTCHER (1998:99) recommends limiting new assessments to fixed time intervals. In practice, adaptations are often only undertaken if price changes exceed 10 percent.

104 On the contrary, one could also state that if the present generation fails to set aside enough money for reconstruction, it will consume the values of something whose cost will continue to be paid by taxpayers of next generations.

tion and technical advances, some authors have argued that the present generation of users pays more than the actual costs encountered through their use (FRIEDL 1996). This is regarded by some as a violation of the equivalence principle and an excessive double burdening of current users. Others refute this argument by stating that the financial means generated through depreciation based on the replacement cost valuation method are merely in line with the wear and tear of assets caused by the current user generation through their current use.

Debates also relate to how contributions from the public budget to investments in water infrastructure should be dealt with. A long-term dependency on public resources is only avoided and self-sufficiency achieved if this contribution to investment is recovered appropriately from users. Otherwise a financing gap will prevail, which either has to be compensated for by repeated transfers from the public budget, direct contributions from users at the time of reinvestment (which may entail sharp increases in prices), or external finance.

Finally, if large investments are partly financed through direct financial contributions from users, and these are not depreciated, there will also be a financing gap at the time of reinvestment. Depreciating user contributions would, however, lead to double-payment by the present generation (STEENBOCK 1984, BROD 1988:82): first through their direct contribution to the initial investment, and second through their user fees entailing a contribution to future renewals. However, this argument is again refuted by others (cf. BÖTTCHER 1998) who claim that there are two different "services" being paid for, the first one of gaining access to the infrastructural system (the initial investment), and the second one of using the service that is being offered. These issues are taken up again at greater length in Chapter 3. This brief discussion of differences in accounting procedures already points at important difficulties for comparative assessments of cost-recovery.[105]

Depreciation under an economic cost approach

The economic cost approach is primarily intended to estimate effective resource use, i.e. all factor uses in production of a good or service should be valued at their scarcity values at the time of use (BALS AND NÖLKE 1990). The Article 9 objective of a sustainable provision of water services is best taken account of by the maintenance of asset value approach with depreciation based

105 Cf. HOLLÄNDER ET AL. (2008) for an illustration of the effect of different accounting procedures on the level of capital costs.

on the replacement values of existing assets.[106] Only under this approach, can assets be maintained at constant service levels and the value of infrastructure remains constant over time despite inflation and technical change.[107]

As already highlighted above, the financing structure of the initial infrastructure investment should also be considered in determining the appropriate methodology for depreciation. If a contribution from public budgets were granted for lowering the financial burden on users, it would either have to be recovered from users through depreciation (in this way, assets are transferred into the equity of the service provider) in order avoid a financing gap at the time of renewal, or a comparable contribution either from users or from the public budget would have to be made:

- If the public transfer is renewed, it must be made explicit in the cost-recovery assessment by reducing the degree of internal cost-recovery accordingly (cf. Chapter 3.2).
- If the required financial means are generated through direct contributions by users at the time of reinvestment, these have to be made explicit on the revenue side of the cost-recovery assessment. From a social and economic perspective, a smoothing of the budgetary expenses required from users is to be preferred.

The adopted financing structure depends on the approach to cost-recovery, as outlined in Section 2.2.1 above.

A further issue requiring attention is the appropriate timeframe over which assets are depreciated, i.e. the expected useful life of existing infrastructural assets.[108] Generally, the useful physical life is distinguished from the useful eco-

---

106 This classification is reinforced by Annex III of the WFD, which states that the cost-recovery assessment should "*tak[e] account of <u>long term forecasts of supply and demand for water</u> in the river basin district and, where necessary: (1) estimates of the volume, prices and costs associated with water services, and (2) <u>estimates of relevant investment including forecasts of such investments</u>;*" (emphasis added). Hence Article 9 is not only concerned with refinancing historic investments (SCHMUTZER 2006:231), but rather takes a forward-looking perspective on future investment needs.

107 This assumes, however, that depreciation cash flow is solely employed towards infrastructure maintenance and renewal and not used for financing investments in other areas, which would again lead to a financing gap at the time of reinvestment.

108 Next to the useful physical lifetime (i.e. the timeframe in which an asset is able to physically fulfil the required functions) and the useful economic lifetime (i.e. the timeframe from the assets purchase up to the point where replacement makes sense under economic considerations, as maintenance costs become too expensive and a replacement of the asset become profitable), some authors additionally distinguish the useful procedural lifetime (i.e. the

nomic life. While there is extensive technical guidance on the appropriate use duration for the assets required for the provision for water services (cf. for the German context publications by e.g. LAWA, DWA, DVGW), variations in the anticipated lifetime of assets may lead to significant differences in capital costs (YARD 2004, SCHULZE 1994). Generally, under the economic cost approach, the timeframe should be determined by the economic lifetime of assets as effective resource use forms the basis for cost accounting. Estimation errors in the timeframe should therefore be adjusted in consecutive accounting periods, whenever appropriate (BÖTTCHER 1998).

Another aspect discussed in the literature, which may lead to different depreciation bases, is whether returns on the reserves set aside for renewals should be subtracted from the sum to be depreciated (REIF 2002). For the sake of simplicity it is assumed here that all capital inflows are directly reinvested into the infrastructural asset base (cf. section on opportunity costs of capital below). However, there may be different national rules concerning the handling of interest on reserves, with differences accruing e.g. to equity and foreign capital. The treatment of reserves has thus to be made explicit in the assessment of cost-recovery (cf. Chapter 3).

2. Opportunity cost of capital

Opportunity costs of capital constitute an estimate of the rate of return that can be earned on alternative investments. The return on capital is the interest on debt and the return on equity. The so-called weighted average cost of (debt and equity) capital is usually considered an appropriate measure of the rate of return on capital. Opportunity costs of capital are a form of remuneration for the monetary capital provided. In the literature, there is extensive debate on how these opportunity costs should be accounted for. A long-debated issue in public sector economics relates to the adoption of the appropriate interest rate for investments in the water sector, namely, whether the market interest rate or a social discount rate should be applied (MASSARUTTO 2002B, BARRAQUÉ 1999, BALS AND NÖLKE 1990, ZWEHL 1988).[109]

timeframe after which new legislative requirements demand a replacement of assets). Different timeframes for depreciation may influence assessments considerably.

109 Private investors usually require much shorter repayment schedules and thus higher interest rates, in order to achieve risk/return profiles that are comparable with the rest of the economy, while the public sector can, and according to some, should be less pessimistic. This myopic behaviour of private capital markets will mean that capital purchased on the market will have a higher price than its opportunity cost (measured by the social discount rate). (MASSARUTTO 2004A)

Analogous to the discussion above, the rate of return should be calculated on assets based on the replacement cost method, in order to mirror the capital commitment through the consumption of goods (BÖTTCHER 1998). Similarly, rate of return calculations should address the entire asset base under an economic cost approach (BALS AND NÖLKE 1990). These issues are dealt with more specifically in Chapter 3.

Administrative costs

Administrative costs are related to the regulation of water services. According to the WATECO Guidance document (CIS 2003A), this cost category includes the costs of administering a water abstraction licensing system or the costs of monitoring compliance with existing regulations on wastewater disposal. Furthermore, it relates to administering the provision of the water service itself. Detecting administrative costs may not always be straightforward (CIS 2004A), as several actors may be involved directly or indirectly in the process of water services provision. Examples include the following:

- In England and Wales, the costs of the environmental regulator, the Environment Agency, are financed through water abstraction and wastewater discharge licenses that are paid by the privatised water companies (CIS 2004A).
- In Germany, the pluralistic structure of *Länder* legislation (i.e. the Federal States) and the decentralised structure of water supply have produced a wide variety of different administrative arrangements, e.g. relating to water abstraction charges.[110] The revenue from these charges is employed very differently across the Länder. Often revenues are used for reimbursing compensation payments made by the water suppliers to farmers in return for their adherence to stricter environmentally friendly farming practices in water conservation areas than the code of good farming practices already prescribes. These schemes aim at precautionary water protection in water abstraction zones in order to reduce water treatment costs.[111] Administrative

110 Eleven German States operate a system of water abstraction charges, with large differences in the schemes (cf. Section 4.1).

111 In light of Article 9, however, these payments must be rigorously questioned. While they are an effective tool in terms of groundwater protection and legally allowed (also in line with EU legislation, see EUROPEAN COUNCIL 1999), from the polluter-pays perspective of Article 9, they are contradicting WFD objectives (AMMERMÜLLER 2009, HEINZ 2005). Furthermore, Article 5 reports and monitoring results demonstrate that the required reductions in diffuse pollution from agriculture can only be obtained through intensified activities from the agricultural sec-

costs are thus encountered by the Länder administration in managing the water abstraction scheme, by the water supplier in managing its cooperation with local farmers, and again by the water authority, which reimburses costs of voluntary agreements (cf. HOLLÄNDER ET AL. 2008).
- In the case of irrigation, administration costs may be encountered at the different structural levels of water services provision (e.g. in the form of overhead costs of governmental irrigation organisations).[112]
- In the form of financing primary regulatory functions at the respective authorities (ABERNETHY 2005), such as making rules about abstraction, uses, disposal and pollution; supervising the application of a system of water abstraction rights or licenses; or ensuring compliance with rules through monitoring of activities.

Although the relevant figures are in part readily available (e.g. the administrative costs related to the operation of certain economic instruments), administrative costs may often not be straightforward to quantify and the analysis will have to rely on qualitative descriptions or expert judgement in order to gain an understanding of their magnitude.

The following Table 5 provides an indicative overview on potential administrative costs in water services provision.

Table 5: Examples of administrative costs

Administrative cost	**Actors involved**
Monitoring compliance with environmental standards	Environmental regulating agency, water service provider
Monitoring compliance with water service quality (e.g. drinking water standards)	e.g. Drinking water agency
Administration of water abstraction licensing system	Regulating agency, water service provider, farmers
Administration of water pollution / effluent charges	Regulating agency, water service provider
Checking compliance with voluntary agreements or abstraction licences	Water suppliers, farmers, water authority
Overhead costs of governmental irrigation organisations	Authorities, farmers

tor. The traditional interpretation of the PPP for agriculture may thus probably no longer be maintained.

112 As highlighted in OECD (2002:12), irrigation districts are assigned an instrumental role in water policy implementation and water management in some OECD countries such as Spain, Turkey, Mexico and the US (i.e. collecting charges and levies issued by Basin Agencies from farmers and transferring the revenue to the Agency).

Furthermore, administrative costs include the direct costs of administering the provision of water services itself, e.g. the part of personnel costs that is related to administrative tasks, the rent for office space, office supplies, etc.

Other direct costs

The WATECO Guidance document included under this cost component "*costs of productivity losses due to restrictive measures*" such as the loss of agricultural production resulting from the creation of a retention area. These indirect costs related to the provision of water services generally do not accrue to the operator of the service, but rather to an uninvolved third party. Whether these costs have to be included in the assessment of cost-recovery depends on whether compensating payments have been undertaken as a result.

In terms of the proposed assessment in Chapter 3, the WATECO definition of these costs forms part of the cost component '*resource costs*' (cf. Section 2.2.2.3 below). For the purpose of the assessment, the cost category '*other direct costs*' refers instead to costs encountered by water service provides that do not fit into one of the other cost categories of financial costs, such as concession fees paid to municipalities for the right to use public roads for installing the pipe system for water supply or sewerage collection, or other fees.[113]

Taxes and subsidies

Taxes and subsidies qualify as transfers within society and should therefore generally be excluded from the estimation of economic costs. However, it is important to distinguish between general taxes and environmental taxations schemes related to the environmental damages caused by water services' provision: while general taxes need to be deducted from financial costs, environmental taxes may represent internalised environmental costs and as such should not be excluded from the assessment but rather be transferred to the environmental costs category.

Subsidies need to be made explicit in the revenue side of the cost-recovery assessment in order to assess the degree to which cost-recovery of water services is achieved internally and to assess the adequacy of contribution of different sectors (i.e. the degree of cross-subsidies). While subsidies may not necessarily constitute a breach with Article 9 (cf. Section 2.2.1), they have to be made

113 While this cost category serves as a conglomerate for costs that cannot be clearly allocated to another financial cost category, it should be noted that costs stemming from extraordinary events are excluded from the analysis and should not be considered here.

explicit and warrant a justification. Subsidies in water management may take on very different forms in the water sector (cf. KRAEMER 2003), and analyzing subsidy systems is thus a complex task. The OECD (1996) defines subsidies as "*government interventions through direct and indirect payments, price regulations and protective measures to support actions that favour environmentally-unfriendly choices over environmentally-friendly ones*". This definition includes direct subsidies[114] in the form of direct payments by the government to certain water service providers or users, and indirect subsidies in the form of soft loans, tax concessions or allowances, guaranteed minimum prices, discounts or preferential procurement policies and accelerated depreciation allowances. In particular, indirect subsidies are not always easily identifiable. Generally, subsidies are of particular importance in the context of agricultural water services (cf. Section 2.3.4).

Generally, subsidies can have two main objectives: either they are instituted to compensate users for a cost they incur in response to a required action or a prohibition, or subsidies are constructed so as to set the necessary incentives for achieving certain desired, but not required, actions (KRAEMER ET AL. 2003). From an economic perspective, they may lead to inefficient situations, e.g. by channelling resources to problems that are no longer high priority or supporting activities considered counter-productive from an integrated societal perspective (PIETERS 2002). Subsidies can be of a fiscal nature (i.e. paid out of public funds) or can take the form of para-fiscal cross-subsidies through redistribution between regions or among user groups. When certain user groups do not contribute adequately to the costs of water service provision, the resulting revenue gap will have to be filled (i.e. subsidised) by other sources. These sources can be contributed by the public at large (e.g. through general taxation) or by other user groups that pay a larger fraction of the total costs (including resource and environmental costs) than they generate and thus cross-subsidise water service provision to other users. Article 9 explicitly addresses the issue of cross-subsidies among users by asking for the sectors (i.e. households, agriculture and industry) to contribute adequately and in line with the polluter-pays principle to cost-recovery of water services.[115]

---

114 '*Direct subsidies*' may, for example, consist of monetary transfers from the (central or local) government: to the provider of water services in the form of investment subsidies (i.e. capital subsidies leading to lower fixed costs); to the provider of water services for co-financing the operation of the infrastructure (operational subsidies leading to lower variable costs); or to water users (in the form of income transfers, leading to lower prices or charges paid by the user) (CIS 2003A).

115 As WATECO outlines, cross subsidisation may arise among different users (households, agriculture, and industry), different regions (dry and wet, populated or less populated)

2.2.2.2 Environmental costs

According to the WATECO Guidance document, environmental costs *"represent the costs of damage that water uses impose on the environment and ecosystems and those who use the environment [...]. This loss in welfare may encompass lost production or consumption opportunities as well as non-use values [...]"* (CIS 2003A). The Drafting Group Eco II further clarified the definition provided by WATECO by linking it to the concept of total economic value (cf. Chapter 2.1): costs of damage *"imposed on the environment and ecosystems"* arise if non-use values attached to a healthy, functioning ecosystem are reduced or hampered. *"Costs to those who use the environment"* are related to the use values, which are associated with the actual or potential future use of a natural resource (e.g. drinking water, fish consumption, irrigation water) (CIS 2004B).

In order to make these definitions operational, a clear definition and delineation of environmental damage is required. As nearly every use of water resources has an effect on water bodies, a reference frame is needed for determining the significance of the effect. In theory, *'damage'* arises when there is a discrepancy between the status quo and a target point or situation. The target situation can be determined through existing environmental norms or standards or through a past situation, characterised by lower pollution levels and corresponding damage costs, which is to be regained (CIS 2004B). In the case of the WFD, the target situation is defined through the objective of achieving good water status.[116]

Even if an environmentally damaging effect is known, it is often difficult to measure the extent or size of the damage and to translate it into monetary terms (i.e. costs) (cf. Section 2.4.2 on the measurement of environmental costs). Furthermore, linking damages to a specific water use or a certain (group of) water users is inherently difficult, in particular in the case of summation effects, diffuse pollution or environmental damages which relate to historic events (e.g. contaminated sites). Generally, the magnitude of environmental costs is very much site-specific (GÖRLACH AND INTERWIES 2005, MASSARUTTO 2004A,

---

and/or different types of users (rich or poor, small or large users etc.). The accounting for cross-subsidies has already been discussed in Section 2.2.1. and will be taken up again in Section 2.2.3

116 As pointed out by CIS (2004B), residual damage costs may still be present even when a target situation has been reached e.g. because the target level has been set in accordance with politically or socially acceptable limits instead of the environmental demands of a healthy functioning ecosystem. Furthermore, part of the environmental damage may also be unknown at the time the target is defined due to the uncertainties relating to the determination of environmental cause-effect relationships.

CIS 2004B), as the degree to which a certain water service leads to environmental damages relates to the framework conditions of the river basin district in which the service is provided (e.g. the type of water, stream flow and velocity, river depth, changes to the rivers morphology, its hydro-chemical condition, ecological sensitivity and biocoenosis as well as the socio-economic conditions (e.g. population density, agricultural uses etc.)).

According to Article 9, only those environmental costs in aquatic ecosystems or dependent land ecosystems caused by the provision of a particular water service, or which influence the costs of service provision, must be considered in the assessments of cost-recovery. If over-abstraction for drinking water provision or irrigation affects the functioning of wetlands or effluent discharges cause a loss in biodiversity, these costs should be taken into consideration in the cost-recovery assessment and be factored into water pricing systems. Likewise, if diffuse pollution from agriculture causes environmental costs which impose higher financial costs on drinking water service provision, these should be addressed through the sector's adequate contribution.[117]

For the assessment of cost-recovery, it is important to distinguish between environmental costs which have already been internalised and external environmental costs.[118] Internalised environmental costs include costs incurred by the water service provider in order to avoid environmental damages (avoided costs approach, cf. Section 2.4.2), as well as the payments made on existing internalisation instruments such as charges on water abstraction or effluent discharges. These environmental costs already have monetary values attached to them and are included in the cost statements of water service providers. External or residual environmental costs linked to the provision of water services are, however, so far not accounted for and are thus left uncovered. For assessing their relevance and expressing them in economic terms, information on the status quo, i.e. on the physical situation of each water body affected by a water service's provision has to be available. Investigation of physical pollution pathways allows the attribution of environmental damages to the activities that cause them.

To what degree and under what circumstances environmental costs may arise in the provision of water services to households and the agricultural sector is discussed in Sections 2.2.3.1 and 2.2.3.2 respectively. Questions concerning

117 The differentiation and assessment of environmental costs is discussed in detail below.

118 An external cost exists when the following two conditions prevail (PEARCE AND TURNER 1990): (1) an activity by one agent causes a loss of welfare to another agent; and (2) the loss of welfare is uncompensated.

the measurement of environmental costs in theory and in practice within the cost-recovery assessment are taken up in Section 2.4.2.

2.2.2.3 Resource costs

The interpretation of the term "resource costs" differs in the relevant guidance documents from the CIS process. The WATECO guidance document (CIS 2003A) and the Commission communication on pricing policies (EUROPEAN COMMISSION 2000) interpreted resource costs as *"the costs of foregone opportunities that other uses suffer due to the depletion of the resource beyond its natural rate of recharge or recovery"* (emphasis added). In this understanding, resource costs are related to e.g. groundwater over-abstraction beyond the natural rate of recharge. The link to *'other uses'* comprises both current and future uses of the resource. In this understanding, resource costs are thus equivalent to the scarcity value of water resources.[119] Resource costs are hence only present, if there is water scarcity, i.e. if there is not enough water of sufficient quantity or quality to satisfy the demand of all users.[120]

The working groups DG Eco II took a broader view on resource costs by interpreting them as *"an economically inefficient allocation of water, either in terms of water quantity or water quality, over time and across different water use(r)s"* (CIS 2004B, emphasis added). Following this interpretation, resource costs are the result of a misallocation of water. Resource costs thus equal the difference between the economic value in terms of net benefits of present or future water use (e.g. allocation of emission or water abstraction permits) and the economic value in terms of net benefits of the best alternative water use (now or in the future). Resource costs then only arise if an alternative water use generates a higher economic value than the present or foreseen future water use (i.e. the difference between net benefits is negative).[121] Contrary to the definition in the WATECO guidance document, resource costs are therefore not necessarily confined to water resource depletion only (in terms of water quantity or water quality). They arise instead as a result of an (economically) ineffi-

---

119 Scarcity costs typically occur on a seasonal basis, given that water is a renewable resource.

120 As HEINZ (2005) points out, the definition provided by WATECO may cause terminological overlaps to the cost category of environmental costs: for instance, excessive depletion of water resources may lead not only to resource costs, but also to environmental costs (e.g. economic losses due to impaired forests).

121 An economic definition of *'scarce'* underlies this conception, i.e. not scarce in the sense of *'insufficient'* (everyday definition) but rather as *'not unlimited'* (economic definition). Even if there is *'enough'* water, it is not unlimited and scarcity costs may thus arise for nearly all water uses.

cient allocation of water and/or pollution over time and across different water users.

The estimation of use and non-use values provides the necessary basis for the calculation of resource costs according to the DG Eco II definition (CIS 2004B), as knowledge of the value of water in its alternative uses is required. For instance, resource costs arise if ecosystem restoration and conservation – measured through the use and non-use values attached to the water environment by the general public (households) and/or recreation – generate a higher economic value than, for example, current or future water abstraction by agriculture or water pollution by industry. However, it is not always evident which potential alternative uses could occur. Competition for water between uses is not always detected in a straightforward manner, as certain water uses are not undertaken due to known restrictions, although these alternative uses might theoretically produce a higher value.[122]

In the practical implementation process of Member States, the definition of the WATECO guidance document figures most prominently as evaluated in the European research project *'Aquamoney'* (GÖRLACH ET AL. 2007). While the definition provided by the Drafting Group Eco II (CIS 2004B) is considered to be theoretically sound and unambiguous, it is generally criticised for not being well-adapted to practical implementation.

2.2.2.4 Conclusion: the cost approach of the WFD

The WFD aims to establish cost transparency and has chosen an economic cost approach. Three cost categories are to be addressed in the assessment of cost-recovery, namely financial costs, environmental and resource costs. A further definition and concretisation of the sub-components of these cost categories has to be undertaken from the perspective of each water service. Accordingly, the next sections clarify the content of the three cost categories for public drinking water provision and sewerage services (Section 2.2.3.1) as well as for irrigation water services (Section 2.2.3.2).

---

122 It also needs to be considered that while certain uses, such as drinking water provision or irrigation, may be less productive in terms of monetary equivalents generated than the use of water as an input in a production process, they may have a high social value (cf. Section 2.1.2) and hence be assigned a preference in use. Accordingly, water is implicitly valued more highly in its drinking water use. Generally, it is a political decision to prioritise certain water uses over others. For a more detailed discussion, cf. Section 3.2.5.

2.2.3 Cost factors in water services provision

The following section provides a brief introduction to the cost factors relevant to the assessment of cost-recovery, differentiated by provision of water services to households (i.e. water supply and sewerage management) and to the agricultural sector (i.e. irrigation water supply). This section forms a particular basis for the development of the framework for assessing cost-recovery in line with WFD Article 9 that is developed in Chapter 3.[123]

2.2.3.1 Cost factors in the provision of water services to households

As classical network-bounded services, both water supply and wastewater services to households are characterised by a high degree of fixed costs, which amount on average to more than 80 percent of total costs (LEIST 2007, REIF 2002). The pipe system constitutes the largest portion of these costs. The variable costs, which are incurred in proportion to service provision (i.e. the amount of water provided or the amount of sewerage managed), are comparatively small, with a share of about 20 percent of total costs, which is mainly linked to energy consumption. Due to the high percentage of fixed costs, the total costs of water supply and sewerage management can largely be controlled in the early planning phases (HAAS AND FRANZ 1998 in LEIST 2007). The level and the composition of the costs of water supply and similarly of wastewater management are influenced by a large number of framework conditions, such as natural conditions (local water availability, raw water quality, hydro-geological conditions, climate and vegetation, etc.), the population demography and density within the service area as well as the structure and composition of the service area (e.g. number of connected inhabitants, size and composition of the service area, structure of water demand etc.), the established accounting rules on capital costs (cf. Section 2.2.2.1) and legal framework conditions (e.g. quality standards, concession fees for using public territory for laying the pipe network etc.).[124]

---

123 Factors relevant to the cost side of the equation are already introduced here. The discussion of the differences that a particular cost accounting may have, e.g. the way depreciation is accounted for, is, however, postponed in part to Chapter 3, where optimal accounting procedures in line with Article 9 are recommended.

124 A brief discussion of important framework conditions for water services provision is undertaken in Section 2.2.4.

The following paragraphs highlight important cost factors within the cost components of financial costs and environmental and resource costs for water services provision to households.[125]

Financial costs

Operation and maintenance costs

The operation and maintenance of the pipe system takes up the largest part of this cost component in the case of water supply and sewerage services (BOSOLD 1994). These costs are mainly fixed and vary with the size of the service area (i.e. the larger the service area, the higher the percentage of total costs stemming from the pipe system ((MUTSCHMANN AND STIMMELMAYR 1999:700), cf. Section 2.2.4).[126]

Generally, operation and maintenance costs comprise the following elements for water service provision to households (LEIST 2007, CARDONE AND FONSECA 2003):

- Personnel costs (e.g. staff involved in operation, maintenance, routine preventive maintenance, repairs, and construction for minor rehabilitation);
- Energy costs (electricity connections plus usage fees);
- Material costs (such as chemicals for treatment, tools, spare parts, equipment, and administrative supplies);
- Hired and contracted services (e.g. for construction); support costs (such as transport costs, training support, technical assistance);
- Monitoring costs (e.g. water and effluent quality parameters, technical conditions of the system, etc.);
- Costs of measures for water source protection and conservation, which may qualify as internalised environmental and resource costs (and thus have to be extracted for the assessment); and

---

125 The sub-components of administrative costs, other direct costs and taxes and subsidies are not taken up again in detail here, as they have already been illustrated in the examples above.

126 For the assessment of cost-recovery of water services provision, it is important to note that the side-service of fire protection may have an influence on the dimensioning of the pipe network. The associated costs should from a systematic point of view be factored out of the cost-recovery assessment for water services due to the public good nature of the side service (EWRINGMANN 2006). The costs should be borne by society as a whole and not by water service users. However, if drinking water services to households have a connection rate close to 100 percent, the group of water service users and beneficiaries from fire protection is approximately identical.

- Other costs (such as unaccounted for water due to leakage in the system or vandalism);
- General and support expenditures (such as personnel and management services, legal and property management services, research and development, audit services, public and employee relations services, land and property maintenance, storage of materials) (OFWAT 2006, OFWAT 2007).

These can again be subdivided into recurrent or variable cost and fixed cost components.[127] Fixed operating costs include, for example, the installation and maintenance of water meters, the major part of personnel costs as well as predetermined levels of bulk water imports (LEIST 2007, OFWAT 2006, FORST 2001, GROMBACH ET AL. 2000). The following Table 6 provides an indicative overview on operation and maintenance costs in water services provision to households structured according to the services' functional parts.

Capital costs

As outlined above, capital costs constitute a large part of the total costs of water services provision. The pipe system makes up the largest share of investments, followed by other infrastructure assets (BARTSCH 2007). As Section 2.2.2.1 highlighted, capital costs may vary substantially with the chosen accounting method. Generally, this cost component comprises all costs encountered in guaranteeing the serviceability of the assets of a water supply or sanitation system over time. Assets can be grouped into (OFWAT 2006):

- Infrastructure asset (e.g. underground systems, impounding and raw water storage reservoirs, dams and sludge pipelines);
- Non-infrastructural or operational assets (e.g. intake works, pumping stations, treatment works, boreholes, operational land, offices, depots and workshops, residential properties directly connected to supplies, and land around a water supply source held for the purpose of protecting the wholesomeness of water supplies); and
- Other tangible assets (including e.g. non-operational plants, machinery, vehicles and surplus land).

127 E.g. a large part of personnel costs forms part of fixed costs. Through outsourcing, personnel costs may also be substituted with costs of employing third parties.

Table 6: Operation and maintenance costs in water services provision to households

Water service component	Examples of associated operation and maintenance costs
Water supply – water resources and treatment	e.g. operation and maintenance of pumping equipment, aqueducts, reservoirs and associated dam structures; energy costs of pumping; buildings and equipment used for raw water abstraction, conveyance of raw water to treatment facilities; chemicals for raw water treatment and associated sludge and wastewater disposal; bulk purchases of water, costs of voluntary agreements with farmers[128];
Water supply – distribution of treated water through piped system	e.g. operation and maintenance of pumps, buildings and equipment used for the transfer of water from treatment to service reservoirs, service reservoirs and towers, aqueducts and mains (including e.g. actual repairs; mains flushing, disinfection and testing); energy costs of pumping; costs of inspections of e.g. reservoir security and structural and aesthetic conditions of the site; installation, replacement and repair of water meters; monitoring activities (e.g. pressures and water quality); inspection activities (e.g. of new domestic connections);
Sewerage services – collection	e.g. inspection activities (e.g. maintenance examinations and physical inspection of man entry sewers), sewer cleansing and removal of blockages; sewer maintenance and repair including manholes and emergency work; inspection and operational maintenance of pumping stations; energy costs of pumping;
Sewerage services – treatment	e.g. energy costs of sewerage pumping, provisions of preliminary treatment, operation and routine maintenance of all aspects of primary, secondary and tertiary sewerage treatment;
Sewerage services – sludge treatment and disposal	e.g. thickening of sludge by gravity and low energy processes, sludge digestion, mechanical sludge dewatering, air drying of sludge, sludge incineration, costs associated with sludge disposal, i.e. the removal of sludge from the point of collection to its final resting place (including intermediate storage costs);
Customer services[129]	e.g. correspondence with customers; billing processes; meter reading;
Other services	e.g. provision of laboratory services; water and sewerage/effluent quality reviews; provision of operational scientific services;

Source: based on Ofwat (2006), Ofwat (2007), Wetzel (1996).

128 Voluntary agreements with farmers may be considered to be avoided environmental costs, which figure as financial costs in the accounts of drinking water service providers but are caused by other water users, i.e. agriculture. In Chapter 3, these are defined and assessed as Type II EC, cf. Section 3.2.4.

129 While these costs are generally recorded under personnel costs, material costs, etc., they qualify as administrative costs according to the cost classification outlined above and the assessment framework proposed in Chapter 3.

The costs of servicing capital and the opportunity costs of equity should be accounted for. Issues that have to be considered include, among others, the interest payments on debt, the structure of amortization as it relates to depreciation, the economic lifetime of assets, associated bank fees and the level of inflation (cf. Section 3.2.2).

Environmental and resource costs

As Section 2.2.2 highlighted, the incidence of environmental and resource costs is always site-specific and determined by the local conditions under which water service provision is taking place. However, while the degree to which resource costs occur is dependent upon locally relevant alternative water uses (cf. Section 2.2.2.2 and Chapter 2.4. on the measurement of resource costs) and thus cannot be undertaken on a generalised level, environmental costs of water services provision can be sketched as a non-exhaustive list.

The provision of water supply services to households may, in particular, be linked to environmental costs stemming from an impairment of the natural rate of recharge due to overabstraction, with negative impacts on aquatic species, biodiversity and water dependent land ecosystems such as wetlands or forests. Furthermore, morphological changes caused by the infrastructural requirements for water supply (e.g. reservoirs, dams) or the chemicals and materials used for raw water treatment may also give rise to environmental costs (cf. Section 2.4.2 on the measurement of environmental costs).

In the case of sewerage management, environmental costs may take the form of water pollution through inadequately treated discharges or discharges into sensitive waters (i.e. an impairment of the natural rate of recovery), a loss in biodiversity or aquatic species due to polluted discharges, and the environmental costs of sewerage sludge disposal.

These costs are sometimes accounted for and thus internalised into the financial costs of water service providers through specified instruments. In the case of water supply services, these instruments include abstraction licences, (damage related) abstraction charges, or specific taxes on water use. For sewerage service, environmental costs may be internalised through specific taxes on wastewater, pollution discharge licences, or pollution charges, which are dependent upon the pollution load of effluent water.

2.2.3.2 *Cost factors in the provision of water services to agriculture*

Water services to agriculture are understood in this study as comprising water provision for irrigated agriculture (both as self-service and as collective service, see Section 2.4.1). The costs of irrigation water use cannot be easily divided into clear cost categories, particularly, as costs become more complex and difficult to measure under increasing degrees of scarcity (OECD 2002). If water resources are abundant, the attribution of the financial costs of irrigation systems can be straightforward, as they are often privately assumed by farmers. In contrast, in regions with scarce water resources, a long range of additional intermediate costs may be related to the provision of irrigation water services (OECD 2002).The problem of multiple direct and indirect users and beneficiaries besides irrigated farmers further complicates the clear attribution of costs[130] for these services, most notably in large irrigation areas.[131]

Furthermore, historic or current state transfers play an important role in agricultural water services provision, complicating a clear cost accounting. Large water-works have in the past been traditionally realised and financed by the state, in the framework of structural policies aimed to sustain the agricultural economy (MASSARUTTO 2002A), which has often promoted irrigated agriculture and thus encouraged high water consumption.[132] Generally, the composition of the different cost components can vary significantly across areas both within and across countries. The level of costs will also vary depending on the local conditions and the accessibility of water resources (e.g. groundwater use is generally related to higher pumping costs).[133]

---

130 The allocation of costs among different users should be based on the degree to which the users benefit or participate in the service (cf. Section 3.4.2 on '*adequate contribution*').

131 EASTER AND LIU (2005:13) demonstrate that irrigation projects generally serve multiple beneficiaries, as they are often multipurpose projects (e.g. supplying water for households next to irrigation water or promoting flood control and hydropower), or provide for indirect beneficiaries by means of increased agricultural production.
For the European context, OECD (2000:13) summarises that large and complex facilities and institutional arrangements are needed in Member States facing a high degree of seasonal water scarcity, as water has to be stored and transported over large distances and distributed through complex networks.

132 As OECD (2002:12) points out, the role of subsidies may be considered less influential in terms of dollar per cubic meter, but more so in terms of taking the initial decision to become a groundwater user, as the existence of subsidies for covering part of the required investment costs may induce more farmers to undertake the necessary investments.

133 According to OECD (2002), irrigators using groundwater resources usually pay all costs associated with the water use, except for environmental and resource costs. Hence, the costs incurred by farmers using groundwater resources can be taken as an (minimum) approxima-

In general, three levels can be discerned at which costs are encountered: (1) the private level, i.e. the farmer, (2) the district level, i.e. farmer organisations, and (3) the state / authority level. Whether costs are encountered at private or at district level depends largely on the availability of water resources and the irrigation techniques applied. Under good availability of groundwater resources, private costs tend to be more relevant. The level of state costs is estimated by OECD (2002:15) for the Spanish case within the range of 5.1 percent to 32.2 percent of total costs.[134]

The following paragraphs briefly highlight important cost factors within the cost components of financial costs and environmental and resource costs for water services provision to households.[135]

Financial costs

Operation and maintenance costs

Annual maintenance planning for agricultural water service provision is not an easy task, as many irrigation systems are living infrastructural systems, which change over time. While canals and infrastructural systems are expanded in some regions, they are neglected in other areas because they are no longer needed or are too difficult to maintain (CORNISH ET AL. 2004A). Furthermore, determining suitable norms for adequate maintenance is not trivial, as restoring assets to their original state may not be possible, or in many cases, useful (CORNISH ET AL. 2004A). Generally, operating costs can be grouped into the following sub-categories:

- Personnel costs (e.g. for operating equipment and managing irrigation systems at the private and district levels (e.g. water-works used to service farmers or those contributing to regulate river runoff), for undertaking routine preventive maintenance and repairs on pumping equipment and field irrigation equipment at the private level and on multiple-user infrastructures owned by the state or government agencies)[136];

---

tion of the ability-to-pay of farmers operating under similar conditions but using surface water.

134 It should be noted that the costs incurred at the three levels cannot simply be added up at the individual level, as many irrigtaors face either private costs or district costs, etc.

135 Again, the sub-components of administrative costs, other direct costs and taxes and subsidies are not taken up in detail here, as they have already been illustrated with examples above.

136 As OECD (2002) notes, personnel costs *"vary significantly across irrigation techniques, crops, labour market conditions and land-holding structure"*.

- Energy costs (electricity connections plus usage fees, mainly related to pumping),
- Material costs (such as chemicals for treatment, tools, spare parts, equipment, and administrative supplies).

Depending on the organisational structure for irrigation water services these costs are incurred at different levels (i.e. at individual or collective levels).

Capital costs

The exact specification of this cost category for agricultural water services is dependent upon the irrigation techniques applied and the kind of organisational structure under which water service provision is taking place (i.e. at individual or collective levels). Generally, it includes:

- Investment costs associated with the construction of water-works used to service farmers, the building of water reservoirs and canals, the regulation of river runoff and other infrastructural systems required for the provision of irrigation water;
- Investment costs for field irrigation equipment.

These costs generally vary with location, water delivery method, irrigation technology, and season (MASSARUTTO 2002A). For the analysis, the costs of servicing capital and the opportunity costs of equity should be accounted for (cf. Section 3.2.2.2).

Environmental and resource costs

While environmental costs of agricultural water services are just as site-specific as in the case of households, they account on a global level for a large and growing share of water pollution (DINAR 1998).[137] Foremost, agricultural water services may affect the environment through (EASTER AND LIU 2005, OECD 2002, MASSARUTTO 2002A, OECD 2001, EUROPEAN COMMISSION 2000, IEEP 2000):

- Groundwater contamination by pesticides and nitrates due to over-fertilisation;
- Soil salinity and water logging;

137 Next to the natural conditions under which the agricultural water use is taking place (climate, soil condition, etc.) the extent to which agricultural water use leads to environmental costs may be influenced through agricultural policies that promote water consuming crops or subsidisation schemes for the erection of irrigation technologies, etc.

- Overexploitation of water resources, which affects water and non-water-related ecosystem services and functions;
- Soil erosion and the loss of soil fertility;
- Negative impacts upon wetlands, biodiversity, wildlife habitats and landscapes;
- Adverse effects of river regulation and flow diversion required for irrigation purposes.[138]

There is clear evidence that the highest environmental costs stem from self-supply from aquifers, while large transfers of surface water create fewer problems (i.e. related to the risk of low flows and impacts on landscape) that can be more easily tackled (e.g. through abstraction licences) (IEEP 2000). Furthermore, environmental costs linked to irrigation water use are often interdependent, which adds complexity to their measurement.

Similar to the case of households, the assessment of resource costs cannot be undertaken at a generalised level, due to its dependence upon locally available alternative water uses. However, as agricultural water use is particularly important during summer months with low precipitation levels and varies with harvest times, the incidence of resource costs generally varies on a seasonal basis.

2.2.4 Framework conditions and cost drivers

The costs of water service provision are influenced by external framework conditions and cost drivers, which may vary regionally and over time. Relevant framework conditions for water service provision range from natural characteristics of the service area (e.g. availability of water resources, quality of raw water sources, topography), to socio-demographic factors (e.g. size and structure of the service area, population density, number of connected inhabitants/farms and related water use per connection) (cf. HOLLÄNDER ET AL. 2008). Conditions are inherently site-specific and need to be considered as a relevant reference frame to water service provision of an investigation area in order to achieve meaningful results, in particular when it comes to cross-regional or cross-country comparisons (cf. Section 3.3.1). Costs of water service provision

138 As outlined by OECD (2002), water service provision to agriculture may also be linked to external benefits e.g. in terms of landscape preservation. In the case of Japan, irrigation and drainage canals support urban infrastructure by letting in drainage from city areas. In Spain the development of irrigation in isolated or rural and aging communities helps as a side effect to maintain an active population in such locales and to promote agro-food industries.

may differ widely due to these factors, as discussed in-depth by HOLLÄNDER ET AL. (2008, 2009) for the case of drinking water provision.

In addition to these framework conditions for water service provision, the impact of the following two driving forces on the costs of water service provision is widely acknowledged and appropriate policy responses are subject to intensive debate, both in the political arena as well as in the scientific community: demographic change and climate change. Due to their encompassing nature, these factors are now revisited in greater detail.

Demographic change

Over the past decade, population decline induced by demographic change has led to reduced water consumption in many European regions. This trend was often amplified by regional migratory movements, ongoing changes in consumer behaviour with the adoption of water-saving technologies and accompanied by falling commercial and industrial demand for water services (LIBBE 2007). For many providers of water supply and wastewater management services, demographic change has necessitated a change in strategies, both in terms of required adaptations of infrastructural systems (e.g. adapting network capacities, increasing the flexibility of existing infrastructural systems, adjusting medium to long-term capacity planning, adapting maintenance and investment strategies) as well as in terms of a redesign of tariffication schemes to recover the rising fixed costs per cubic meter of water provided (HOFFMEISTER ET AL. 2008, cf. also Section 2.1.4).

For the network-related technical infrastructure, the following functional problems may result from reduced water demand and wastewater flows caused (in part) by demographic change (HOFFMEISTER ET AL. 2008, LIBBE 2007, KOZIOL 2004):

- In the case of water supply, reduced water flow in the pipe network leads to lower flow velocity and in consequence longer retention periods or stagnation in parts of the network, with negative effects on drinking water quality[139] and a risk of bacterial aftergrowth[140];

139 In addition to these network-related effects, reduced abstraction levels may lead to rising water tables with negative impacts on houses (i.e. damp cellars).

140 As KOZIOL (2004) points out, "*this danger* [bacterial aftergrowth] *is particularly great during the temporary fall in consumption during the summer months (school holidays, etc.) and in the case of combined fire-fighting water and public drinking water supply, which requires large pipe diameters.*"

- For sewerage services, reductions in the flow velocity within the sewer system may induce sedimentation within the pipe system (especially in low gradient sections of the network), leading to corrosion and odour problems. As a consequence, maintenance needs of the network may increase and earlier exchanges may become necessary.
- Decreasing amounts of effluent may lead to a reduced efficiency of existing, increasingly oversized sewage treatment plants.

Despite these negative effects, a reduction in overall system capacity is often not feasible, owing to the necessity of guaranteeing security of supply and ensuring fire protection provisions. Furthermore, the infrastructural system cannot be adjusted continuously, but rather in a step-wise process. Due to the water sector's high capital intensity and the long economic lifetime of assets, the possibilities for reducing fixed costs are very limited in the short-run. Instead, efficiency gains may be generated through technical or organizational optimization. If a deconstruction of infrastructure is not feasible or advisable from an economic perspective, the negative effects of system underutilization can be mitigated by regular flushing of the pipe system, affecting operation and maintenance costs. In any case, adaptations of the network have to be closely meshed with measures of urban development in order to avoid inefficient investments.

Against this background, the structure of tariffication schemes and related incentives move to the centre of water service providers' attention. In regions unaffected by water scarcity, setting incentives for further reducing water demand may be considered counter-productive in terms of sustainable water service provision.

Climate change and climate variability

Although the impacts of climate change on water resources vary widely among European regions, three main challenges to the management of water resources can be identified: (a) an increase in extreme precipitation events, (b) a decrease in water availability during summer seasons, and (c) a resulting deterioration of water quality (HOLZWARTH ET AL. 2007). Overall, climate change may necessitate adaptations in water user behaviour as well as in tariffication schemes to temper potential negative effects on sustainable service provision.

For *water supply and wastewater management systems*, the main challenges emerge from an increased intensity and frequency of extreme events. High precipitation levels put sewerage systems under additional pressure, with the hydraulic capacity of networks being exceeded more frequently (HOLZWARTH ET

AL. 2007). In addition, periods of high precipitation levels can lead to increased water pollution from diffuse sources (KRYSANOVA ET AL. 2005),and inundations of sewerage systems, or overflow from wastewater treatment plants could be aggravated (HATTERMANN ET AL. 2008). While increased precipitation and runoff is thus expected to lead to an increase in water availability in northern Europe, which could be beneficial for water supply service provision, regions in southern and central Europe are expected to suffer increasingly from seasonal water stress. In consequence, water supply systems will have to accommodate longer periods of warmer and drier weather and will face the challenge of satisfying consumer demand during periods of intensified water shortages.[141]

In addition to water supply and wastewater services provision, the agricultural sector and thus the provision of *irrigation water services* will also be affected by climatic changes. As OECD (2009) remark, "*farming systems and water resources have been increasingly vulnerable to climate change and climate variability*" and the trend is expected to continue (IPPC 2008).

Like its impact on water supply services, climate change is expected to have a dual effect on irrigated agriculture (OECD 2009), differing from southern and central Europe to northern Europe. Especially in arid regions, agriculture will be affected by decreased precipitation or increased evapotranspiration levels (or both). In central Europe, projected shifts in precipitation patterns are expected to reduce water availability during the vegetation period in summer, leading to an increase in the demand for irrigation water. Rising temperatures and evaporation rates would also aggravate the situation in southern Europe, where the dependency on water for irrigation is high and water resources are already scarce (HOLZWARTH ET AL. 2007). Not only cash crops, but also livestock, could be affected, especially during drought or heat waves (HATTERMANN ET AL. 2008). For farmers, climatic change may translate into significant additional costs, either from increased expenses for water pumping and transportation, or from production losses (HATTERMANN ET AL. 2008). Nevertheless, in regions where higher temperature and/or precipitation levels reduce climatic constraints for agricultural production, the anticipated changes in climate may also turn out to be beneficial (HATTERMANN ET AL. 2008 quoting EISENREICH 2005). In northern Europe, the agricultural sector could benefit from longer vegetation periods and higher crop yields leading to positive opportunities and benefits for agriculture in these

141 Other potential effects on the environment include: reduced water levels affect water quality negatively, as pollutants from point sources will become less diluted. In combination with increased water temperatures and reduced dissolved oxygen levels, this could seriously affect the ecological balance of freshwater systems (HOLZWARTH ET AL. 2007). Low run-off levels would further accelerate this problem, leading to lower flow rates and less mixing of the water body in streams (HATTERMANN ET AL. 2008).

regions, e.g. as new land may become available for farming (HATTERMANN ET AL 2008). These developments will, however, be mirrored by land lost for farming in other areas, e.g. in water-stressed regions due to desertification and salination processes as well as by production losses. In addition, northern regions may face an increase in nutrient losses and erosion due to increased precipitation with negative impacts on soil fertility as well as eutrophication of aquatic ecosystems.

For agriculture in flood-prone areas, increased damages may result from flood events. Higher precipitation rates in combination with higher temperatures can also lead to production losses due to an increase of certain plant diseases (HATTERMANN ET AL 2008). Despite these challenges, agriculture is expected to have a high potential for adaptation to climate change, e.g. by diversification and adjusted management schemes (i.e. through shifts to less water-demanding crops and shifts in cropping seasons) (HOLZWARTH ET AL. 2007).

At the water management level, an integration of policy spheres will become even more important (agriculture, water management, land use planning, flood protection, nature conservation). For the European context, the WFD, the Floods Directive and the Marine Strategy Directive offer an umbrella for community-wide activities. In addition, funding schemes from the Common Agricultural Policy could be targeted at further improving the adaptive capacity of European agricultural systems (HOLZWARTH ET AL. 2007). To frame its activities, the European Commission launched a Green Paper on *'Adaptation to Climate Change'*, and a Communication on *'Water Scarcity and Droughts'* in July 2007, which is based on a current assessment on the water scarcity problem, and addresses measures needed to complement existing policy responses.

Article 9 of the WFD is also expected to offer guidance in the adaptation process to climate change due to its second objective of *'setting incentives for efficient resource use'*. Pricing policies are considered a key element in adaptation strategies (HOLZWARTH ET AL. 2007). Against the background of the notable regional variations of the effects of climate change across Europe, it will be important to exploit pricing schemes in a differentiated and well-adapted manner. Overall, climate change can be expected to prompt adaptations in tariffication schemes in order to better accommodate the new challenges posed to service providers in this context.

2.3 Water pricing as a tool for achieving cost-recovery and beyond

With its requirement to establish by 2010 water pricing policies that *"provide adequate incentives for users to use water resources efficiently"* the WFD assigns water pricing a prominent role in its overall implementation process. In water policy-making across Europe and beyond, the objectives attached to water pricing as a policy tool have undergone significant changes over the years. A vast body of literature attests to the constant interest given in academics to the different facets of water pricing. Water pricing was foremost conceived as a tool for achieving an efficient allocation of water resources among different and competing uses. As public budgets became more and more restricted, water pricing was increasingly regarded as an appropriate financing mechanism. Justifications for the introduction or reform of water charges centred on the financial need for cost-recovery to fund further projects, relieve state finances, extend water services provision to remote areas and ensure the physical integrity of and continued benefits from irrigation schemes (MOLLE AND BERKHOFF 2007) and household water services.

The Dublin Conference (1992) marked a turning point in the debate (MOLLE AND BERKHOFF 2007) and may be regarded as a landmark shift to the economic dimension of water use (see Chapter 2.1). While the emphasis of water management practices had before centred on supply management tools, the focus now shifted towards demand management with the aim of better employing available resources. In the 1990s water prices, and more generally economic incentives, came to be seen as key policy tools endowed with the potential to achieve multiple objectives.[142] Tied in with this development, ecological objectives, which made water resources conservation a critical issue, moved into the focus, and the debate on water pricing became more oriented towards incentive setting. This shift in policy objectives was reinforced by an increasing intersectoral competition for water. In response, water prices were regarded as a

142 From the 1990s onwards, a growing body of literature explored the theoretical principles and potential impacts of water pricing and water markets. JOHANSSON (2000) interpreted water pricing in a World Bank report as a *"primary means...to improve water allocations and to encourage conservation"*, while JONES (2003) stated for the OECD that *"anything scarce and in demand commands a price"*, and that consequently *"water pricing is increasingly seen as an acceptable instrument of public policy"*. The Asian Development Bank highlighted in a report on its water strategy from 2001 the *"needs to promote efficiencies in water use by supporting demand management, including water pricing"* (for an overview on the evolution of the water pricing debate, cf. MOLLE AND BERKHOFF 2007).

means for promoting desirable environmental objectives, while maximising water productivity and aggregate economic welfare (MOLLE AND BERKHOFF 2007).

In practice, however, prices in the water sector are rarely encountered in microeconomic terms, but rather as a complex mixture of financial instruments (such as tariffs, specific earmarked taxes and charges, general taxes) along the value chain (MASSARUTTO 2004A). Water pricing is thus only one tool among others for achieving cost-recovery. As the analysis of Article 9 in Chapter 1 as well as the investigation of the WFD cost-recovery understanding in Section 2.2.1 made clear, water pricing in accordance with the polluter-pays principle along with internalisation instruments for environmental and resource costs is seen as the primary mechanism for achieving cost-recovery in the WFD context. Furthermore, an emphasis is set on the incentivising function of prices, which may in part contradict the cost-recovery objective (cf. Chapter 1.4).

The following sections first provide a justification for water prices, both from an economic as well as from an environmental perspective. Subsequently, an introduction is given to prominent water pricing models, before specificities of pricing water services to households and the agricultural sector are investigated in greater detail.

2.3.1 Potential objectives and limits of water pricing

Considering the diverse nature of water as outlined along the three dimensions in Chapter 2.1 above, why should it be priced? Would this not attest to the predominance of the economic dimension of water? As the investigations below shows, water pricing can be defended from an economic and from an environmental perspective alike.

Generally, three rationales are invoked in justifying water pricing (HECHT AND WERBECK 2006, EWRINGMANN 2006, KRAEMER ET AL. 2003, HANSJÜRGENS AND MESSNER 2002):

- Fostering allocative efficiency by informing users about the (full) economic costs of their water use and distributing water to its most productive uses. Furthermore, in this function, water prices promote an efficient use of financial means spent by public and private actors for water services infrastructure (*allocation and information function*).[143]

143 By including the full opportunity costs of water resources in water prices, resources are allocated to their most efficient use from a micro-economic point of view (given that certain market conditions are met, which can in the case of water be drawn into question (no free

- Guaranteeing financial stability of the water sector through cost-covering water service charges (*revenue generation and financing function*). Through water prices revenue is generated for sustaining the service independent of the public budget and financing further environmental investments to prevent or mitigate environmental and resource costs.
- Promoting water conservation objectives and ecologically sustainable water management through incentive-setting (*environmental incentive function*). In this context, prices are seen as policy instruments for inducing behavioural changes among consumers and water services providers in accordance with exogenously given environmental policy objectives, i.e. avoiding overuse and depletion of '*nature capital*' by curbing water demand though pricing. This rationale is often also linked to the polluter-pays principle.

The first two rationales relate to the economic perspective on water prices, while the third rationale reflects the environmental perspective. Both are discussed in the following paragraphs.

2.3.1.1 Justifying pricing for water services: an economic perspective

The allocation and information function of water prices

From a microeconomic perspective, water prices lead to an efficient allocation[144] of water resources, if each quantity of water generates the same value at the margin in all alternative uses.[145] Meeting this condition implies that collective welfare is theoretically maximised and cannot be improved further by a reallocation of water resources among different uses (*Pareto optimality*). In order for water prices to fulfil this allocative function, water users must be informed about the true cost of their water use, including the degree of scarcity and related welfare implications. Under this condition, water prices fulfil a sig-

---

movement of good / services possible, no free entry and exit into the market, etc., cf. Section 2.1.4 and Section 2.3.1.1 below).

144 TIETENBERG (1996) defines a set of four basic characteristics promoting efficiency: (1) Universality – a condition in which all resources are privately owned, and all rights are fully specified; (2) Exclusivity – all benefits and costs resulting from resource ownership and use accrue solely to the owner, either directly or through sale to others; (3) Transferability – rights to resources are completely transferable between resource owners in voluntary exchange; (4) Enforceability – rights are protected from involuntary seizure or encroachment by others.

145 If water generated the same value in all alternative uses at the margin, collective welfare is theoretically maximised and cannot be improved further by a reallocation of water resources among different uses.

nalling function and allow water users to make informed decisions. Ideally, water is then only employed for a certain water use, if the benefits from this use exceed or are equal to the total costs associated it.

Two important prerequisites are associated with the allocative and informative function of prices. First, all costs associated with the use of water resources have to be internalised into water prices (HECHT AND WERBECK 2006, HANSJÜRGENS AND MESSNER 2002). Second, a well-functioning market has to be operational in order to allow a free allocation of water resources to the most productive uses. Both prerequisites are difficult to meet in practice. While environmental costs are not easily accounted for (cf. Section 2.4.2) and therefore often left unconsidered in water pricing, the physical and network characteristics of water services hamper the establishment of a functioning market.[146]

In addition to these prerequisites, an efficient allocation of water resources in microeconomic terms may raise equity concerns (cf. Section 2.1.2), as only those who are willing (and able) to pay the equilibrium price will be using water. Accordingly, individuals with restricted means to pay for essential survival water uses may be disadvantaged in comparison with rich individuals, who may have a higher willingness to pay for non-essential uses (e.g. swimming pools, garden watering, etc.). Under social considerations, the price charged on users should therefore not exceed a critical limit value, which excludes those who cannot pay (BARRAQUÉ 1999).[147]

For the agricultural sector, the empirical evidence on the impact of water prices on the efficiency of water use is murky (CORNISH ET AL. 2004). For households, GARCÍA-VALINÃS (2005) point out that many empirical studies have analysed efficiency issues (GARCIA AND REYNAULD 2004, CASTRO ET AL. 2002, TIMMINS 2002, RENZETTI 1992), while only few have approached the distributional aspects of water pricing (RENZETTI 2000, HALL AND HANEMANN 1996).

In terms of Article 9, the allocative function of water prices is referred to in Objective 2, *"setting incentives for efficient resource use"* and in the need for basing resource allocation decisions on the total economic costs of water

---

146 Water cannot be easily moved to where it is demanded (e.g. due to high transportation costs, bulkiness and the risk of qualitative deterioration), cf. Section 2.1.4 and 2.1.1.

147 In addressing this point MOHAMED AND SAVENIJE (2000) note that efficiency consideration should focus on the socio-economic (rather than only the economic) net benefits to society of water use. Thus if one user is short of his basic water needs, the social willingness to pay will rise considerably (even if the economic ability to pay is zero) such that efficiency requires prioritizing the allocation of water needed for meeting the basic needs of that user. According to the authors, basic needs (i.e. drinking water) and public health needs (i.e. sanitation) should therefore be given an infinite value of willingness-to-pay (i.e. marginal benefits) in the demand curves of water.

resource use by considering both financial and environmental and resource costs in water pricing.

The financing function of water prices

A second economic justification for water prices is provided by their financing function. Through pricing, the necessary financial means can be generated for ensuring the financial viability and the sustainable provision of water services, independent of volatile public budgets and political will. While the initial investment in water service infrastructure has in most European Member States been undertaken with significant contributions from the public budget, subsidies both for investment as well as for the operation and maintenance of water supply and sewerage services have been significantly reduced since the 1980s (KRAEMER AND BUCK 1997, KRAEMER 2003).[148] The reduction in public support for water services was facilitated by increasingly restricted public finances and a growing consensus that the provision of water services should adhere to the principle of cost-recovery (OECD 1987, MOLLE AND BERKHOFF 2007).

As in the household sector, the rationale of financing service provision through contributions from users gained importance in the irrigation sector (MOLLE AND BERKHOFF 2007). However, here the focus was less on refinancing the initial project investment costs, but rather on being able to finance the long term maintenance of the capital infrastructure and the (critical) nature capital. In this understanding of cost-recovery, termed '*sustainability cost*' or '*quasi-full cost-recovery*' by TARDIEU AND PRÉFOL (2002), there might well be a public contribution for the initial investment, provided that later on, users will be able to continue financing operation and depreciation of the system in order to prevent its value from deteriorating, thus preventing the dependence on new public contributions in the future. In general, the cost definition applied (e.g. the degree to which initial investment costs or environmental and resource costs are considered)[149] determines the level of water prices under the financing function.

The financing function of water services is referred to implicitly in Objective 1 of Article 9 "*achieving cost-recovery of water services*", as well as in the underly-

---

148 OECD (2003) provides an overview on the incidence of subsidies in the water sector.

149 Under the financing function of water prices, the role of environmental and resource costs is often limited to the extent that they affect the future costs of providing the service. If lower groundwater tables result in higher pumping costs, or purification costs increase due to lower water quality, the environmental costs of water use affect the (future) costs of service provision. Similarly, the merit good characteristics of water services (e.g. public health benefits) are often neglected by this view on water prices. (MOHAMED AND SAVENIJE 2000)

ing objectives, *"sustainability of water services provision"* and *"transparency of financial flows"* (cf. Chapter 1.4).

2.3.1.2 Justifying pricing for water services: an environmental perspective

The incentive function of water prices

From an environmental perspective, water prices are considered an effective policy tool for inducing environmentally friendly behaviour and for reaching environmental policy objectives in an efficient manner (MOLLE AND BERKHOFF 2007, KAIKA 2003, AVIS ET AL. 2000, EUROPEAN COMMISSION 2000, DE MOOR AND CALAMI 1997). The rationale for this use of prices is based on the theory of environmental taxes, evolving from the work of Pigou in the 1930s to the theoretical underpinnings provided by BAUMOL AND OATES (1988).[150] Using prices for stimulating environmentally friendly behaviour is based on the idea that water users react negatively to price increases and therefore reduce their water consumption when charged higher prices. Thus, setting prices at an appropriate level may trigger water conservation and curb down excessive abstraction of water from ecosystems.

In theory, water prices have the potential to be more cost-efficient than traditional command-and-control instruments, as polluters are given more flexibility with regard to the ways in which they achieve the objective.[151] The incentive function can, however, only develop its potential if rates are set sufficiently high to stimulate a change in actors' behaviour (i.e. to invest in emission abatement or in less water consuming irrigation techniques). Theoretically, water prices can be structured in such a way that they align actor incentives with water policy objectives, promote the implementation of the polluter-pays principle and simultaneously generate the necessary financial means for counteractions (i.e. by burdening the responsible parties with pollution-related costs). As a prerequisite, water prices must vary in relation to the quantity of water consumed or the quantity and quality of discharges, making water metering a prerequisite. [152]

---

150 While PIGOU proposed to determine the level of a tax in such a way as to internalise externalities completely, thereby fostering an optimal allocation of environmental resources, BAUMOL AND OATES recognised the difficulties inherent in measuring externalities precisely and therefore advocated the use of taxes even if their target level has been determined exogenously (cf. also the discussion in MASSARUTTO 2004A).

151 The incentive function of water prices may also promote innovation and competition by placing established practices under scrutiny (ENVIRONMENTAL LAW INSTITUTE 1999).

152 GREEN (2000) and OECD 1999A point out that the cost of metering should be considered as a transaction cost, which can only be justified if the value of water saved is greater than the

However, whether water prices actually lead to behavioural change and promote more efficient water resources use is dependent upon the shape of the demand curve: the more elastic the demand curve, the more effective is the price signal (cf. Figure 11 below).[153] Economic theory suggests that residential water demand should be price inelastic for three reasons (RENWICK AND ARCHIBALD 1998): (1) there exist no close substitutes for water in most of its uses; (2) the amount of money spent on water is generally a relatively small share of the typical household budget; and (3) water is frequently demanded jointly with some other complementary good.

Empirical investigations have shown that the elasticity of water changes along the demand curve (cf. Figure 11). If water is very abundant and used wastefully in non-essential uses, higher water prices lead to a more efficient and less wasteful use, i.e. the demand for water is elastic with respect to price. However, considering that water has no substitutes in its most important uses, the demand for water becomes price inelastic with increasing scarcity of water supplies (i.e. a one percent increase in price will lead to a less than one percent decrease in demand) and perfectly inelastic under extreme water scarcity (DALY AND FARLEY 2004:197).[154] Accordingly, reference to a demand curve as being inelastic or elastic must always be made for a specific range of prices. In addition, MASSARUTTO (2005) points out that technical or infrastructural characteristics may also influence the elasticity of water demand, as they may influence

---

cost of metering. The OECD adds that, by promoting improved public water infrastructure, metering can make it easier to spread the health benefits associated with improved water quality to a wider segment of the population. Low price elasticities of water demand are found in the literature, but paradoxically the introduction of metering is reported to have allowed significant reductions in water demand in most case studies analysed in the literature, leading to reductions in the range of 10–40 percent of initial demand. But, as OECD (2009) highlights, the choice of metering may also induce a positive selection among water service users, with only those connecting to metering devices expecting to profit from it.

153 The price elasticity of water demand signals the percentage change in demand for water that is to be expected from a percentage change in water price.

154 As water is classified as an inferior good, income elasticity should be low as well. However, different effects may play a role. For example, higher income may lead to higher value uses (e.g. part of consumption of tap water is replaced by mineral water) or higher water use (due to higher consumptive applications such as swimming pools, etc.). BAUMANN ET AL. (1998) and DALHUISEN ET AL. (2003) report that income correlates positively with household water consumption, although not in clear ways. MARTINEZ-ESPINEIRA AND NAUGES (2004) find that water bills do not seem to be a determining factor in explaining water use, if water bills account only for a very small proportion of disposable household income. DOMENE AND SAURI (2006) report in reference to LOH AND COGHLAN (2003) that indoor water consumption remains relatively stable for different income categories, while larger variations are linked to outdoor uses, as documented by SYME ET AL. (2004).

the adaptive capacity of users, both in the case of households (i.e. water-saving appliances, retrofitting of water-using fixtures)[155] and agriculture (i.e. change in crop type, adaptation of irrigation equipment). For very low prices the effect of a price change on consumption is negligible, as consumption is rather determined by technical system characteristics than by the price of water. For high price levels, elasticity tends to be lower as a certain minimum consumption cannot be substituted and is thus indispensible.[156]

Figure 11: Elasticity of water demand along the demand curve

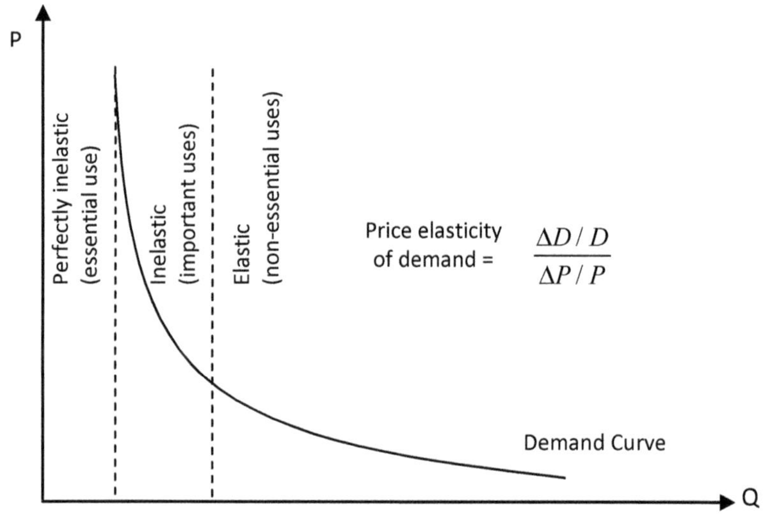

Source: Daly and Farley (2004:197).

In the literature, water demand elasticity values are generally reported to be quite low.[157] For drinking water demand by households, OECD (2009) concludes

155 Based on an evaluation of studies, DOMENE AND SAURI (2006) conclude that the adoption of domestic water-saving devices (e.g. water-saving appliances for toilets, showerheads and faucets) and garden irrigation technologies may have a considerable negative impact on water consumption levels. In reference to GATERSLEBEN ET AL. (2002), the authors conclude that water use seems to "*relate more to socio-demographic variables (household size and income) that influence specific behaviours, than to attitudinal variables*", cf. also Section 2.2.4.

156 Studies by GÓMEZ-LIMÓN AND BERBEL (2000) and VARELA-ORTEGA ET AL. (1998) demonstrate that while demand for irrigation water is inelastic for low water prices, higher prices result in an elastic interval on the water demand curve.

157 Only limited studies on water demand elasticities are so far available for the European context.

that *"estimates found in European research in the last 35 years consistently report a low responsiveness of demand to marginal price changes"*, with a 10 percent price increase leading to an expected decrease in water demand of 1 percent to 2.5 percent. While the evidence from studies is scattered, DALHUISEN ET AL. (2000) conclude from a meta-analysis of empirical or modelling studies on water price elasticities for residential water use that (1) the vast majority of studies indicates that both price and income elasticities of water demand are smaller than 1 (in absolute terms), but significantly different from zero;[158] and (2) that price elasticities depend on the season[159] and the exact use for which water is demanded (demand for in-house vs. outdoor uses[160], industrial vs. domestic vs. agricultural demand, etc.).[161] Furthermore, water demand tends to be more sensitive to water prices in the long-run than in the short-run, because over longer time periods, capital investments are not fixed (Olmstead and Stavins 2008).[162]

For the case of agriculture, CORNISH ET AL. (2004A) highlight a number of influencing factors on the price elasticity of water demand: it is shaped by the availability and relative cost of alternative water sources (i.e. groundwater vs. surface water supplies); it depends on the crop value, with elasticity being higher for low value crops[163]; the relative cost of water as an input to the production

158 DALHUISEN ET AL. (2000) and OLMSTEAD AND STAVINS (2008) find an effect on demand for water of reporting the price of water on the bill.

159 Winter demand was reported in several studies as less sensitive to price changes (cf. RENZETTI 1992, HOWE 1982, CARVER AND BOLAND 1980).

160 Outdoor uses are found to be more sensitive to prices (RENWICK AND GREEN 2000).

161 As DALHUISEN ET AL. (2000:7) point out, several difficulties relate to the estimation of elasticities. First, there is no general agreement upon estimation procedures. While estimation by ordinary least squares is generally acknowledged to yield biased estimates, there is no generally agreed upon alternative available. Second, the available studies on water demand elasticities differ in the detail in which various categories of demand have been distinguished (e.g. indoor and outdoor demand). Third, the authors note that studies may differ considerably with regard to location, time span, and variables that have been controlled for (like temperature).

162 Furthermore, as OLMSTEAD ET AL. (2007) outline the pricing structure may also affect price elasticity, with higher elasticities under increasing-block prices (in which the marginal volumetric water price increases with consumption) than under uniform volumetric prices (cf. Section 2.3.2.5). OECD (2009) also reports on larger effects linked to changes in tariff structures.

163 As RPA (2000) demonstrates, the value added of agricultural water use is highly crop-specific and can shift markedly in response to product market conditions. GÓMEZ-LIMÓN AND RIESGO (2004) observe that actual behaviour patterns vary significantly when specific groups of irrigators (clusters with homogeneous decision criteria) are studied. MASSARUTTO (2005) points out that higher water prices may also lead to higher water consumption in agriculture, if farmers shift to higher-value crops which consume more water. Raising water prices will then

process is decisive, with high production costs and a low relative share of water leading to low elasticities due to limited incentives for changing irrigation methods; it is influenced by application rates (i.e. where farmers are applying excessive amounts of water, there is scope for conservation without the necessity to change irrigation methods), the season and the phase of cultivation; it is dependent upon the ability to change to other less water-intensive crops or crop varieties and determined by the ability to change to more efficient irrigation technology. In addition, the price range for which the elasticity of water demand is measured may also play a role, with higher price ranges leading to higher elasticities (CORNISH ET AL. 2004).

Due to the specificities of irrigation water demand, HOYT (1984) argues, water prices will provide only limited incentives for water conservation in the short-run, as switching to more water-efficient technologies or management practices takes time. In cases where efficient, high-technology, on-farm water management is already in place, the potential effect of water prices on irrigation water demand is further reduced (CORNISH ET AL. 2004A).[164] Thus the short-term elasticity of water demand in agriculture may be very low compared to the long-term elasticity, depending on the irrigation technologies in place.

In a recent study, CHOHIN-KUPER ET AL. (2003) show that the use of water prices as a demand management instrument is far more limited in the agricultural sector, and that prices alone may not suffice to elicit significant changes in the behaviour of farmers, while DINAR (2000) and JOHANSSON (2000) provide some evidence to the contrary. Overall, there is more compelling evidence on changes in cropping patterns and technology, but these changes are typically caused by a variety of interacting factors, making it difficult to discern the effect of water pricing (MOLLE AND BERKHOFF 2007). In any case, for households and agriculture, price elasticities must be interpreted in the context from which they have been derived.

In Article 9, the incentive function of water prices is addressed through Objective 2, '*setting incentives for efficient resource use*' as well as Objective 3, '*making the polluter pay*' (cf. Chapter 1.4). It is important to highlight that the economic justifications for water pricing, i.e. the financing function and the allocative efficiency function of prices, do not have to coincide with water pric-

---

not reduce overall water consumption, but stimulate an increase in water productivity, i.e. the value obtained per unit of water consumed.

164 VARELA-ORTEGA ET AL. (1998) conclude from a study of three regions in Spain that the technical endowment in an agricultural district has a major effect on its response to water pricing. The higher the technical proficiency, the less responsive the district is to water pricing. They conclude that "*equivalent water charges would create widespread effects on water savings, farm income and collected government revenue across regions and districts*".

es as incentivising instruments. For setting incentives, water prices only have to vary with the quantity consumed, without necessarily meeting cost-recovery levels, as the success of an incentivising instrument can be determined by the extent to which initial revenues from it fall as behaviour changes (SPECK AND EKINS 2000).

2.3.2 Operationalising water pricing: an introduction to water pricing models

Opponents of pricing for water services highlight the low elasticity of water demand (cf. Section 2.3.1.1), the transaction costs linked to the administration of a charging system, the public good components of water resource uses and equity issues, which they find neglected in most pricing schemes. These issues can, however, be reduced through the design of the pricing structure. According to the OECD (1999B:17), pricing structures should be designed by considering allocative (economic) efficiency; equity; financial requirements for ensuring a sustainable service provision; public health aspects promoted through water services provision; environmental efficiency; consumer acceptability and understanding; and administrative costs (cf. Sections 2.3.1.2 and 2.3.1.1).

There is an extensive body of literature on different pricing models, testifying to the fact that for some economists, *"the question [is] no longer the desirability or possibility of using price regulation but a mere technical debate on how to determine the 'optimal price"* (MOLLE AND BERKHOFF 2007). In the following section, water pricing models which are most prominent in the literature are briefly reviewed, namely marginal cost pricing, Ramsey pricing, two-part pricing, block-rate pricing and peak-load pricing.[165]

2.3.2.1 Marginal cost pricing

Marginal cost pricing was first proposed by JULES DUPUIT in 1854.[166] In the 1930s and 1940s, the popular consensus held that marginal cost pricing constituted

---

165 Cf. RUSSELL AND SHIN (1996) for an extensive review of pricing theories.

166 As COASE (1970) points out, DUPUIT (1854) did not insist that a good's price should be made equal to its marginal cost, but rather indicated the loss that would be suffered if prices diverged from marginal costs. However, as DUPUIT also indicated the loss that would be suffered if prices *did* equal marginal costs, he did not take explicit sides in the debate and left the question open. Based on an investigation of DUPUIT's work, EKELUND (1968) concludes that Dupuit did not develop the short-run marginal cost principle in full, which he instead attributes to Hotelling (1938).

the optimal pricing rule for public utilities, even for those industries, where marginal cost is below average costs, necessitating a government subsidy for sustaining production (COASE 1970).[167]

According to standard microeconomic theory, the price of any good (under perfect competition) should reflect the marginal cost of its production. Prices signal a good's economic scarcity only if the consumption of an additional unit is priced at the additional cost caused by this last unit (cf. Section 2.3.1.1 on the allocative and informative function of prices). Under the condition that all relevant cost categories have been considered in the process, marginal cost pricing would lead to an overall improvement of society's welfare, as a good is only consumed up to the point where the utility obtained from consuming the additional unit is equal to the disutility created by its production. Proponents of marginal cost pricing for water resources point out that any divergence of prices from marginal costs leads to distorted price signals, and thus to an inefficient resource use from the perspective of society as a whole. In this understanding, only water prices equal to marginal cost allow consumers to adequately compare the benefits obtained with the costs (including the opportunity costs of foregone alternative uses) imposed by the consumption decision. A unit price, which differs from marginal costs, would then lead to consumption levels that are either too high (for prices below marginal cost) or too low (for prices above marginal cost) in relation to the socially optimal level of consumption (MONTEIRO 2005).

The suitability of marginal cost pricing for public utilities in general or for water services in particular, has been a continuous topic of debate in the literature (CHAMBOULEYRON 2003, Green 2003, CHÁVEZ AND QUIROGA 2002, MERRET 1997, RUSSELL AND SHIN 1996, KIM 1995, ZARNIKAU 1994, DANDY ET AL. 1984, BAUMOL ET AL. 1982, FELDMAN ET AL. 1981, TURVEY 1976, RIORDAN 1971, COASE 1970, HIRSHLEIFER ET AL. 1960, COASE 1946). While marginal cost pricing is regarded by some economists as the universally preferred pricing method, others highlight the problems linked to its implementation in the case of water services. The following arguments are put forward by critics of the universal application of the marginal cost pricing rule.

First, even among proponents of marginal cost pricing there is an ongoing debate whether short-run or long-run marginal costs should serve as the basis

167 In 1938 HOTELLING put the argument as follows: *"A free bridge costs no more to construct than a toll bridge, and costs less to operate; but society, which must pay the costs in some way or other, gets far more benefit from the bridge if it is free, since in this case, it will be more used. Charging a toll, however small, causes some people to waste time and money in going around by longer but cheaper ways, and prevents others from crossing. The higher the toll, the greater is the damage done in this way."*

for price setting. For the water sector, the difference between short-run and long-run marginal costs is particularly high, as the sector is characterised by high capital intensity (see Section 2.1.4 and Section 2.2.2). Accordingly, basing prices on long-run or on short-run marginal costs produces very different results. For new investment decisions the right signals can only be provided by long-run marginal costs.[168] However, in those cases where the infrastructural system for water service provision has already been constructed and demand is less than installed capacity, the cost of serving additional users is very low (cf. Section 2.1.4). As the fixed costs have already been incurred, they should not enter into the decision-making process and thus prices should be based upon short-run marginal costs.[169] Hence, the marginal cost principle does not have a straightforward method of implementation, when, at least in the short-run, capacity of operation is fixed, and demand varies over time (for example, by season as in the case of agriculture).[170]

Linked to the first argument is a second problem highlighted notably by COASE (1970), who posited that what happens at the margin is only one factor which matters in the formulation of a policy recommendation. If in addition to economic efficiency other aspects, such as the long-run financial viability of service provision, are to be considered, marginal cost pricing may not offer a suitable solution. For the case of industries, which are characterised by decreasing average costs (cf. Section 2.1.4), a pricing strategy based on short-run marginal costs will lead to negative revenue (BAUMOL ET AL. 1982).[171] Thus it is incompatible with an internal cost-recovery approach, as it only allows for recovering part of the operational costs, without any consideration given to

---

168 If long-run marginal cost pricing is applied, it may lead to less than optimal economic operations, because of the lack of incentive to achieve cost efficiency (CHETTRI AND VENKATESAN 1983:290).

169 In the late 1970s the deregulation economist ALFRED KAHN made a case for short-run marginal cost pricing for the airline industry by outlining the negative effects of an airplane taking off with empty seats as long as additional customers would be willing to pay for the short-run marginal costs associated with having them on board.

170 CHÁVEZ AND QUIROGA (2002) suggest that in those cases price should be set at levels that equal short-run marginal costs and to decide on the desirability of plant size expansion by comparing long-run marginal cost (i.e. including operation costs as well as cost of expanding capacity) to willingness-to-pay during high season demand. If demand in high season is greater than long-run marginal cost, expanding capacity leads to welfare improvements.

171 According to GARCIA AND REYNAUD (2004) *"marginal cost pricing does not reveal whether it is worth incurring the fixed costs. Marginal cost pricing requires that the public authority decides for consumers whether they are ready or not to pay for covering total costs. Unfortunately, in many cases, the public authority may lack information to correctly implement this decision."*

future capacity expansion or the replacement of infrastructure (CHÁVEZ AND QUIROGA 2002, MASSARUTTO 2004A). While collective welfare is maximised, choosing a price equal to marginal costs prevents the service provider from meeting the necessary revenue requirements for guaranteeing a sustainable operation (SHAW 2005, CHÁVEZ AND QUIROGA 2002), rendering subsidies a pre-requisite.[172] These, however, lead to distorted price signals for those goods or services from which the required tax revenue is collected.[173] Second-best solutions are then required if fixed costs are to be recovered (see below).

Third, the measurement of marginal cost can also be problematic (RUSSELL AND SHIN 1996) and is often linked to high costs, which may outweigh any efficiency gain that would in theory result from the use of the marginal cost approach (GREEN 2003:95). For the water sector, metering of water use is a prerequisite for the application of marginal cost pricing (signal to users, incentive to safe water). Metering, however, is expensive and can be regarded as a transaction cost.[174] Opponents of the marginal cost approach stress that that the cost of metering can only be justified if the value of water saved is greater than the cost of metering and water users react to price signals (cf. Section 2.3.1.2). Accordingly, the marginal cost approach is an optimal pricing rule only if the monitoring of individual consumption is not difficult or costly and demand is elastic enough.

Pertaining to the difficulties attached to the implementation of marginal cost pricing and its dependence upon subsidies for guaranteeing a sustainable service provision, there is little evidence of real-world applications of marginal cost pricing for water services provision (RENZETTI 2000).

---

172 As BAUMOL AND BRADFORD (1970) point out, if the necessary revenue is generated by means of an excise tax *"this is obviously a decision to make some prices depart from marginal costs after all. If it is obtained by an income tax, it is the price of labour, which is forced away from its marginal costs. Any tax, except a Pigouvian poll tax, which might perhaps be more felicitously called an "inescapable tax" will unavoidably affect some price."* Hence, one is in effect dealing with a problem in the area of second best.

173 An exception to this rule is given by Pigouvian taxation.

174 As GREEN (2003) posits in references to publications by OFWAT, meters require replacement and renovation within about a seven-year cycle. For the context of England and Wales, OFWAT calculated the additional costs of metering domestic consumers as £29.60 per year as compared to an average bill for water of £112 and £236 for water and sewerage combined. Therefore, whether or not metering is economically justified depends primarily upon the consumption figures, the marginal cost of water and the reduction in demand expected to occur as a result of metering.

2.3.2.2 *Ramsey pricing*

The main disadvantage in applying marginal cost pricing to the water sector (or any other natural monopoly sector) is that the operator will be dependent upon subsidies. This problem is addressed by a pricing framework which was first proposed by RAMSEY in 1927 and further formalized in the context of public sector pricing by BOITEUX (1956, 1971) and FELDSTEIN (1972), which aims to achieve cost-recovery for public utility service providers. Ramsey pricing is based on discriminating water users by their demand elasticity. Different groups of customers are charged the marginal costs of their consumption plus a quota of the fixed cost that is inversely proportional to their demand elasticity (BAUMOL AND BRADFORD 1970[175], MITCHELL AND HANEMANN 1994). Thus, water users with a less elastic demand are charged higher prices for their consumption. Since the Ramsey pricing rule derives from maximizing consumer plus producer surplus subject to a cost-recovery constraint, this method produces inefficient (i.e. *second best*) outcomes. As water is not a physical final good, demand characteristics may vary across time, customer classes, space and for the various uses of the provided water services (KIM 1995). [176] Hence, different groups of users can normally be easily identified. However, Ramsey pricing may not satisfy the equity criterion because low-consumption and low-income consumers are sometimes characterized by a very price inelastic demand (GARCÍA-VALIÑAS 2005).[177]

Probably because of its complexity, this method has not been adopted by industry (LOEHMAN 2004).[178] Although its design theoretically combines a striving

175 Under non-significant cross-price effects, the results of that model can be summarized by the following equation: (pi – c) / pi = δ * 1/ηii . This formula is known as Ramsey-Boiteux prices, Ramsey prices or the inverse price-elasticity rule. The parameters pi and c, denote, respectively, the price and marginal cost of ith good: η_{ii} is the elasticity of demand of ith good with respect to its own price; and δ is a constant, the so-called Ramsey number, which allows the margins over marginal costs to be adjusted in all markets. If the assumption of zero cross-elasticities is not met, the results lose some of their simplicity but remain valid, as shown in BAUMOL AND BRADFORD (1970).

176 KIM (1995) estimates marginal costs for a cross section of U.S. water suppliers and combines these estimates with demand elasticities to simulate second-best prices for water services provision, His results show that while the existing price structure is quite different from the one suggested by marginal costs, it does not depart substantially from the second-best optimum.

177 FELDSTEIN'S proposal (FELDSTEIN 1972) considers RAMSEY-BOITEUX pricing adjusted by a factor that reduces the regressivity associated to that kind of tariffs. The model proposes using the marginal utility of income to weight individual surpluses.

178 Furthermore, the required price discrimination between different groups of customers may also not be legally accepted.

for allocative efficiency with the requirement for costs to break even (i.e. the *financing function* of prices), in practical settings the two-part tariff approach is more widely used.

2.3.2.3 Two-part pricing

In 1946, a two-part tariff design was first proposed by COASE, which consisted of a share of fixed costs plus marginal costs, in order to address the problem of insufficient revenue-generation for achieving cost-recovery under decreasing marginal costs (COASE 1946, 1970). In practice, two part tariffs are often composed of a lump-sum connection fee and a per-unit charge approaching marginal cost (cf. Sections 2.3.3 and 2.3.4). Due to the fixed tariff component, two-part tariffs allow for meeting both efficiency[179] and revenue generating objectives. While this tariff design became well established in residential water service pricing, critics highlight that the fixed-cost allocation is arbitrary, and hence the scheme may not be individually rational, i.e. a user may be better off with self-supply (LOEHMAN 2004). Furthermore, the scheme may have efficiency problems (VOHRA 1990) in response to which BROWN AND SIBLEY (1986) proposed that '*personalized*' fixed cost shares should be set such that users would not choose to leave the system. In practice, the fixed charge is often based on meter size which may be independent of demand level.

2.3.2.4 Block-rate pricing

Another response to the perceived equity problem inherent in Ramsey pricing is a price structure of increasing or decreasing block rates (DIAKITE ET AL. 2009, HARRIS ET AL. 2002). Like Ramsey pricing and two-part pricing, this approach is based on the principle that the actual consumer should bear the full cost of maintaining the provision of the service as argued by COASE (1946).[180] Under an increasing block structure, water use per billing period is divided into a number of discrete blocks, for which consecutively higher prices are established (i.e. high-volume users are charged the marginal costs of water, while lower volume

179 It is often suggested that the extent to which a two-part tariff can exploit consumer surplus, depends on the income elasticity of demand for the service under investigation. But as GABOR (1956) shows, the relationship between income elasticity and the efficiency of two-part tariffs is rather complex.

180 Increasing or decreasing block-rate pricing can solve the problem of excess profits and losses generated by the marginal pricing in cases of economies and diseconomies of scale respectively (DIAKITE ET AL. 2009).

rates are set so that full cost-recovery is achieved).[181] This pricing structure allows for promoting economic efficiency in increasing returns to scale utilities (HARRIS ET AL. 2002), while simultaneously recovering the full costs of service provision. Equity concerns can be addressed by charging the first block at a relatively low per unit price and cross-subsidising small (and presumably poor customers) through returns obtained from customers in high consumption blocks.[182] In addition, increasing block rates are able to promote water conservation objectives (LIU ET AL. 2003), as they discourage excessive water use (i.e. under the assumption that water demand is more price elastic in high blocks).[183]

However, this pricing scheme may also be linked to a number of drawbacks: the determination of the number of blocks and the associated price increases may engender rent-seeking behaviour (HALL AND HANEMANN 1996 in LOEHMAN 2004, BOLAND AND WHITTINGTON 2000); the revenue of the water utility may become more variable as consumers may switch from one price block to another by changing their water consumption (DIAKITE ET AL. 2009); and the pricing schemes may score more poorly than other schemes in terms of simplicity, transparency and implementation. Mostly due to its potential in addressing equity concerns, the use of the increasing block-rate system is widespread in developing countries (LIU ET AL. 2003, BOLAND AND WHITTINGTON 2000), while in the developed context its potential for promoting water conservation objectives has made it a recommended approach.

2.3.2.5 Peak-load pricing

The peak-load pricing approach is based on the observations that demand is not evenly distributed over time but fluctuates with the season of the year and that capacity must be sized to meet maximum or peak demands (ARBUÉS ET AL. 2003, HARRIS ET AL 2002). Accordingly, a considerable part of the fixed costs incurred in the provision of water services stems from meeting peak demand from households and agricultural water users in dry summer months. In response, the theory of peak-load pricing was introduced and advanced by BOITEUX (1949), STEINER (1957) and most notably WILLIAMSON (1966). It promoted economic efficiency by informing customers of the added cost of their peak period use (WARD 2007). Ideally, peak-load pricing stimulates a demand reduction in peak periods

181 For the (original) argumentation on decreasing blocks, cf. WILLIG (1978).

182 Furthermore, GABOR (1956:32) highlights that if *"the rate of consumption is subject to wide seasonal variations, block tariffs appear to be more equitable, even if the total annual cost to the consumer is the same as under the two-part system"*.

183 The evidence on the price elasticity of demand under block tariffication systems is however mixed, cf. ARBUÉS ET AL. (2003) for a discussion.

and thus contributes to lower capacity expansion costs over time (BAKKER ET AL. 2003). By levying a larger share of the costs of service provision on peak period users, who cause overall higher fixed costs due to the required capacity expansions, this pricing approach problematises equity considerations (ARBUES ET AL. 2003).[184] Finally, peak-load pricing may also promote water conservation objectives, by providing incentives for reducing water demand during periods of water scarcity.[185] Problems with the application of peak-load pricing relate in particular to determining the time differentiated pricing schedule that results in optimal capacity (HARRIS ET AL. 2002).[186] In line with the approach's composition, peak-load pricing is mainly applied in areas suffering from water scarcity in dry summer months, and practiced foremost in pricing water for agricultural uses (cf. Section 2.3.4).

2.3.2.6 Summary Overview

The following Table 7 briefly summarises the main characteristics of the six pricing models introduced in this section along the justifications for water pricing: (1) allocative efficiency, (2) financing, and (3) incentive setting for water conservation (environmental perspective).

Table 7: Summary overview on water pricing models

	Marginal cost pricing	**Ramsey pricing**	**Two-part pricing**	**Block-rate pricing**	**Peak-load pricing**
Allocative function	yes	no	yes	yes	yes
Financing function	not under declining average costs	yes	yes	yes	yes
Incentive function for water conservation	yes	no	yes	yes	yes

184 However, as BEECHER ET AL. (1994:63) mention, the final effect of a seasonal pricing scheme is not certain, as it might only reduce average but not peak demand and thus fail to reduce the need for capacity expansion.

185 LYMAN (1992) found peak demand more elastic than off-peak demand and showed that the elasticity of peak-demand affects off-peak demand when consumers purchase more water-efficient appliances.

186 As HARRIS ET AL. (2002) point out, additional complications related to shifting peaks (PRESSMAN 1970), diverse production technologies (CREW AND KLEINDORFER 1976) and possible storage (NGUYEN 1976) are less important in the water sector, as there is mainly a seasonal peak (peaks of daily demand and reserves, e.g. for fire protection are systematically provided for).

2.3.3 Specificities of user groups: pricing water services for households

While introductory indications of the use of water pricing models in the household sector have already been provided in Section 2.3.2, the following paragraphs complement the information base with a brief summary of the results of a recent review of water pricing practices in OECD countries (OECD 2009).

In this study, the leading OECD investigation of water pricing policies and practices in OECD countries from 1999 has been updated to determine whether new trends emerge and to expand the analytical frame to non-OECD countries. In addition to an OECD survey (undertaken in 2007–2008), the study draws on the results of an investigation of pricing levels and structures undertaken by Global Water Intelligence in 2007 and 2008, which covered more than 150 cities in OECD countries, extending to Eastern Europe and emerging economies. In addition, data collected by the International Water Association (IWA) were integrated into the OECD assessment and further amended by expert studies on specific issues (e.g. related to the investigation of demand elasticities and the nature of social tariff structures).

The OECD (2009) report concludes that there is a continued decline in the use of decreasing block tariffs and flat-fee systems for household tariffs. Instead, two-part tariffs consisting of a fixed charge and a variable tariff component, i.e. a uniform or increasing block volumetric charge, are more often employed. The study reports that charges for wastewater service provision are increasingly separated from water supply charges. In addition, the degree to which wastewater charges are based on the actual costs encountered in the services' provision is on the rise, leading to increases in the prices charged for wastewater service provision. Overall, the study finds that the response of household water service users to marginal price changes is small. It finds a more significant (while potentially temporary) effect of tariff structure changes, in particular if a shift from flat to volumetric rates is undertaken. The study also confirms that social concerns continue to be an important element in the design of tariffication schemes and are addressed either through innovative tariff structures or parallel income-support mechanisms.

2.3.4 Specificities of user groups: pricing water services for agriculture

In the pricing of water services for the agricultural sector, four main approaches are found in practice, namely: (1) area-based pricing, (2) volumetric pricing,

(3) two-part pricing, and (4) block-rate pricing.[187] These are briefly introduced in the following paragraphs and reviewed with regard to their efficiency, financing, equity and incentive-setting prerequisites.[188]

2.3.4.1 Area-based pricing

The most widely used pricing scheme for irrigation water use is a fixed per hectare charge (DUDU AND CHUMI 2008, EASTER AND LIU 2005). Area-based charges, which are generally calculated according to the average cost pricing rule[189], are directly dependent on the size of the irrigated area (or on farm size, CORNISH ET AL. 2004A) and have no link either to the volume of water supplied or any indirect measure of consumption. As EASTER AND LIU (2005) point out, *"pure area-based pricing is appropriate in places where water is not scarce, where crops are not varied, and where meter installation is difficult or costly"*. Extensions of area based pricing include (EASTER AND LIU 2005):

- *Crop-based pricing*, i.e. fixed per hectare charges varying by crop type, often with higher rates for intensive water-consuming crops and artificially low rates for staple foods, thus encouraging water conservation and promoting social policy objectives (HELLEGERS AND PERRY 2004). Crop-based pricing may set particular incentives for changing cropping patterns, when irrigation charges are sufficient for making less water-consuming crops relatively more profitable (CORNISH ET AL. 2004A, CHOHIN-KUPER ET AL. 2003).
- *Area-irrigated based pricing*, i.e. per hectare charges differentiated by water delivery costs for different irrigation methods, with lower rates for gravity-based irrigation systems and higher rates for pump irrigation.
- *Area-season–based charges*, i.e. with higher water prices per hectare charged during dry and water scarce seasons, characterised by peak demands, and lower prices during off-peak wet seasons, when water is relatively plentiful. This pricing scheme may contribute to limit the number of hectares irrigated in peak-seasons and thus help to curb infrastructure expansion (cf. Section 2.3.2.6).

---

187 Cf. JOHANSSON ET AL. (2002) for a review of other pricing methods for allocating water resources and related evidence from case studies world-wide.

188 It should be noted that in the case of agriculture, water pricing schemes may be instituted at the level of farmers (direct service charge), the district level (contribution to covering irrigation district's costs) and the state level (e.g. environmental taxes).

189 Area-based charges may also form part of a two-part tariffication system (cf. Section 2.3.4.3 below), in which the area-based charge is designed to cover the fixed costs of service provision.

Another pricing scheme, which is so far only marginally applied, is *area-technology–based pricing*. Under this scheme, farmers using water-saving technology (e.g. drip and sprinkler irrigation instead of flood irrigation) pay lower water prices per hectare.

Area-based pricing schemes generate a predictable revenue stream and allow for the achievement of cost-recovery objectives through an easy-to-implement and transparent pricing scheme (HELLEGERS AND PERRY 2004, TSUR AND DINAR 1997).[190] Depending on the scheme's exact specification, only data about the farm size, the irrigated area, the crop choice or the irrigation technique applied is required, which is likely to be available or easy to collect in most cases (DUDU AND CHUMI 2008).[191] The transaction costs of the scheme are generally low, as no volumetric measurement with related equipment prerequisites is needed. Due to their non-volumetric measurement, simple per hectare water prices have been identified by some authors as *"the most adverse incentive affecting irrigation across OECD countries"* (OECD 2002). Since per area prices are fixed costs for farmers and hence the marginal cost of water application is zero, there is no further effect on water consumption through this pricing scheme once the irrigation area decision has been made. Thus, the scheme is likely to promote an inefficient allocation of water (DUDU AND CHUMI 2008) and will not reallocate water to more productive uses or sectors.

Even though area-based pricing provides no incentives to farmers for reducing water use per hectare, it is still widely applied in irrigation districts throughout the world due to the simplicity of its implementation (EASTER AND LIU 2005). Nevertheless, there has been a trend away from pure area-based pricing systems with most of the recently designed systems adopting new features.

2.3.4.2 Volumetric Pricing

Under the volumetric pricing approach for irrigation water, water prices are based on the actual amount of water delivered, rendering accurate metering of water use a prerequisite. Per unit charges can either be based on the marginal cost of service provision, promoting an economically efficient allocation of water (cf. Section 2.3.2.1), or on the average cost approach, enabling viable

190 However, as EASTER AND LIU (2005) point out, defining organisation and maintenance costs are important *"because the water supply entity may have an incentive to inflate the costs charged to farmers"*.

191 If the area-irrigated based approach is used, however, this data will need to be estimated for each period or season, as during wet seasons less area will be irrigated than during hot summer months (EASTER AND LIU 2005).

service financing without subsidies (cf. Section 2.3.2.2). Volumetric pricing can be applied even if the total quantity of water available to the user is set as a quota, which aims at efficient water allocation (Hellegers and Perry 2004). This pricing approach sets incentives for rationale water use and encourages water saving, provided the charges are significant in relation to the value of water and the cost of management. As long as the net incremental productivity of water exceeds the water price, demand will increase. The threshold at which demand is significantly affected by volumetric charges depends on the elasticity of water demand, the productivity of water, the proportion of land devoted to permanently irrigated crops and the irrigation technology applied (De Fraiture and Perry 2002, Garrido 2001), among other factors. Due to the need for metering and the related need for management (and control infrastructure), the implementation costs of this pricing approach can be significant (i.e. in the case of a large number of small users) (Hellegers and Perry 2004).[192] Furthermore, if the volumetric charges are based on marginal costs, this pricing approach does not allow full cost-recovery under decreasing average costs (e.g. in the case of large canal systems).[193] Due to its drawbacks, volumetric pricing is generally only applied in conjunction with another system, i.e. as one component of a two-part pricing system.

2.3.4.3 Two-part pricing

For irrigation water services, the two-part pricing scheme combines a volumetric pricing component with a fixed charge, which is often based on the size of the area irrigated (Easter and Liu 2005). As in the case of households, this combined approach promotes the cost-recovery objective of water pricing and ensures a certain revenue flow regardless of how much water is available and delivered, while simultaneously setting incentives for reduced water use through the volumetric tariff component. On the negative side, this dual pricing scheme is less transparent for farmers, requires metering and monitoring capacity and involves administrative costs as outlined above.

---

192 If metering is established at higher levels, the pricing approach loses its incentive function for the individual farmer. As a solution to this metering-level problem, Mohamed and Savinije (2000) propose that water user associations negotiate with farmers about the appropriate use and payment for each farmer.

193 As Easter and Liu (2005) report, the marginal costs of irrigation water services may exceed average costs in the case of pump irrigation using groundwater leading to revenue over-collection.

2.3.4.4 *Block-rate pricing*

Block rate pricing schemes for irrigation water use may promote stability of demand by varying the volumetric price depending on total water demand (HELLEGERS AND PERRY 2004). By charging a low price (even below organisational and maintenance costs) for the first block, this scheme may effectively address equity concerns and support subsistence farming at low costs. In order to compensate for the loss in revenue and to promote cost-recovery, a cross-subsidisation is, however, required through subsequent block rates, which need to be set high enough to ensure the recovery of fixed costs (TSUR ET AL. 2002). Block-rate prices are charged either to the individual farmer or group of farmers on a volumetric basis, thus requiring adequate metering and monitoring equipment. If the overall price level and price differentials between blocks are sufficiently high, block-rate pricing may effectively influence the extent of irrigation water use (DINAR AND MODY 2004, CHOHIN-KUPER ET AL. 2003). According to EASTER AND LIU (2005), this pricing scheme is appropriate to use in cases where water is scarce, farm incomes are low, and water charges are high relative to net farm income. However, as CHOHIN-KUPER ET AL. (2003) point out, this scheme is rarely applied in practice to irrigation water use.

2.3.4.5 *Selected evidence on water pricing in agriculture*

The empirical evidence on irrigation water pricing confirms the leading assumption that water prices for agricultural water uses tend to be too low for recovering the costs of service provision.[194] Furthermore, as the discussion on the elasticity of water demand in agriculture in Section 2.3.1.2 already highlighted, the impact of water prices on the demand for irrigation water is influenced by a number of conditions and surrounding factors. As VARELA-ORTEGA ET AL (1998) highlight, the impact of water pricing schemes depends strongly on regional, structural and institutional conditions. The researchers simulated the behaviour of farmers and their responses to different water pricing scenarios and found that equivalent water charges may create widespread effects on water savings, farm income and collected government revenue across regions and districts. ARROJO (2001) reports that the way farmers react to prices is significantly influenced by the available alternatives, i.e. by the possibilities to switch to less

---

194 According to OECD (1999B), *"marginal cost charging is rarely encountered in irrigation water prices. More commonly, irrigation prices are intended only to make farmers responsible for all variable costs of supplying water, whereas part or all the fixed costs are covered by public agencies, at tax-payers' cost."*

water-consuming crops and to practice extensification. Similarly, VARELA-ORTEGA ET AL. (1998) show for Spanish cases that farmers react to higher water prices first with a change to less water-intensive crop types, before exiting irrigation altogether by switching to rain-fed crops. Furthermore, the study found that the technical endowment in an agricultural district has a major effect on its response to water pricing. The higher the technical proficiency, the less responsive the district is to water pricing. By inducing crop changes, higher water prices may also lead to higher water consumption in agriculture. As MASSARUTTO (2005) points out, farmers may shift to higher-value crops which consume more water. Raising water prices will then not reduce overall water consumption, but stimulate an increase in water productivity, i.e. the value obtained per unit of water consumed. DE FRAITURE AND PERRY (2002) conclude that to induce a reduction in demand, considerable price increases are required (either in general levels of charging, or through more complex multi- level charges), which may often find no political acceptance. As MARTINEZ AND ALBIAC (2004) conclude, water pricing may also fail in curbing water demand, as the number of illegal and thus uncontrolled wells is too high in some countries, and the required price levels, which would have to be reached for affecting water demand, would be unacceptably high. Similarly, BERBEL AND GÓMEZ-LIMÓN (2000) highlight for the Spanish context that water prices will only affect demand for water at very high levels, and will then be associated with stark impacts on farm income (i.e. a reduction in farm income of about 40%), which may significantly affect rural development and employment. Due to these ambiguous effects of water prices on water demand and sustainability, several studies have emphasized the importance of a policy mix for turning water pricing into an effective tool in reaching water policy objectives (HELLEGERS 2002, DINAR AND MODY 2003).

2.4 WFD Article 9 and its implications for cost accounting and pricing of water services

The following sections are intended to complement the basis for the methodological framework developed in Chapter 3 on assessing cost-recovery in line with Article 9. They provide definitions on key Article 9 concepts, namely:

- The delineation and requirements related to water services and water uses (Section 2.4.1); and
- The definition and measurement of environmental and resource costs (Section 2.4.2).

2.4.1 The definition of water uses and water services and related implications

As earlier discussions have shown, the definition and delineation of water services and water uses has important implications for the implementation of Article 9 and the required financing mechanisms for water services provision, i.e. the degree to which an internal cost-recovery, including environmental and resource costs and in line with the polluter-pays principle is demanded by the Directive (cf. Section 1.4.4 and 2.2.1.2).[195] The WFD defines water services in Article 2(38) as

> *"All services which provide for households, public institutions or any economic activity:*
> *(a) Abstraction, impoundment, storage, treatment and distribution of surface or groundwater;*
> *(b) Wastewater collection and treatment facilities which subsequently discharge into surface water"*

As discussed in Section 1.4.1, Article 9 asks to *take account of the principle of recovery of the costs of water services, including environmental and resource costs [...] in accordance in particular with the polluter pays principle*. Although Article 2(38) includes a list of activities that have to be undertaken for qualifying a service as a *'water service'* in terms of the Directive, the practical definition of water services has raised extensive discussions in the practical implementation process. Foremost, it was discussed, whether all of these activities have to undertaken simultaneously, or whether it is sufficient if some of these activities are met for qualifying a service as a *'water service'*. While the Guidance documents from the CIS process promoted a broad interpretation of water services[196], which includes all services having a significant impact on water status, their legally-non-binding status and the relatively vague formulations still left room for interpretation by Member States, as the Article 5 reports attest (cf. Figure A2.2, Annex). While water supply and wastewater management services

195 Some activities with no significant impact on water status are neither water services nor water uses. Clearly, this distinction cannot be made systematic as it is based on the analysis undertaken in accordance to Article 5 and Annex II, e.g. in some cases, fishing will have no impact on water status, but over-fishing has a significant impact on the ecology of a river and water status.

196 CIS (2004A): *"the appropriate water services to include in the assessment are those that are identified in the pressures and impacts study as having a significant impact on the status of water bodies."*
The Wateco Guidance Document (CIS 2003A) states that: *"To achieve maximum transparency, to ensure equitable and effective treatment vis-à-vis the internalisation of environmental and resource costs, and to preserve competition between economic sectors, water services should, where necessary, include both services provided by third parties and self-services."*

for the different sectors were approved as water services by responsible authorities, service providers and environmental NGOs alike, the inclusion of additional activities under the definition of water services was a matter of extensive debate. In particular, services for the following water uses, which were identified in most Article 5 reports as having a significant impact on water status, were discussed (WWF AND EEA 2006, EUROPEAN COMMISSION 2007):

- Flood protection, impacting on water status through the construction and maintenance of dykes (i.e. impoundment and storage),
- Navigation, necessitating the dredging of rivers and specific infrastructure built (i.e. impoundment),
- Hydropower, requiring the building and operation of dams (i.e. impoundment and water storage); and
- Agriculture, via drainage and other infrastructures (i.e. abstraction, impoundment and storage).

As an analysis of the Article 5 reports reveals (WWF AND EEA 2006), most Member States have, however, opted for a narrow definition of water services, sticking to those services that fulfil *all* the requirements listed in Article 2(38) simultaneously. Thus, the majority of Member States restricted its definition of water services solely to water supply and wastewater management (including self-services, if significant, cf. Figure A2.2, Annex). Of the investigated Article 5 reports, only river basin districts in France, the Netherlands, Portugal, Spain, Latvia and Romania opted for applying a less narrow interpretation by covering additional activities with a significant impact on water status, such as navigation, hydropower and/or flood protection. Self-supply of water for irrigation purposes and industrial activities has generally been included in the definition of water services, if the service provision classifies as *significant* in terms of its impact on water status. Furthermore, collective irrigation water supply qualifies in those countries, where it is a significant water use.[197]

In May 2006, environmental NGOs filed a complaint with the EUROPEAN COMMISSION against the narrow definition of water services applied by eleven Member States, contesting that only a broader approach to water services may promote the Directive's implementation, in particular regarding the polluter-pays principle. In their response, Member States referred to the origins of Article 9 and highlighting that while it was originally constructed in a broader manner, the amendments added through the negotiation process led to a reduced application scope, addressing water supply and wastewater services only. Another

197 For the German context, self-supply for irrigation purposes is only considered as a water service, if it has a significant impact on water status.

strand of argumentation supportive of a narrow definition of water services, builds on the formulation of Article 2 and its listing of different, consecutive activities, which only fully apply to water supply and wastewater services. As the European Commission favoured in its reply a broad definition of water services, the conflicting interpretations of Article 2 remain to date to be resolved.[198]

In the course of this study, the methodology for assessing cost-recovery in line with Article 9 is exemplified for the following water services:

(1) Drinking water supply to household,
(2) Irrigation water supply[199], and
(3) Wastewater services provided to households.

The definition of the term "water use" is defined in Article 2(39) of the Directive:

> *"Water use means water services together with any other activity identified under Article 5 and Annex II having a significant impact on the status of water."*

By contrast to the approach taken for water services, Article 2 does not specify aa list of water uses to be considered (CIS 2003A). Basically, only the activities that cause significant impacts on water bodies and therefore pose a risk to achieving good status are covered by the definition of water uses. Experience shows that navigation, hydropower generation, agriculture, tourism, and industrial activities are important water uses which may cause significant impacts and therefore have to be taken into consideration. The following Section 2.4.2 provides an overview on the potential impact of these activities on water status, due to which they may qualify – depending on the exact specifications within each river basin district – as water uses.

For the group of water uses, disaggregated into at least industry, households and agriculture, Article 9 merely requires *"an adequate contribution [...] to the recovery of the costs of water services,[...] taking account of the polluter pays principle."* Hence, while water services have to adhere to the principle of cost-recovery, water uses only have to contribute *adequately* to this cost-recovery, i.e. in line with their share in the costs of service provision and the polluter-pays principle (cf. Figure 12). Should agricultural cultivation practices e.g. lead to higher water treatment costs for drinking water provision, these additional costs should be borne by the agricultural sector, i.e. the polluter.[200] In practice,

---

198 A decision is still pending (status November 2009).

199 Drainage services are touched upon in the assessment as far as their interrelation to irrigation water supply necessitates an integrated analysis.

200 Not considering the potential exceptions outlined in Article 9(3) and 9(4).

determining the polluter may, however, not always be straightforward due to summation effects or diffuse pollution, which can often not be traced back to an identifiable polluter without considerable monitoring. Furthermore, the traditional interpretation of the polluter-pays principle for certain sectors may also pose difficulties to the Article's practical implementation, most notably in the case of agriculture.[201] Section 3.3.3 outlines how the adequacy of contributions from different water uses to the cost-recovery of the water services addressed in this study can be assessed.

Figure 12: The requirements of Article 9 for water services and water uses

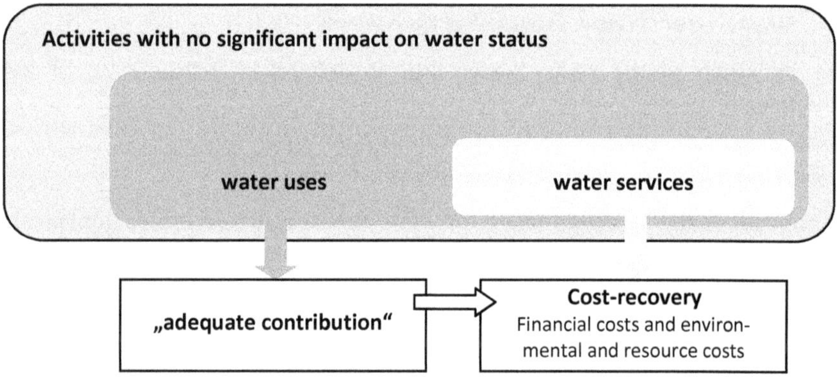

Source: extended based on CIS (2003a).

The definition of water services also determines the degree to which environmental and resource costs have to be considered in the implementation process of Article 9: they only have to be included in the assessment as far as they are encountered due to the provision of a designated water services or are

201 Traditionally, the scope of the polluter-pays principle in the agricultural sector has been determined by the code of good farming practice. On its meeting on 15 November 1999 the European Council (*13078/99*) concluded that farmers themselves have to bear the costs of complying with the code of good farming practice. Should they, however, take on additional activities for protecting the environment which extend beyond this code, Member States are allowed to compensate farmers accordingly. Hence, the definition of the code of good farming practice simultaneously determines the scope of application of the polluter-pays principle in agriculture. If this interpretation remains valid for WFD implementation, the agricultural sector could theoretically only be held liable for e.g. additional water treatment costs, as long as these costs are encountered due to an insufficient implementation of the code of good farming practice.

caused by other water uses and lead to higher costs of water service provision.[202]

2.4.2 Environmental and resource costs: substantiating the concept

While generic definitions on environmental and resource costs have already been provided upfront in this Chapter, the following section discusses the role of environmental and resource costs within the overall cost-recovery assessment and presents theoretical approaches for their measurement.

2.4.2.1 Measuring environmental and resource costs: theoretical concepts and related difficulties

According to the definition provided by the CIS Working Group DG Eco II (CIS 2004B), the measurement of resource costs entails the valuation of water in alternative uses (cf. Section 2.2.2.3). Resource costs may arise, for example, if ecosystem restoration and conservation – measured through the use and non-use values attached to the water environment by the general public (households) and/or recreation – generate a higher economic value than for example current or future water abstraction by agriculture. Hence, an economic valuation of the use and non-use values associated with the alternative use – here, ecosystem restoration and conservation – is required for assessing the magnitude of resource costs in the provision of the current water service '*irrigation water supply to agriculture*'. Several difficulties are linked to this approach: first, alternative uses may not always be evident (e.g. due to traditionally determined water use patterns which discourage alternative uses altogether); second, the information on use values and in particular non-use values associated with alternative water uses is not readily available. While the economic toolbox provides a variety of approaches for the measurement of both use and non-use values, they are all linked to important difficulties (cf. Box 4 below).

Similar difficulties are linked to the assessment of environmental costs. Generally, two distinct approaches are considered for measuring environmental costs, namely the *environmental damage cost approach*, also referred to as the

202 To give an example, only that part of diffuse pollution caused by agriculture which leads to higher costs for drinking water provision (or other designated water services) has to be recovered from polluters – additional pollution, which does not affect the costs of water services provision is out of the scope of Article 9. The same argumentation applies to resource costs.

benefit based approach, and the *damage avoidance cost or cost-based approach*:

- Under the benefit-based approach, the benefits that are foregone if a certain environmental damage occurs are estimated as an approximation to environmental costs.
- Under the cost-based approach, instead the cost of measures that are necessary for preventing a certain environmental damage from occurring, or for rectifying an existing environmental damage are considered as a proxy of environmental costs.

The cost-based approach links up to the cost-effectiveness analysis and the river basin management plans, which contain the measures needed for reaching good water status and thus for protecting waters from or rectifying existing environmental costs. The required information base for applying the cost-based approach is thus partly collected within the WFD implementation process. However, the programmes of measures address *all* environmental damage costs to the aquatic environment, irrespective of whether they are linked to the provision of water services or not. Only the latter are however of relevance to the cost-recovery assessment. In addition, the cost-based approach cannot cover non-use values and may thus only approximate a lower limit of environmental costs.

The benefit-based approach instead requires new information to be collected, namely a valuation of foregone benefits, which could also be employed for assessing of resource costs. Theoretically, the economic toolbox offers numerous approaches for the assessment of benefits (cf. Box 4), which are however all linked to specific drawbacks and are time consuming in their execution. As many Member States lack experience in benefit valuation procedures and are reluctant to base their political decisions on the results obtained, benefit valuation studies have so far not been widely employed for assessing environmental costs in practice. They would however be required for assessing the resource costs associated with many alternative uses, if their monetisation is aimed at.

Box 4: Economic valuation of benefits: revealed and stated preference methods

Revealed preference methods

Revealed preference (or market-based) methods draw on market information and behaviour for eliciting the economic value of associated non-market impacts, i.e. by observing actual behaviour and, in particular, purchases made in actual markets (PEARCE ET AL. 2006). As formulated by RUSSELL (2001), these methods seek to quantify the market *'footprint'* of non-market goods. Among the revealed preference methods, the hedonic pricing method, the travel cost approach and the avoided cost method are most prominent in the literature and hence briefly reviewed.[203]

Hedonic pricing method (HPM)[204]

The HPM is based on the assumption that the price of a market good is a function of its attributes (IVM 1998), some of which are intangible environmental goods. By trading these market goods, consumers simultaneously express their values for the intangible environmental goods (PEARCE ET AL. 2006). The value of an intangible environmental good can thus be deduced by observing behaviour in the market for a related good. For example, other things remaining constant, the benefits stemming from an improved watershed area may be deduced from increased land prices (VENKATACHALAM 2004B). Although it has been widely used, the application of the HCM may be hindered by several factors: first, market goods often contain several intangible characteristics, which cannot be easily separated and thus measured independently (i.e. non-market characteristics often display multi-collinearity) (PEARCE ET AL. 2006); second, the assumption that individuals have full information about the intangible environmental attributes of a market good, so that their preferences can reflect the value of the environmental attribute under consideration may not be true (VENKATACHALAM 2004A); third, like all revealed preference methods, the HCM does not allow for capturing non-use values.

Travel cost method (TCM)

The TCM is based on the assumption that the time and money individuals spend on travelling to a specific recreational site (e.g. parks, woodland, beaches, lakes etc.), gives an indication of their willingness-to-pay for visits to this site (PEARCE ET AL. 2006). More specifically, the TCM estimates a demand function that explains the number of visits to an area as a function of the total costs incurred in visiting the area such as travel costs (e.g. train ticket, petrol expenses, wear and tear and depreciation of the vehicle), travel time (valued at the appropriate opportunity costs, i.e. the individual's wage as an indication of the shadow price of time), possible entrance fees and others such as the costs

---

203 For a recent review of these methods, cf. BOYLE (2003).

204 See PALMQUIST (1991) for an in-depth discussion of the economic and econometric aspects of hedonic methods.

for fishing gear when valuing recreational fishing at a watercourse) (IVM 1998). Thus, the TCM recognises that individuals use a number of factor inputs for experiencing a recreational site and hence that travel and the recreational area are (weak) complements (PEARCE ET AL. 2006). The TCM can also be used of estimating the value of a change in the recreational site, e.g. an improvement of water quality. The TCM is often regarded as one of the most effective approaches in valuing recreation services (BOCKSTAEL ET AL. 1991, WARD AND LOOMIS 1986), despite some problems linked to its application: first, some of the cost items are themselves intangible (e.g. the time spent travelling); second, travel itself can also be regarded as valuable by individuals or they may undertake multi-purpose trips and visit several sites (e.g. international tourists), making it difficult to discern and attribute a certain portion of the costs to the recreational value of one site (PEARCE ET AL. 2006); third, the TCM is only capable of measuring the use value of recreational sites; fourth, substitute sites may further complicate the estimation of recreational value based on the TCM (MARKANDYA ET AL. 2002, IVM 1998).

Avoided cost method (ACM)

The ACM uses information on averting behaviour, i.e. on measures undertaken by individuals for insulating themselves from a non-market bad (PEARCE ET AL. 2006). The costs of these measures, both direct and indirect in terms of time or of restrictions imposed on actions the individual would otherwise undertake, act as the relevant proxy for estimating the economic value of not encountering a certain non-market good. The ACM is mainly used for estimating the economic value of environmental damage costs by looking at the amount of expenditure required for either averting a particular environmental damage or restoring a damage that has already taken place (VENKATACHALAM 2004B). In terms of the WFD, the (non-disproportionate) costs of measures may be considered as an approximation of the value of reaching good water status. The application of the ACM may raise a number of complications: first, expenditures on defensive or rectifying measures generally provide only a lower bound estimate of the value of the (avoided) impact; second, measures may be of a multi-purpose nature, thus making it attribute the costs appropriately. Third, again this method is not capable of incorporating non-use values.

The main advantage of revealed preferences methods lies in the fact that they rely on observed behaviour and thus on actual decisions made by individuals. However, they are not able to include non-use values in their value estimates.

Stated preference methods

Stated preference methods offer a direct survey approach to estimating individual or household preferences. As this technique can also be employed for assessing impacts on non-market goods, it is of particular importance to the WFD implementation context. The following sections briefly review the contingent valuation method, choice modelling and the benefit transfer method.

Contingent valuation method (CVM)

The CVM aims to trace out the willingness-to-pay (WTP) distribution for a population of economic agents for a proposed change in a good (CARSON ET AL. 2001) by constructing a hypothetical market in a survey context.[205] Typically, individuals are asked their WTP to ensure a welfare gain from a change in the provision of an environmental amenity, or their willingness-to-accept (WTA) compensation for enduring a welfare loss from the reduced level of the environmental amenity (CUMMINGS ET AL. 1986, MITCHELL AND CARSON 1989, PORTNEY 1994, VENKATACHALAM 2004A, CARSON ET AL. 2003). The term *'contingent'* refers to the fact that values revealed by respondents are contingent upon the particular scenario described in the survey (PORTNEY 1994, CARSON ET AL. 2003). In exact economic terms, the CVM estimates the Hicksian consumer surplus due to a change in the provision of good (BATEMAN AND TURNER 1993, VENKATACHALAM 2004A).[206] It is generally recognized that only the CVM is applicable to the estimation of non-market economic values of natural resources (MITCHELL AND CARSON 1989, BATEMAN AND TURNER 1992, SMITH 1993, CARSON ET AL. 2001, CARSON ET AL. 2003). As a stated-preference method, contingent valuation circumvents the absence of markets for natural resource services by presenting consumers with a choice situation in which they have the opportunity to buy or sell the services in question (CARSON ET AL. 2003).[207] Although some economists (e.g.

---

205 For a detailed description and assessment of the CV method cf. MITCHELL AND Carson (1989).

206 HICKS (1946) classified the consumer surplus measure into two different categories, namely the compensating variation and the equivalent variation. The compensating variation refers to the amount of money income that has to be given up by the consumer to attain the increased level of utility (I.e. WTP measure). The equivalent variation refers to the amount of compensation required to be provided to the individual so that she could attain an improved utility level in case the provision of the public good does not take place (i.e. WTA) (Venkatachalam 2004).

207 As indicated by Carson et al. (2001), the contingent valuation method is not the only technique capable of capturing passive use values. Other techniques of the class of *'constructed markets'* (CARSON 1991) include actual referenda (DEACON AND SHAPIRO 1975) and simulated

DESVOUSGES ET AL. 1993B) continue to question the effectiveness of the CVM and in particular its aptness for measuring non-use values, the discussion has rather shifted towards guidelines for conducting such studies (see e.g. ARROW ET AL. 1993, CARSON ET AL. 1999).[208] The method has been frequently applied in the US and less so in Europe.[209] CVM gained further in popularity after non-use values had been recognised as important components of total economic value in the environmental economics literature. Nevertheless, the number of studies addressing non-use values in particular is still restricted, especially for the European context.

Choice modelling (CM)

The CM approach addresses the fact that environmental impacts are often multidimensional. Proposed projects or policy actions therefore often lead to multiple changes, whose assessment requires distinct valuations (PEARCE ET AL. 2006). CM approaches, such as choice experiments, in which respondents are asked to choose their most preferred option from a choice set of at least two options, are increasingly used for valuing multidimensional environmental problems.[210] However, CM studies encounter largely the same problems as CV studies. As there are only few studies so far that compare the findings of reliability and validity of CV and CM studies, there is no conclusive evidence so far (PEARCE ET AL. 2006).

---

markets (BISHOP AND HEBERLEIN 1979). Nevertheless, the contingent valuation method is by far the most frequently applied technique.

208 The contingent valuation method was first described by CIRIACY-WANTRUP in 1947. He investigated the benefits of preventing soil erosion and observed that several of the favourable effects were non-market goods (VENKATACHALAM 2004). CIRIACY-WANTRUP proposed as one possible way of estimating these benefits to elicit the individuals' willingness to pay for these benefits through a survey method (see PORTNEY, 1994, HANEMANN 1994). Although CIRIACY-WANTRUP conceptually developed an approach for addressing the problem, he never pursued his idea further. The first contingent valuation survey was carried out in 1963, when ROBERT DAVIS interviewed hunters and wilderness lovers in order to estimate the benefits of outdoor recreation in the Main Woods (USA).

209 CARSON ET AL. (1995) produced a bibliography of 1,674 contingent valuation studies and papers.

210 Other forms of choice modelling include contingent/conjoint rating, pared comparisons and contingent ranking.

Benefit transfer (BT)

The BT approach comprises the transposition of monetary environmental value estimates gathered at one site (i.e. the study site) through market-based or non-market-based economic valuation techniques to another site (i.e. the policy site) (BROUWER AND SPANINKS 1999). The main motivation for using BT is to economise on study costs and implementation time by using previous research results. Thus, BT has the potential of quickly informing decision-making. Generally accepted practical guidelines for conducting BT are still missing. However, PEARCE ET AL. (2006) outlined the following elements as constituting best practices: First, the original studies, on which the BT is to be based, must themselves be accurate and of high quality. Second, in conducting a benefits transfer, the study and policy sites must be similar in terms of population and population characteristics. Third, the change in the provision of the good being valued at the two sites also should be similar. The validity and reliability of environmental value transfer remain, however, contested.[211] The main argument put forward by critics of BT us that BT introduces subjectivity and greater uncertainty into appraisals in that analysts must make a number of additional assumptions and judgements to those contained in original studies (PEARCE ET AL. 2006).[212]

Hence, while the benefit-based approach is more in line with standard environmental economic procedures, the cost-based approach is more pragmatic as the necessary data is more readily available (GÖRLACH AND INTERWIES 2005) and is in part generated by the ongoing WFD implementation process. Therefore this approach received more attention in the implementation process so far.[213]

In general, the assessment of environmental costs is undertaken from an anthropocentric viewpoint, i.e. environmental and resource costs arise if and only if a specific change in a water body results in a welfare loss to human beings, who depend on water and use water in some form (CIS 2004B). Welfare losses may however occur at different scale (e.g. relating to individual persons,

211 See a special issue of Ecological Economics in April 1998 (25, 1).

212 For a broader review of benefit transfer studies cf. BROUWER (2000) and ROSENBERGER AND LOOMIS (2003).

213 Notable exceptions are Sweden, which compiled a database on valuation studies within the WFD implementation process and the United Kingdom, which has a longstanding tradition in employing economic tools in environmental policy making and can draw on a number of studies. The European research project *'Aquamoney'* (www.aquamoney.org) may provide further insights and support the use of economic valuation studies in WFD implementation through appropriate guidance for policy-makers.

sectors or society as a whole), which is determined by the nature of the damage, i.e. whether the physical ecosystem damage occurs at a local, regional, national or transnational level (CIS 2004B). In general, the magnitude of environmental and resource costs is site-specific and dependent upon a number of surrounding factors (e.g. weather conditions, season, soil type, etc.). The same activity may therefore give rise to widely divergent environmental and resource costs.

The following section discusses, to which environmental and resource costs different water services may give rise to and how these can theoretically be attributed or traced back to specific polluters. A detailed plan of procedure for the assessment of environmental and resource costs is deferred to Section 3.3.3.

2.4.2.2 Determining environmental and resource costs for the purpose of cost-recovery assessments

In the assessment of cost-recovery, only those environmental and resource costs have to be considered that are linked to the provision of water services. While environmental costs tend to be very site-specific, the following Table 8 provides an indicative overview on environmental costs that may be linked to the provision of different water services.

An extensive analysis and guidance on procedures for estimating environmental and resource costs within the assessment of cost-recovery in line with WFD Article 9 is deferred to Sections 3.3.3 and 3.3.5.

Once environmental and resource costs have been estimated, an assessment of cost-recovery necessitates to evaluate, how these costs are currently financed, i.e. to which degree the polluter (or beneficiary) is paying, which share of the costs is borne by other water users and to which extent costs are externalised on the environment or future generations. In the WFD implementation process the results from the analysis of pressures and impacts can provide important insights for attributing environmental costs to responsible damaging activities. However, several caveats remain, e.g. the results from the IMPRESS analysis may not be available at the required scale; singling out the effect of water services alone may not be trivial; or historic environmental damages may no longer be traceable to a specific activity. The appropriate methodological procedures for attributing costs to different water uses and for assessing their adequacy are developed in Section 3.3.3.

Table 8: Environmental costs of water services

Water service	Potential environmental costs
Drinking water supply to households	Overabstraction (impairment of natural rate of recharge with negative impacts e.g. on aquatic species, biodiversity and water dependent land ecosystems such as wetlands or forests) Morphological changes (e.g. reservoirs, dams)
Sewerage management (wastewater collection and treatment)	Water pollution through inadequately treated discharges Water pollution through discharges into sensitive waters (impairment of natural rate of recovery) Degradation of soil and water quality through sewerage sludge Loss of biodiversity through polluted discharges Loss of aquatic species through polluted discharges
Irrigation water supply services[214]	Overabstraction of groundwater Soil salinity and water logging Diffuse pollution of surface and groundwater (through nitrates, pesticides, plant protecting agents) Land erosion Eutrophication Loss of soil fertility (secondary impact: reduced productivity of agriculture) Negative impacts on wetlands Loss of biodiversity Adverse effects of river regulation and flow diversion
Hydropower*	Morphological changes through impoundments (e.g. dams) Reduced fish population / diversity due to loss of passability (→ secondary impact on leisure / industrial fishing) Loss of biodiversity
Navigation*	Morphological changes through impoundments, dredging, etc. Reduced fish population / diversity due to loss of passability (→ secondary impact on leisure / industrial fishing) Negative impacts on wetlands Loss of biodiversity
Flood management*	Morphological changes through impoundments (e.g. dykes, dams) Reduced fish population / diversity due to loss of passability (→ secondary impact on leisure / industrial fishing) Negative impacts on wetlands Loss of biodiversity

* *The classification of hydropower, navigation and flood protection as water services is still disputed (cf. Section 2.4.1 above). Furthermore, they are not at the focus of this study.*

In the Article 5 reports for 2004, most Member States have not undertaken an in-depth assessment of environmental and resource costs related to differ-

214 The Impacts can be caused due to farm level practices, field level water application systems, water distribution systems, water supply or drainage systems or large reservoirs or tanks.

ent water services, but merely provided concretised definitions[215] for the national / river basin context and reported on existing instruments for their internalisation. While partly being related to time and capacity constraints, this procedure also reveals that the concept is politically difficult to communicate and Member States lack experience in applying the relevant techniques and in conducting the required assessments.[216]

2.5 Difficulties related to a European comparison of cost-recovery and water pricing

As the discussion in Section 2.2.1 revealed, the understanding of the concept of cost-recovery varies greatly between and even within European Member States. Hence, without common definitions of cost categories to be considered in the assessment and of the way to account for different revenue generating mechanisms, comparisons of cost-recovery are inherently flawed. Presently, some studies consider cost-recovery to be achieved if only operation and maintenance costs are recovered, while others address the entirety of financial costs or attempt to include environmental and resource costs, e.g. through existing internalisation instruments. Nevertheless, there is no agreed upon terminology on what should be included in the analysis, which also allows taking account of national specificities.

Present cross-country (and even within country) rankings of water prices generally have very low information content. This is linked to a number of factors: First, the underlying service that is being compared may differ significantly across European Member States. In some Member States, water prices also account for the costs of general water management activities (e.g. resource protection measures unrelated to water services provision), flood protection

---

215 Within the European research project *'Aquamoney'* (GÖRLACH ET AL. 2007), an investigation was undertaken among European Member States on the definition of environmental and resource costs applied within the implementation process. Of the twelve countries, which provided the required data for the assessment, seven applied the Wateco definitions of environmental and resource costs, three Member States developed their own definitions and one Member State used both the Wateco definitions as well as the definitions provided by the Working Group DG Eco II (CIS 2004B). In addition, the analysis revealed that a number of Member States do not differentiate between *'environmental costs'* and *'resource costs'*.

216 As GÖRLACH AND INTERWIES (2005) point out, most Member States attempt to balance their efforts spent on estimating environmental and resource costs with the actual use of the (additional) information obtained. These attempts can be well-linked to Annex III of the WFD, which refers to the need to maintain proportionality in data collection.

services, rainwater management of public places and roads, or fire protection services etc. Second, due to the high fixed costs of the water sector, the entire tariffication scheme has to be compared. Looking only at cubic meter prices and the basic fixed charge, as is often done in price comparisons, may neglect important additional financing schemes, both endogenous (e.g. investment contributions, connection fees) as well as exogenous (e.g. direct subsidies for water services provision), impacting on the level of cubic meter prices and basic fixed charges. To circumvent problems related to differences in tariffication structures, often typical consumer profiles are constructed (e.g. 150 m³ for a four-person household) and compared. Yet, these are often likewise flawed. To take Germany, for example, daily water consumption per inhabitant varies between 85 and 143 litres per day (cf. Section 4.2.1). Hence, there are distinct regional disparities present with regard to what is a '*typical*' consumption profile. Due to the high proportion of fixed costs, the service provider has to augment its cubic meter prices with decreasing water consumption in order to reach cost-recovery. High water prices per cubic meter therefore cannot be taken as a signal of higher expenses per household for the service that is provided (i.e. if consumption is lower, households may pay the same yearly bill as households facing lower cubic meter prices with higher consumption levels). Water price comparisons are additionally hampered even at the local level, by differences in the local framework conditions for water service provision (e.g. the natural and socio-structural conditions, cf. HOLLÄNDER ET AL. 2009, 2008, BARTSCH 2007), as well as by differences in water service quality.

Nevertheless, there is a demand for water price comparisons, which are regularly conducted both at international (e.g. by the OECD or the United Nations) or the European level (e.g. METROPOLITAN CONSULTING 2006, UBA 1998), as well as at regional level within Member States. Due to the water sector's monopolistic structure, there is an increased need for transparency in price setting. However, without '*getting the basics right*', comparisons of water prices cannot succeed in establishing transparency and facilitating adaptations in water policy to fully exploit the potential of water pricing in terms of establishing sustainability of service provision and a fair cost allocation across user groups and generations. To this end, a methodological framework has to be conceptualised that unambiguously delineates the service to be compared (i.e. what is to be considered in the analysis); provides a clear understanding and definition of the cost categories that should be considered for a given water service and how these costs should be accounted for; and allows for documenting which financing schemes are in place for the provision of the water services and whether these are of an endogenous or exogenous nature.

In the long-run, the implementation of WFD Article 9 may facilitate well-grounded comparisons of water service financing across Europe, as its implementation process necessitates a common understanding on cost categories and related financing mechanisms. If these implementation challenges are addressed through a methodological framework, which is capable of addressing national specificities, more meaningful water price comparisons could result as a by-product. The following Chapter 3 proposes a step-wise methodology for assessing the implementation of the different objectives of WFD Article 9 in an integrated way. The main drawbacks for water price comparisons as outlined above have to be addressed in the process. By offering an assessment framework for costs and revenue generating mechanisms that accommodates national and regional specificities, the proposed methodology offers one stepping stone for a further development of a framework for cross-country water price comparisons.

3 Methodology for a comparative assessment of cost-recovery in line with WFD Article 9

"The adoption of common definitions for key cost variables would facilitate the comparison between costs and prices [...] for different water services, uses and countries."

European Commission (2000a)

This chapter first recapitulates the analytical prerequisites for developing the assessment framework, by describing the *'ideal path'* to cost-recovery in line with WFD Article 9. Next to a cost accounting system, which delivers definitions and accounting guidelines in line with Article 9 on each cost component to be considered in the assessment, the chapter provides a complementary revenue accounting framework. Furthermore, a pragmatic procedure for an area-wide integration of environmental and resource costs in the assessment of cost-recovery is presented. The chapter also addresses the question which framework conditions are decisive for a comparative assessment of cost-recovery in different EU Member States and through which indicators they can be effectively measured. Next, it proposes an assessment scheme for justifying exemptions from achieving cost-recovery in line with Article 9 and provides guidelines for analysing the implementation of the Article's additional objectives. The chapter concludes with an overall plan of procedure.

3.1 Analytical prerequisites: the ideal path to cost-recovery according to WFD Article 9

As the discussion of possible interpretations of the cost-recovery concept along with the analysis of Article 9 objectives revealed, the WFD promotes a strong internal approach to cost-recovery (cf. Section 2.2.1.2). This is documented by the role designated to the polluter-pays principle in achieving cost-recovery, the requirement of an adequate contribution from different sectors and the reliance on an economic cost approach (i.e. the inclusion of environmental and resource costs). Hence, the Directive aims at a high degree of internal financial and environmental and resource costs recovery, with a use-/pollution-dependent allocation of costs onto different user groups and no costs being transferred to the next generation (cf. case 1, Figure 13). However, an iterative approximation to this objective is accepted (i.e. starting from a *'sustainability cost'* or *'quasi-full cost-recovery'* approach) and divergences may be justified, given that certain conditions are met (i.e. in geographical, social, climatic or economic terms, cf. Section 3.3.2). Furthermore, in case that the public good components of water services cannot be fully extracted for the assessment, they may also justify the adoption of alternative financing mechanisms (i.e. for these components). During the first implementation cycle, *transparency on how costs are currently recovered* and a *full cost accounting approach* (i.e. which costs are encountered and by whom) are hence at the centre of attention.

The following graph takes up the discussion of possible approaches to cost-recovery from Section 2.2.1.1 and allows for depicting different pathways to implementing the cost-recovery objective of Article 9.

Figure 13: Depicting the 'ideal path' to cost-recovery according to Article 9

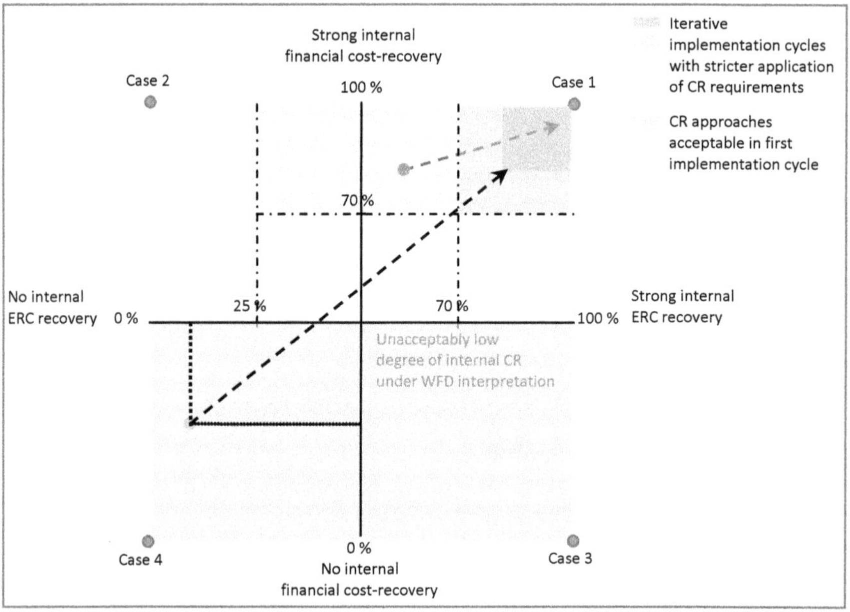

Note: The 'cases' refer to the description of cost-recovery approaches in Section 2.2.1.1, Box 3. (CR: cost-recovery, ERC: environmental and resource costs)

The x-axis in Figure 13 delineates the approach to environmental and resource cost-recovery, ranging from no internal environmental and resource cost-recovery, i.e. a full financing through public budgets, to a 100 percent internalisation of environmental and resource costs and hence a full recovery from water service users. Analogously, the y-axis depicts the degree to which financial costs are recovered from water service users. The Figure takes up the four polar cases described in Section 2.2.1.1, Box 3. As outlined in the graphical representation, case 4 and case 3 cannot be reconciled with the WFD approach to cost-recovery. Ultimately, case 1 in the far upper right corner of Figure 13 is to be adopted by all Member States in order to comply with the WFD approach to cost-recovery (cf. above, unless exemptions can be invoked). The iterative approximation, which may be followed in implementing the requirements, is outlined by the grey areas. The lower-bound threshold value for compliance up to 2010 has been set at 70 percent of financial cost-recovery in the graphical representation above in reference to UNNERSTALL (2005:12) and at 25 percent of environmental and resource costs recovery.[217] While different threshold values may be determined and defended on a political level, it should be noted that any acceptable cost-recovery quadrant will become smaller as implementation progresses.

In order to place an investigation area (be it a service provision area, a river basin district, or a Member State) within this graph and to map out its divergence from the ideal cost-recovery approach depicted by *case 1*, a clear understanding of what the recovery of financial and environmental and resource cost-recovery entails and which guidelines apply to their accounting in the understanding of Article 9 is required. To this end a number of analytical prerequisites need to be developed in the following sections.

First, the financial cost categories, which are required for the assessment, have to be defined along with guidelines for their accounting, in order to allow for a comparable cost basis (Section 3.2.2) and to make divergences (i.e. due to national or local particularities) transparent. Subsequently, the cost accounting framework needs to be complemented by a revenue accounting matrix, which documents all endogenous (i.e. direct contributions from users of the investigated water service) and exogenous (i.e. public transfers) recovery mechanisms in place (Section 3.2.3). In combining these two strands of the assessment, it can be analysed how financial costs are currently recovered, i.e. which cost-recovery path is currently followed (y-axis, Figure 13) and in how far there is a

217 UNNERSTALL (2005) does not differentiate between financial and environmental and resource cost-recovery, but refers to an overall cost-recovery level of 70 percent. Hence, Figure 13 depicts in this understanding minimum levels of cost-recovery to be achieved by 2010.

divergence from the stylized ideal path (i.e. which cost categories are accounted for differently than proposed by the respective guidelines). Section 3.2.4 amends the analysis with an investigation of environmental costs and their recovery. To this end, a differentiation between environmental costs in aquatic ecosystems or dependent land ecosystems which are *caused by the provision of a particular water service* (Type I EC) and environmental costs that are *caused by other water uses* (Type II EC), leading to higher (financial) costs of water service provision[218], is undertaken. Subsequently, Section 3.2.5 investigates the role of resource costs, and the impact of unaccounted for resource costs in particular, on the overall degree of cost-recovery in water service provision. Taken together, Section 3.2.4 and 3.2.5 then allow for depicting an investigation area's position on the x-axis of Figure 13.

For the purpose of this study, the DG Eco II definition of resource costs has been adopted, according to which resource costs emanate from an "*economically inefficient* allocation *of water, either in terms of water quantity or water quality, over time and across different water use(r)s*" (CIS 2004B). Once the different assessment components have been established, Chapter 3.3 maps out the current overall implementation pathways to Article 9 objective 1 (i.e. recovery of financial and environmental and resource costs, cf. Section 1.4.1). To this end, relevant background information on the investigation area is first collected on the basis of selected indicators, in order to allow for a meaningful interpretation of results. Then, the role of Article 9 exemptions (i.e. social, environmental, economic or geographic and climatic conditions) in explaining divergences from the ideal Article 9 implementation pathway is assessed. Finally, Chapter 3.4 outlines relevant aspects and indicators for an assessment of the complementary objectives or Article 9. In concluding, Chapter 3.5 summarises the entire assessment procedures in a concise '*plan of procedure*'.

Through the assessment framework, which outlines guidelines for a stylized ideal Article 9 cost-recovery pathway, transparency on cost and revenue categories and applied modalities is enhanced; financing mechanisms become visible; and cost as well as revenue structures are revealed. Divergences from the objectives can be documented and checked for their compliance with the exemption regulations provided by Article 9, based on an integrated analysis, supported by a core set of supplementary investigatory questions and indicators. Hence, *transparency of how costs are currently recovered* and a *full cost*

---

218 e.g. in the form of higher raw water treatment costs for drinking water provision or higher water abstraction costs due to a lowering of water tables caused by other water using activities.

accounting approach (i.e. which costs are encountered and by whom) are at the centre of attention.

3.2 Cost-recovery in line with WFD Article 9: the basic assessment framework

The basic assessment framework proposed in this study for analysing the implementation of *objective 1* of Article 9 consists of four distinct elements. The first analytical step (Section 3.2.2) defines and delineates the financial cost categories that are relevant for the assessment[219] and proposes guidelines for their measurement by addressing, among others, the following questions:

- Which cost categories have to be considered in the assessment and which cost factors need to be included in these categories;
- Which share do the different cost categories contribute to the total costs of water service provision (i.e. to which degree can they impact on total costs; what is the potential impact of a divergence from the accounting guidelines for the different cost categories);
- Which share of the cost categories is fixed and which share is variable;
- To which degree is the magnitude of cost categories dependent upon other water uses;
- By which (external) factors can they be influenced (e.g. geographical conditions) and which trends may be important (e.g. demographic changes).

The second analytical step (Section 3.2.3) proposes a complementary framework that takes stock of the revenue side of the equation. It addresses for instance the following questions:

- Through which mechanisms are water services currently financed;
- In how far do financing mechanisms mirror a polluter/user pays understanding;
- Which rationale is invoked for public contributions (e.g. is public financing limited to the public good components of water services' provision or to extended side-services).

219 Chapter 2.2 already provided an introduction to the different cost categories relevant to the assessment of cost-recovery for the water services under consideration in this study, i.e. water supply and wastewater services provided to households and irrigation water services for the agricultural sector.

Figure 14: Overview on cost categories for the assessment of cost-recovery

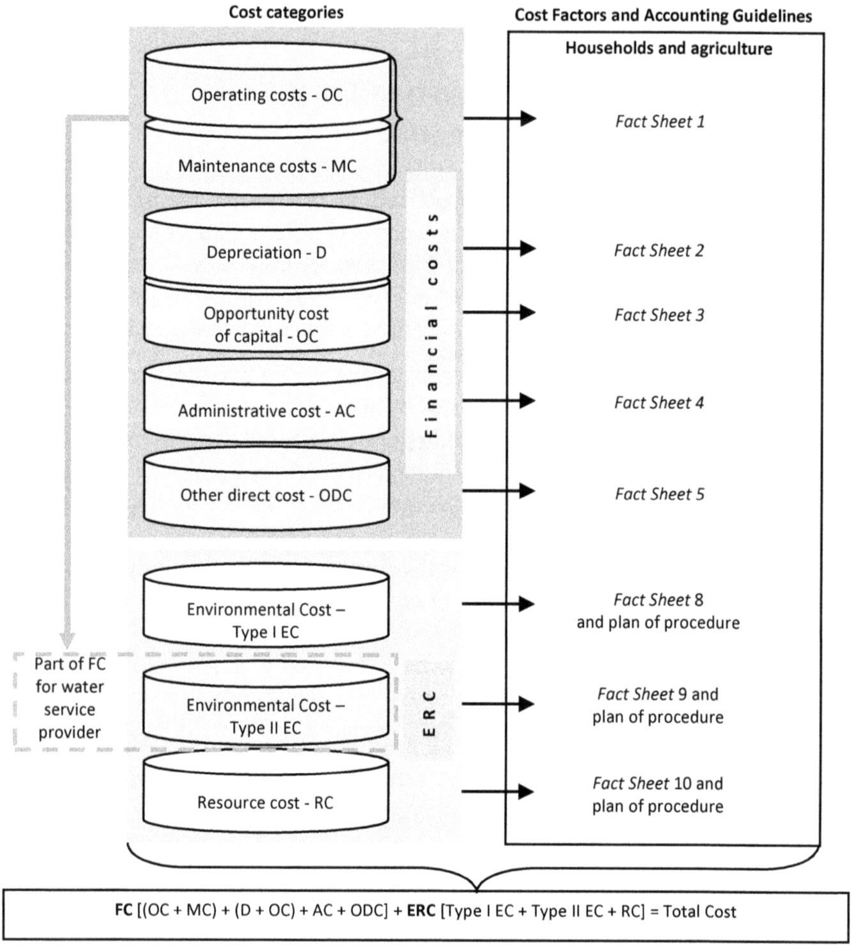

On the basis of these two strands of investigation, the cost-recovery pathway to financial costs (y-axis in Figure 13) can be determined (Step 3), which combines the assessment of financial costs, with the respective refinancing mechanisms and outlines divergences from the stylized ideal path.

In a fourth analytical step, an investigation of environmental costs is undertaken, which differentiates between existing internalisation schemes and residual costs, i.e. those environmental costs that are not yet addressed through existing measures and instruments. A cost-based approach is chosen for the measurement of residual environmental costs, drawing on information compiled in other WFD implementation steps, most importantly the POM as part of the RBMP.

Subsequently, the relevance of resource costs is assessed in a fifth analytical step (Section 3.2.5). Through these complementary assessments the evidence on the relevance of both environmental and resource costs can be used for determining the adjusted overall cost-recovery degree.

3.2.1 Definition and delineation of parameters for the cost assessment

Figure 14 summarises the cost categories to be included in the assessment of cost-recovery from Chapter 2 and links up to the upcoming assessment steps.

A number of requirements for cost accounting could already be deduced from the investigation of the objectives of Article 9 presented in Chapter 1 and the subsequent overview on theoretical underpinnings presented in Chapter 2. These are taken up accordingly in the accounting guidelines for the respective cost categories below. Furthermore Box 5 outlines a number of general guidelines for the cost-recovery assessment.

The general steps of the assessment do not vary for different water services. Yet, where particularities of household or agricultural water services are concerned, these are highlighted and taken up in the illustration cases in Chapter 4. Furthermore, the specificities of water services will be of greater importance in Sections 3.2.4 (assessment of environmental costs) and 3.2.5 (assessment of resource costs). References are taken up as appropriate.

Box 5: General guidelines for the cost assessment

The approaches for cost accounting in water services provision differ markedly both within and across European Member States (Maestu 2005, Strosser 2007). As the cost accounting framework developed in this study is intended to be applicable at all relevant analytical scales for WFD implementation, it has to be capable of serving as an umbrella approach for national and regional divergences. At the same time, it needs to establish a set of basic rules that can be adhered to in any application case. The comparative framework developed within the next sections does not seek to alleviate national or regional specificities in cost accounting. Instead it aims to provide a universal *'translation key'*, which allows for meaningful comparisons of Article 9 implementation between different national or regional systems of cost accounting.

The following **general guidelines** apply to the cost assessment of all cost categories:

- In most cases, the required information on costs for water service provision will not be reported in line with the division of cost categories proposed for the Article 9 assessment. The structured *Fact Sheets* allow for documenting in how far the proposed system of cost attribution could be applied, where difficulties were encountered, and in how far divergences impact on the overall cost allocation.[220] In this way, divergences and national specificities can be made transparent and a meaningful comparison can be undertaken in later stages of the assessment.
- Depending on the scale of investigation, reporting on the implementation of Article 9 may require up-scaling of information, i.e. aggregating information from different service areas to the river basin district level. Imprecision in the attribution of costs to the proposed cost categories may be amplified by this procedure. A transparent documentation of all assessment steps allow for tracing back cost allocation errors and evaluating the overall divergence from accounting rules.
- Costs, which are not linked to the provision of a specific water service (e.g. cost for fire protection), need to be excluded from the assessment. Where cost accounts do not allow for the separation of these costs, expert judgement will be required for evaluating their impact on the overall cost assessment (cf. Box 7).
- If cost accounts are consolidated across multiple services or cover multiple uses (e.g. dams that provide water for households and agriculture), expert judgement may be required for disaggregating the required cost information (CIS 2004B).
- Data should be retrieved as far as possible from annual accounts of water service providers and federal, regional and municipal financial statistics. Furthermore, data from environmental agencies as well as scientific research projects may offer additional insights, in particular on the incidence of environmental costs. Expert opinion regarding the quality of data should be reported on in the *Fact Sheets*.

220 The error margin caused by an imprecise cost attribution also depends on the share in total costs of the cost category (e.g. personnel costs make up for a considerable share in the range of 20 percent of total costs of drinking water service provision. In comparison, material costs constitute a rather small share of approximately 5 percent in total costs. Hence, misallocation errors related to '*personnel costs*' will impact more on the overall result, than misallocations related to '*material costs*'.

- For the assessment of residual environmental costs, information on the costs of programmes of measures is required. Next to monitoring data, expert judgement may be required for attributing the relevant share of costs of measures to specific water services.

Source: extended based on CIS (2004b).

3.2.2 Accounting for financial costs of water service provision

As introduced in Chapter 2, financial costs are subdivided for the analysis into operation and maintenance costs, capital costs consisting of depreciation and opportunity costs of capital, administrative costs and other direct costs.[221] To allow for an integrated cost-recovery assessment framework, these sub-categories of financial costs have to be defined more specifically along with guidelines for their accounting, in order to allow for a comparable cost basis. To this end, the following section provides concretisations for each cost category of financial costs on the following aspects in the form of structured *Fact Sheets*:

- The **data requirements** for each cost category, i.e. which data need to be compiled and which cost subcomponents should be accounted for;
- Whether costs are of a **fixed** nature (i.e. independent of the level of service provided) or a **variable** nature (i.e. dependent upon the degree to which a service customer demands a service). This information serves as an input to the upcoming sections that assess other Article 9 objectives, namely the implementation of the polluter-pays principle (*objective 3*), the adequacy of a water use's contribution (*objective 3*) and the provision of adequate incentives through the applied financing scheme (*objective 2*);
- The **quality of data** (ranging from A1 (highly reliable data source and high accuracy of data) to B3 (unreliable data source and low accuracy of data)) [222] and related uncertainties. This information is case study specific and cannot

221 As a reminder, taxes are excluded from the cost-recovery assessment: Taxes qualify as transfers within society and should therefore generally be excluded from the estimation of economic costs. However, it is important to distinguish between general taxes and environmental taxations schemes related to the environmental damages caused by water services' provision: while general taxes need to be deducted from financial costs, environmental taxes may represent internalised environmental costs and as such should not be excluded from the assessment but rather be considered in the assessment of environmental costs recovery (cf. Section 3.2.4).

222 This classification system for data quality is based on OVERATH AND MERKEL (2004) and draws on the IWA assessment framework.

be provided generically. It allows for outlining on which cost components data uncertainties will be highest and where it will be most difficult to obtain comparability of results. It constitutes an important input for an interpretation of results (and for comparisons).

- The **scale of data availability**, its periodicity and data confidentiality issues. Problems of data (dis)aggregation depend in part on the scale of investigation. The information provided assists in the identification of potential bottlenecks for later comparisons.
- **Accounting guidelines** in line with the Article 9 economic cost approach. This part of the *Fact Sheets* specifies distinct guidelines on how data should be compiled, aggregated or recalculated and which calculation procedures should be applied in the assessment.
- The **influence of divergence** from the accounting rule on the overall cost assessment, which is dependent upon the share of a certain cost category in total costs.
- Whether the magnitude of costs is dependent upon particular **driving forces**, e.g. other water uses (i.e. the degree to which material costs for raw water treatment are linked to water pollution caused by agriculture). This information serves as an input to the assessment of the implementation of the polluter-pays principle and the adequate contribution of water uses to the cost-recovery of water services (*objective 3*).
- Potential **difficulties for the cost assessment**. This part of the *Fact Sheet* outlines common difficulties encountered in gathering or calculating the required information/data for a certain cost category.
- **Evaluation of compliance** with the cost accounting framework (through expert judgement) along two criteria:
  (1) Have all costs been accounted for (ranging from a) 'full cost accounting attained', to b) 'not all costs could be accounted for' (complemented by an assessment of the extent of unaccounted costs and the reasons for it).
  (2) Was an unambiguous cost attribution to the different cost components feasible (ranging from a) 'full cost attribution achieved on the basis of existing data'; b) 'minor divergences and mixture of cost components'; to c) 'strong mix-up of cost components and categories, clear attribution of costs not feasible on the basis of existing data').

The *Fact Sheets* aim to increase the transparency of the cost assessment by stating assumptions clearly and showing on which basis the assessment of cost-recovery is grounded. They cannot propose an exhaustive list of all factors, but rather make the general approach as clear as possible and provide the framework for the analysis, while simultaneously allowing for national particularities. As outlined above, they allow for documenting divergences from the proposed ideal Article 9 accounting approach, along with national specificities (i.e. why is accounting done differently).

3.2.2.1 Operation and maintenance costs

The cost category *'operation and maintenance costs'* is generally well documented. Problems for the assessment are, however, often linked to the fact that costs cannot be clearly disaggregated and attributed to a specific water service. For example, if water supply and wastewater services are delivered by the same entity, personnel costs have to be attributed to each service. Moreover, those personnel costs, which are linked to administration, have to be subtracted and accounted for under *'administrative costs'* below (cf. *Fact Sheet* 4). Separating costs linked to the provision of side-services, such as fire-protection or drainage, may also pose considerable difficulties and may require expert judgement to pinpoint the nature of the influence on the cost assessment.

Fact Sheet 1: Operation and maintenance costs

Cost category: Operation and maintenance costs (OM_j)	
Data requirements	Information required on the following sub-components: – Personnel costs – Energy costs – Material costs – Hired and contracted services – Monitoring costs Other operating costs (e.g. unaccounted for water) Cf. Section 2.2.3.1 for a description of the cost components and Table 6 for an overview on operation and maintenance costs along the value chain of water service provision.
Fixed or variable (if both: approximate % division)	→ sub-division into variable and fixed cost components (e.g. for personnel costs) should be indicated, wherever possible.

Cost category: Operation and maintenance costs (OM_j)

Quality of data	Case study specific (generally well documented data). Classification along the following dimensions: A1 (reliable data source, accurate data) to B2 (unreliable data source, low accuracy of data)
Scale of data availability and periodicity of update	Service area. Generally updated for each accounting period.
Confidentiality	Low. Data mostly available from published reports.
Accounting guidelines in line with Article 9 economic cost approach	Cost information on the outlined sub-categories needs to be attributed to the specific water services under investigation. Include only data for the relevant investigation year.
Influence of divergence from accounting rule [depends upon magnitude of cost category in total costs]	Impact can be significant due to share in total costs, e.g. in the case of personnel costs (cf. Chapter 2).
Influencing factors / driving forces for cost category	Indicate whether the magnitude of this cost category is significantly influenced by other water uses (e.g. the degree to which material costs for raw water treatment are linked to water pollution caused by agriculture), and if yes, which cost components are most likely to be affected for the different services.
Potential difficulties for the assessment	Potential overlap with other services and with the cost category 'administrative costs' (e.g. personnel costs, monitoring costs, cf. Fact Sheet 4). Separating Type II EC (e.g. higher material costs for water treatment, or higher pumping costs caused by other water uses). Subtracting costs associated with side services (e.g. fire protection).

Cost category: Operation and maintenance costs (OM_j)	
Evaluation Full cost accounting and attribution to cost categories – *Expert judgement* –	Case study specific. Guidelines: (1) Have all costs been accounted for (ranging from a) 'full cost accounting attained', to b) 'not all costs could be accounted for'. (2) Was an unambiguous cost attribution to the different cost components feasible (ranging from a) 'full cost attribution achieved on the basis of existing data'; b) 'minor divergences and mixture of cost components'; to c) 'strong mix-up of cost components and categories, clear attribution of costs not feasible on the basis of existing data').

While cost allocation may not be trivial, unbundling costs and attributing them to a particular activity is often demanded by benchmarking systems (i.e. which depict specific processes of service provision) or national regulating authorities as well. Difficulties encountered in the clear attribution of costs to specific categories are documented in the *Fact Sheet* under *'evaluation'*, which specifies to which degree the available information could be allocated in line with the proposed framework. This information serves as an input to the overall assessment in Chapter 3.3.

The column *'influencing factors'* documents whether the magnitude of operation and maintenance costs is significantly influenced by other water uses, and indicates the cost components that are affected. This information provides the basis for two subsequent analytical steps, namely (1) the assessment of Type II EC, i.e. those environmental costs caused by other water uses that induce higher costs of water service provision (cf. Section 3.2.4.4), and (2) the investigation of the adequacy of contribution from water uses to the cost-recovery of water services (cf. Section 3.4.2).

3.2.2.2 Capital costs

'Capital costs' are treated very differently in cost assessments throughout and even within European Member States. Different rules are applied to both components of capital costs included in the assessment in line with the WATECO guidance document (CIS 2003A), namely depreciation and opportunity costs of capital (cf. Chapter 2.2). The application of different cost accounting rules, e.g. regarding the valuation of assets, the depreciation period chosen or the interest

Cost category: Capital costs **Cost component: Depreciation (Dj_t)**	
Data requirements	Explicit information on the basis for depreciation: – Approach applied for defining the asset base: historic costs of assets, revaluated historic costs of assets, replacement value of existing assets, etc. – Basis for depreciation (linear, digressive, etc.) – Timeframe (years) for depreciation (and underlying concept, i.e. economic or technical lifetime approach) – Inflation rate considered – Information on the basis for depreciation (i.e. have direct contributions from the public budget or from users been subtracted or not) – Information on the treatment of reserves (i.e. is the money available at the time of reinvestment)
Fixed or variable (if both: approximate % division)	Fixed (short to medium run)
Quality of data	Case study specific; high relevance for assessing the implementation of *objective 4*. Classification along the following dimensions: A1 (reliable data source, accurate data) to B2 (unreliable data source, low accuracy of data)
Scale of data availability and periodicity of update	At service area level: through primary data collection from service providers; in part extractable from company records. Aggregated data from industry profiles / technical associations. Updated for each accounting cycle.
Confidentiality	Partly confidential.

Cost category: Capital costs
Cost component: Depreciation (Dj_t)

Accounting guidelines in line with Article 9 economic cost approach	The following guidelines for the calculation of depreciation can be deducted from the discussions in Chapter 2.2: – Application of the maintenance of asset value approach with depreciation calculated as the replacement cost of existing assets; – Contributions from the public budget or from users should not be deducted from the depreciation base (i.e. contributions either from users or from the public budget have to be made explicit). – Linear basis for depreciation. – Timeframe determined by the actual use duration of assets. – Assets that are no longer in use for service provision or will not be renewed are to be excluded from the depreciation base.
Influence of divergence from accounting rule	Impact can be significant due to the high contribution to total costs (→ high capital intensity of the water sector).
Influencing factors / driving forces for cost category	Demographic change and/or climate change may lead to stranded assets. Costs may be underestimated if the share of the costs that is borne through public contributions or direct user contributions is not taken into consideration. Certain water service user groups may necessitate special investments (e.g. mining industry, industrial productions requiring specialised treatment of wastewater) and thus contribute over-proportionately to this cost category.

Cost category: Capital costs
Cost component: Depreciation (Dj_t)

Potential difficulties for the assessment	Aggregation problems may be particularly relevant, if data have to be combined from different service areas with different underlying approaches to depreciation. There can be overlap between '*depreciation*' and '*operation and maintenance costs*' depending on the classification of investments.
Evaluation Full cost accounting and attribution to cost categories *– Expert judgement –*	Case study specific. Guidelines: (1) Have all costs been accounted for (ranging from a) '*full cost accounting attained*', to b) '*not all costs could be accounted for*'. (2) Was an unambiguous cost attribution to the different cost components feasible (ranging from a) 'full cost attribution achieved on the basis of existing data'; b) 'minor divergences and mixture of cost components'; to c) 'strong mix-up of cost components and categories, clear attribution of costs not feasible on the basis of existing data'). Outline reasons for divergences from the Article 9 accounting rules (e.g. national accounting standards mandate a historic cost approach to depreciation) and the nature of the divergence (e.g. calculating depreciation on the basis of historic costs leads to an understatement of this cost category and thus to uncovered capital costs in terms of Article 9).

Cost category: Capital costs
Cost component: Opportunity cost of capital (OC_{jt})

Data requirements	– Necessary operating capital – Amount and share of equity – Rate of return applied on equity – Dividends paid – Amount and share of debt capital – Interest payment on debt capital
Fixed or variable (if both: approximate % division)	Fixed
Quality of data	Case study specific; high relevance for assessing the implementation of *objective 4*. Classification along the following dimensions: A1 (reliable data source, accurate data) B2 (unreliable data source, low accuracy of data).
Scale of data availability and periodicity of update	At service area level: through primary data collection from service providers; in part extractable from company records. Aggregated data from industry profiles / technical associations. Updated for each accounting cycle.
Confidentiality	Partly confidential
Accounting guidelines in line with Article 9 economic cost approach	– Preservation of substance is being aimed at: rate of return calculation should address the entire asset base under an economic cost approach, i.e. both own and foreign capital. – Rate of return/interest rate may be differentiated for equity and debt capital → reasons for differentiation should be detailed. – Similar to depreciation, contributions from user or the public budget should not be deducted.
Influence of divergence from accounting rule	High impact due to high share in overall costs.

Cost category: Capital costs **Cost component: Opportunity cost of capital (OC_{it})**	
Influencing factors / driving forces for cost category	Development in inflation and rate of return/interest rate.
Potential difficulties for the assessment	Aggregation problems may be particularly relevant, if data have to be combined from different service areas, using different underlying principles for the calculation of opportunity costs of capital.
Evaluation Full cost accounting and attribution to cost categories *– Expert judgement –*	Case study specific. Guidelines: (1) Have all costs been accounted for (ranging from a) 'full cost accounting attained', to b) 'not all costs could be accounted for'. (2) Was an unambiguous cost attribution to the different cost components feasible (ranging from a) 'full cost attribution achieved on the basis of existing data'; b) 'minor divergences and mixture of cost components'; to c) 'strong mix-up of cost components and categories, clear attribution of costs not feasible on the basis of existing data'). Outline reasons for divergences from the Article 9 accounting rules (e.g. national accounting standards prohibit return on equity) and the nature of the divergence (e.g. if the opportunity cost of equity is excluded from the analysis, capital costs are underestimated in terms of Article 9, with potential effects on the sustainability of water service provision (*objective 4*)).

rate applied to equity and debt capital, may lead to differences in the magnitude of this cost category of several 100 per cent (cf. HOLLÄNDER ET AL. 2008, SCHULZE 1994). As capital costs make up for a considerable share in total costs, the application of different accounting procedures for this cost category may significantly impact on the overall result. Hence, for any meaningful comparison of costs and later cost-recovery rates, it is important to make the underlying accounting rules on capital costs explicit.

As discussed in Section 2.2.2.1, the objectives of Article 9 necessitate adhering to the principle of asset maintenance in real terms instead of capital maintenance. As sustainability of service provision is aimed at, Article 9 asks for a calculation of depreciation on the basis of the replacement value of existing assets.[223] If national accounting procedures differ, these differences have to be documented in the following *Fact Sheets* in order to set the basis for comparison. To give an example, calculating depreciation on the basis of historic costs leads to an underestimation of *'capital costs'* under an economic cost approach and thus results in an overestimated cost-recovery rate in terms of Article 9, i.e. there are uncovered costs in terms of a long term sustainability of service provision.[224]

The following *Fact Sheets* outline the proposed accounting rules for the cost components *'depreciation'* and *'opportunity costs of capital'*.

3.2.2.3 Administrative costs

The cost component *'administrative costs'* accounts for all costs related to the provision of water services and their regulation. Costs to be considered in this category range from the direct costs of water service administration (e.g. billing customers) to the costs of monitoring compliance with existing regulations on water service provision (e.g. sewerage sludge disposal) or administering a licensing systems (e.g. water abstraction licences). In many cases, administrative costs will not be recorded separately but included in other cost components, most notably operation and maintenance costs, Type I EC and resource costs. Furthermore, even if administrative costs are reported on, disaggregating costs for a particular water service may be problematic (i.e. due to multiple uses, overlap with other (side) services, etc.).

---

223 Cf. discussions in Chapter 2.3.4 on influencing factors such as demographic change, which may warrant a divergence from this approach, i.e. as it is a replication of the existing asset base in the future may under certain circumstances not be the sustainable approach. These aspects may be accommodated through the *Fact Sheet* under *'Influencing factors / driving forces for cost category'*.

224 Alternatively, it has to be justified, why the historic cost approach is more appropriate in a certain application case, e.g. due to reduced demand for water service in the future caused by demographic changes.

Cost category: Administrative costs
Cost component: Administrative costs (AC_j and AC_{RB})

Data requirements	Different levels (service provider, governmental agency, river basin (district) level, regional irrigation district, etc.): – Direct costs of water service administration – Costs of issuing / administering permits / licences (→ potential overlap with Type I EC) – Costs of monitoring compliance with existing regulations on water service provision (e.g. sewerage sludge disposal, water abstraction licences) – Administrative costs of water service provision at river basin level – Administration of co-operation with other water users, e.g. costs of administering cooperative agreements between water suppliers and farmers / industry (→ cf. Section 3.2.5) – Costs of monitoring compliance with environmental standards (→ administrative cost part of environmental costs assessment)
Fixed or variable (if both: approximate % division)	Mostly fixed
Quality of data	Case study specific. Classification along the following dimensions: A1 (reliable data source, accurate data) to B2 (unreliable data source, low accuracy of data)
Scale of data availability and periodicity of update	Dependent upon the type of administrative cost, e.g. at level of the service area, administrative levels, river basin (district) level, regional irrigation district level, etc.

Cost category: Administrative costs **Cost component: Administrative costs (AC_j and AC_{RB})**	
Confidentiality	May be an issue at a local scale (e.g. data on abstraction licences).
Accounting guidelines in line with Article 9 economic cost approach	Account for all administrative costs that are related to the provision of water services and their regulation. Separate administrative costs for specific water services; make cost attribution criteria explicit.
Influence of divergence from accounting rule [depends upon magnitude of cost category in total costs]	Low
Influencing factors / driving forces for cost category	Case study specific
Potential difficulties for the assessment	Data availability may be problematic for scales higher than the individual water service provider, as information on administrative costs is seldom collected through general statistics. Disaggregation of administrative costs for a particular water service may be problematic (i.e. due to multiple uses, overlap with other (side) services, etc.). Potential overlap with other cost categories: e.g. billing customers, etc. → potential overlap with *operation and maintenance costs;* costs of issuing / administering permits / licences → potential overlap with Type I EC. Disaggregation of administrative costs from other cost components, e.g. operation and maintenance, Type I EC, resource costs.

Cost category: Administrative costs
Cost component: Administrative costs (AC_j and AC_{RB})

Evaluation Full cost accounting and attribution to cost categories *– Expert judgement –*	Case study specific. Guidelines: (1) Have all costs been accounted for (ranging from a) 'full cost accounting attained', to b) 'not all costs could be accounted for'. (2) Was an unambiguous cost attribution to the different cost components feasible (ranging from a) 'full cost attribution achieved on the basis of existing data'; b) 'minor divergences and mixture of cost components'; to c) 'strong mix-up of cost components and categories, clear attribution of costs not feasible on the basis of existing data').

3.2.2.4 Other direct costs

The cost category *'other direct costs'* compiles information on those costs that are not covered through other categories of financial costs, but are incurred in the provision of water services. For example, municipalities may levy fees on water service providers for using public roads or squares for the construction of the pipe network. It is important to report exactly on the nature of the costs that are included under this cost component (i.e. the main aim of the cost accounting framework is to establish transparency). While this cost category collects information on all cost factors that cannot be clearly allocated to another financial cost category, it should be noted that costs stemming from extraordinary events are in general excluded from the analysis and should also not be considered here.

3.2.2.5 Summary

Section 3.2.2 provided guidelines for assessing financial costs of water service provision structured according to *Fact Sheets* for each sub-category of financial costs. On this basis, divergences from the ideal Article 9 accounting approach can be outlined, along with national specificities (i.e. why is accounting done differently).

Cost category: Other direct costs	
Data requirements	All financial costs linked to the provision of water services that cannot be attributed to another cost category. E.g.: concession fees and other fees paid to municipalities, etc.
Fixed or variable (if both: approximate % division)	Case study specific, depending upon the costs included.
Quality of data	Data Quality: Classification along the following dimensions: A1 (reliable data source, accurate data) to B2 (unreliable data source, low accuracy of data)
Scale of data availability and periodicity of update	Case study specific, depending upon the costs included.
Confidentiality	Case study specific, depending upon the costs included.
Accounting guidelines in line with Article 9 economic cost approach	Describe underlying mechanism for determining '*other direct costs*' (i.e. what is the cost category composed of). Make potential overlap with other cost categories explicit. Include only data for the relevant investigation year.
Influence of divergence from accounting rule [depends upon magnitude of cost category in total costs]	Generally, limited impact due to small share in total costs (nevertheless, case study specific).
Influencing factors / driving forces for cost category	Case study specific, depending upon the costs included.
Potential difficulties for the assessment	Vague definitions. Cost attribution. Risk of double counting, if costs cannot be clearly allocated.

Cost category: Other direct costs

Evaluation Full cost accounting and attribution to cost categories *– Expert judgement –*	Case study specific. Guidelines: (1) Have all costs been accounted for (ranging from a) 'full cost accounting attained', to b) 'not all costs could be accounted for'. (2) Was an unambiguous cost attribution to the different cost components feasible (ranging from a) 'full cost attribution achieved on the basis of existing data'; b) 'minor divergences and mixture of cost components'; to c) 'strong mix-up of cost components and categories, clear attribution of costs not feasible on the basis of existing data').

3.2.3 Accounting for revenues in water service provision

In a next analytical step towards assessing cost-recovery of water services, the accounting framework for financial costs outlined in the preceding Section 3.2.2 is amended by a complementary revenue accounting framework, which takes account of all recovery mechanisms in place. It allows for investigating the nature and composition of different revenue streams. The investigation is centred on the rationale that costs are always recovered, but not necessarily from users/polluters (cf. Chapter 2.2). Only in those cases, where costs are burdened onto the next generation (e.g. through public debt or depreciated natural or artificial capital), costs may be considered as uncovered. However, subsidies financed through general taxation may cause many potential distortions in the allocation of resources through distorted incentives, which should not be neglected in the analysis. As a reminder, Article 9 aims at a strong internal cost-recovery approach, i.e. a financing of water services through user/polluter contributions (cf. *'case 1'* in Figure 13 above). Hence, the investigation of revenue streams provides the necessary complementary information for assessing divergences from the strict internal cost-recovery approach intended by Article 9.

The revenue accounting framework focuses on ensuring transparency with regard to different revenue streams by documenting who contributes to which revenue stream; and by which mechanism the level of contribution is determined. Furthermore, the revenue assessment sets part of the required information base for evaluating the adequacy of contribution of water uses and the implementation of the polluter-pays principle (*objective 3*). In addition, basic information on the incentive properties of revenue generation mechanism is compiled. Thus, the following analysis contributes particularly to the later assessment of Article 9 *objective 2* (i.e. the adequacy of incentives for efficient resource use).

Similar to the cost accounting framework above, the exact definition of the service under consideration in the assessment is decisive. For example, revenues generated through side-services, such as fire protection or rainwater management, should be excluded as far as possible from the assessment. If their separation is not feasible on the basis of available data, a qualitative description as to their percentage in revenues should be included based on expert judgement (cf. Box 7).

Revenue generation mechanisms designed as internalisation instruments for environmental costs are studied in unison in the next assessment step, cf. Section 3.2.4.[225]

3.2.3.1 Contributions from users: endogenous revenue components

The revenue accounting framework first addresses all endogenous revenue components, i.e. all direct contributions from users of the investigated water service. For each endogenous revenue component, structured *Fact Sheets* compile information on:

- The payment basis (e.g. per household/connected inhabitant, hectare/crop type, meter size, etc.; obligatory or on demand)
- The structure of charges (e.g. number of blocks, linear, increasing/ decreasing, etc.)
- Collection mechanisms and payment schedules (e.g. one-off payment, weekly, monthly, yearly)
- Variation factors (e.g. season, time-of-day, off-peak)
- Exemptions and eligibility criteria (e.g. prolific families, small farms)
- Links to cost categories (e.g. capital costs of new investments, administrative costs, etc.)

In order to achieve transparency of financial flows and to conduct a meaningful assessment of financial cost-recovery, not only the actual figures are relevant (i.e. how much revenue is obtained from the different revenue categories), but rather, the set-up of the revenue generation mechanisms. For later phases of

225 A clear separation may not always be possible. But *Fact Sheets* allow for documenting revenue attribution difficulties, similar to cost *Fact Sheets* above.

Fact Sheet 6: Revenues collected through endogenous revenue components

Revenue source	Description of the structure of the tarification scheme / element						Data			Evaluation
	Payment basis (e.g. per household, per con-nected inhabitant, per hectare, per crop type, per meter size, pipe size, etc.; obligatory or on demand only)	**Structure of the charge** (e.g. number of blocks, linear, increasing / decreasing, etc.)	**Collection mechanism and payment schedule** (e.g. one-off payment, weekly, monthly, yearly)	**Variation factors** (e.g. season, time-of-day, off-peak)	**Exemptions and eligibility criteria** (e.g. for prolific families, small farms)	**Link to cost categories** (e.g. capital costs of new investments, administrative costs, etc.)	**Rates** (e.g. Euro/m^3, Euro/ha, etc.)	**Revenue collection** (Euro/year and % contribution to overall revenue)	**Data quality** (reliability and accuracy of reported data, rated A1 – B3)	**Problems in reporting**
Connection fee										
Variable charge										
Fixed charge										
Metering charge										
Investment contribution										
Direct taxes										
Other fees / charges										

the assessment, a clear understanding of how the scheme is built up; where there are linkages to cost categories; where exemption regulations are in place and where a differentiation between user groups is undertaken (e.g. special charges or exemptions for certain water uses or customer groups) is indispensable. Furthermore, the degree to which cost categories factor into revenue generation is crucial (i.e. by fixing the revenue objective to be obtained from a certain revenue stream or by introducing a seasonal variation in tarification systems).

3.2.3.2 Exogenous revenue components

Next to endogenous revenue sources through direct contributions from water service users, the role of public transfers in the financing of water service provision has to be investigated. As introduced in Chapter 2.2, the following exogenous revenue components may play a role in water service financing and thus have to be taken into consideration for the assessment:

- Transfers obtained from ear-marked budgets (such as equalisation funds, water taxes)[226];
- Investment subsidies obtained from municipalities or other administrative levels;
- Direct subsidies paid by public authorities for the operation and maintenance costs of water services provision;
- Subsidies obtained from transfer schemes at river basin (district) level[227];
- Subsidies obtained from European transfer schemes for water service provision (e.g. European Structural Fund, Common Agricultural Policy, etc.).

226 There may be an overlap to the assessment of environmental costs, cf. Section 3.2.4 below. For example, France operates a transfer programme for municipal water supply and wastewater services provision, which is funded through earmarked taxes collected at river basin level (i.e. charges for water abstraction or wastewater discharges that are levied by the water agencies). The transfer scheme aims to establish affordability between urban and rural areas and is mainly used for furthering the achievement of environmental objectives within the basin and are paid by water service users through the tariffication scheme (OECD 2009). In effect, they function as regional cross-subsidies for internalizing environmental costs within a river basin.

227 Intra-basin transfers may be considered a smaller hindrance to the implementation of Article 9 objectives than inter-basin transfers. For the implementation of the third objective of Article 9 (Implementation of the polluter-pays principle and the adequacy of contribution) it should however be studied, whether transfers take place within or across customer groups.

Subsidies in the water sector can take various forms. The focus at this point of the analysis is on direct subsidies (e.g. contributions to investments) or indirect subsidies (e.g. low-interest grants, soft loans) distributed by different administrative levels. The role of intra-generational subsidies between user groups (e.g. households and agricultural water users) or between regions (e.g. rural and urban areas) is studied at a later point of the investigation (cf. Section 3.2.4).

In order to investigate the impact of subsidisation schemes on the attainment of Article 9 objectives, the following information has to be provided on each subsidisation scheme (cf. *Fact Sheet* 7 below):

- The eligibility criteria
- The annual amount of payment and payment schedule (lump-sum or recurrent payment);
- Whether it is a historic transfer that still influence the cost structure or an ongoing subsidisation schemes;
- The link to cost categories (e.g. operation and maintenance, capital costs);[228]
- The treatment of transfers / subsidies in depreciation;
- The reliability and accuracy of the reported data (ranging from A1 to B2).

Furthermore, as an input to the assessment of the adequacy of the contribution of different water uses and the implementation of the polluter-pays principle (objective 3), differentiations between user groups are also investigated (e.g. subsidization schemes that apply only to certain water service users (e.g. small farms)).

3.2.3.3 Other revenue components (non-water service related)

Besides direct contributions from users of a water service and transfers from public or ear-marked redistributive budgets, additional revenue sources, which are not water service related, may be present and have to be accounted for. These include for example revenues from activities sold on the market (e.g. services sold to other operators). Furthermore, revenues obtained for other service components may be used for cross-subsidising the provision of a particular water service. For example, an undertaking that provides next to drinking water supply also gas provision to an overlapping group of customers may use

228 Public transfers aimed at rectifying environmental damages caused by the provision of water services are not considered here, but accounted for under Step 4 below.

Fact Sheet 7: Revenues collected through exogenous revenue components

Subsidisation scheme	Description of the structure of the subsidisation			Data			Evaluation
	Eligibility criteria	**Payment mechanism** (e.g. direct financial transfer, reduced interest rate, etc.)	**Payment schedule** (e.g. one-off payment, yearly contribution, recurring contribution at time of investment, historic or ongoing)	**Link to cost categories** (e.g. operation and maintenance; investments, etc.)	**Revenue collection** (Amount of payment)	**Data quality** (reliability and accuracy of reported data, rated A1 – B3)	**Problems in reporting**
Transfers from ear-marked budgets							
Subsidies from municipal level							
Subsidies from river basin (district) level							
Subsidies from other administrative levels							
Subsidies obtained from European transfer schemes							
Other							

revenues from one branch to cross-subsidise activities of the complementary branch of its activities.[229]

Similar to the treatment of extraordinary costs above, extraordinary revenue sources, e.g. from sales of assets, are excluded from the assessment.

3.2.4 Assessing environmental costs and their recovery

In terms of Article 9, two subgroups of environmental costs, which were defined in Chapter 2, are of interest for the implementation process (cf. Figure 15 below). First, those environmental costs in aquatic ecosystems or dependent land ecosystems which are *caused by the provision of a particular water service* have to be considered (Type I EC). They are directly relevant to the cost-recovery assessment and can either have already been internalised through measures conducted by the water service provider in order to avoid environmental damages (i.e. water resource protection costs) or through payments made on existing internalisation instruments, requiring further investigations as to their magnitude and relevance. In part, information on the cost component *'Type I EC'* is available from the river basin management plans, which contain the measures needed for reaching good water status and thus for protecting waters and related land ecosystems from or rectifying existing damages. However, as the programmes of measures address *all* environmental damage costs to the aquatic environment, irrespective of whether they are linked to the provision of water services or not, their separation may not be trivial. As an alternative, benefit based approaches can be used for their measurement, whose application is in practice also linked to notable difficulties (cf. Chapter 2.4).

The second subgroup of environmental costs relevant to the assessment is *caused by other water uses* (Type II EC) and leads to higher (financial) costs of water service provision (e.g. in the form of higher raw water treatment costs for drinking water provision or higher water abstraction costs due to a lowering of water tables caused by other water using activities).[230] They are addressed in Article 9 through the *'adequate contribution'* by other water uses to the cost-

229 This fact poses in particular problems for cost accounting above, as some sub-services such as billing customers may be provided in unison for both branches and costs may not be accounted for separately.

230 e.g. additional pumping costs in the provision of water supply services due to mining activities

Figure 15: Environmental costs

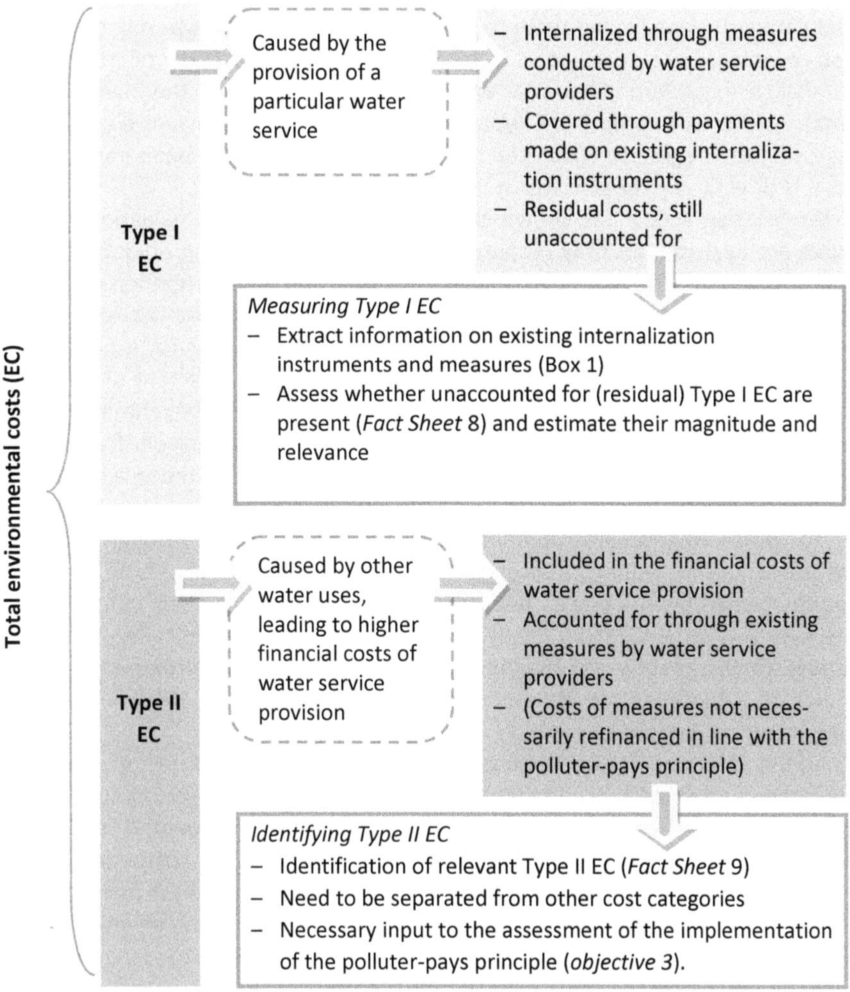

recovery of water services (*objective 3*, cf. Section 1.4.3. and Section 3.4.2).[231] For this subgroup of environmental costs the problem relates less to their over-

231 As a reminder, the concept of an '*adequate contribution*' of water uses to the cost-recovery of water services extends beyond Type II EC and also includes a use-related share of financial costs (cf. Chapter 2.2).

all measurement as they are mostly included in the financial costs of water service provision, but more to extracting and separating them from the other cost categories and linking them to specific water uses (i.e. answering the question of which share of the costs of treating raw water is actually related to the parallel (or historic) industrial or agricultural use of water in the investigation area). As the cost component Type II EC constitutes an important element for assessing the degree to which the polluter-pays principle has been implemented, it is directly relevant to later steps of the analysis.

While Chapter 2.4 has shown that appropriate economic valuation procedures are available for fully monetising and incorporating environmental costs into decision making processes, the present data and information base in most European Member States will only allow for a comprehensive assessment at case study level (GÖRLACH ET AL. 2007). Given limited resources for the relevant assessments and severe time constraints, an area-wide assessment of environmental costs for the first implementation process (and likely beyond) may thus only be achieved through alternative procedures.[232] The assessment framework developed in this study proposes a pragmatic approach for the evaluation of environmental costs for the purpose of the cost-recovery assessment. It assesses to which degree environmental costs are internalised through existing measures and instruments; determines the relevance and magnitude of unaccounted environmental costs and enables a judgement on their influence on the overall divergence from the ideal Article 9 path to cost-recovery. This procedure for estimating the magnitude and relevance of environmental costs draws on a methodology developed by AMMERMÜLLER ET AL. (2008) for a non-monetary cost-benefit assessment.

As the following Figure 16 displays, the proposed procedure for assessing environmental costs is divided into five distinct steps. Step 1 to 3 focus on the identification and estimation of both internalised and unaccounted for Type I EC. Step 4 and 5 aim to make Type II EC transparent for the cost assessment. The proposed procedure accommodates information in different formats and may progressively and iteratively move to a fully monetised environmental costs assessment.

232 CIS (2004A) pointed out that *"in the absence of data to underpin monetary estimation of the environmental and resource costs qualitative description may suffice."*

Figure 16: Step-wise approach for the assessment of environmental costs

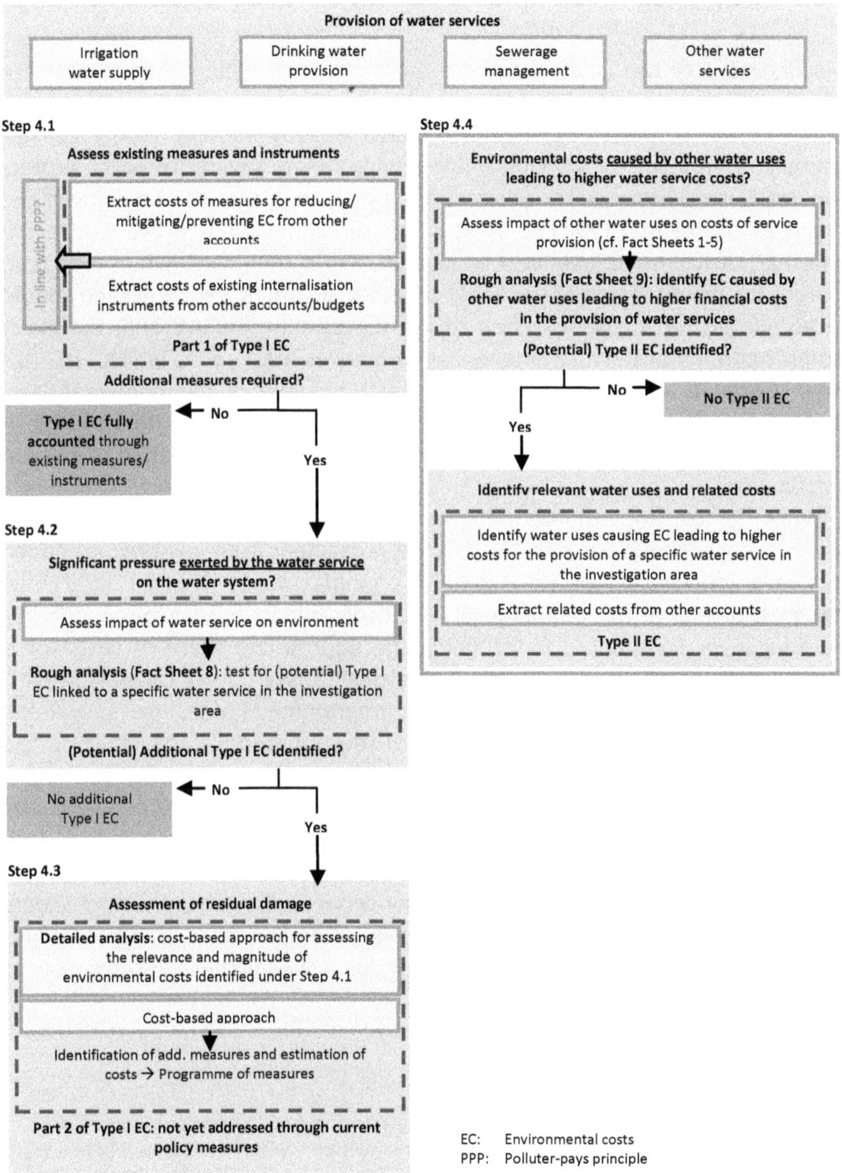

3.2.4.1 Step 4.1: Taking stock and investigating present internalisation mechanisms

The first step of the analysis focuses on those environmental costs that are directly linked to the provision of water services, i.e. Type I EC. It is assessed in how far existing measures and instruments already address Type I EC. For example, if providers of drinking water services finance water resource protection measures within water abstraction zones, the costs of these measures qualify as internalised Type I EC. Next to measures that aim to prevent or rectify Type I EC, instruments may also be operational in the investigation area (e.g. charges for the discharge of sewerage or water abstraction charges) and should be accounted for in the first step of the assessment. To this end, the exact modus operandi of the instruments has to be evaluated, i.e. to which activities the instruments apply, whether they vary with the level of environmental damage (e.g. by levying a higher fee for the discharge of sewerage into more sensitive water bodies) etc.[233] Box 6 provides guidelines for the evaluation of existing instruments.

While the relevant cost information on measures and instruments paid for by water service providers directly can be extracted from existing accounts, measures addressing Type I EC may also be initiated and financed differently, e.g. at RBD level or at national level (e.g. programmes at river basin level for restoring river continuity). Determining the appropriate share of costs for a particular water service may thus not be straightforward in all cases. However, as these measures should form part of the Programme of Measures, knowledge on both the costs and the effectiveness of these measures, as well as on the pollution pathways that cause the need for measures should become more solid over the implementation process and allow for a substantiated cost attribution to different water services. During the first implementation cycle expert judgement may be used for assessing existing schemes and identifying and attributing Type I EC to different water services. Similar to the *Fact Sheets* for financial costs, it is important to make allocation difficulties transparent and to indicate uncertainties in order to allow for a meaningful interpretation of results.

---

233 Furthermore, for later phases of the assessment (cf. Section 3.3), it needs to be evaluated, in how far the instruments adhere to the polluter-pays principle. Thus it is not only important to evaluate, whether measures or instruments are put in place for addressing Type I environmental costs, but also to investigate whether they are levied on the polluters/users of the water service, or why there is a divergence and make this divergence transparent.

Box 6: Evaluation of existing internalization instruments

For each existing internalisation instrument, the following information should be compiled:

- Since when is the instrument operational and why has it been instituted?
- What does the instrument aim at (e.g. internalising environmental damage costs of polluted discharges into surface water bodies)?
- Which activities are addressed by it?
- What is the fee/charge levied on?
- Is the level of the fee/charge determined by environmental targets or rather politically?
- Does the level of the fee/charge vary? With which parameters (e.g. seasonally, according to pollution load, etc.)?
- Are exemption regulations in effect (e.g. exempting certain sectors or activities from the fee/charge)?
- Is the instrument effective in reaching its objectives?

In addition, the following information is required for later phases of the evaluation:

- Does the instrument set incentives for efficient resource use (*objective 2*)?
- Are the revenues of the instrument earmarked? How are they employed (*objective 1 and objective 3*)?

3.2.4.2 Step 4.2: Rough analysis for identifying potential additional Type I EC (residual damage)

Once existing measures and instruments have been analysed, remaining and unaccounted for environmental costs are addressed in a second analytical step. As the principle of proportionality is given priority in Annex III of the Directive in reference to the economic assessments, a detailed analysis and in-depth investigation of environmental costs should be narrowed down as far as possible to relevant aspects. Hence, a rough overview analysis is conducted first (cf. *Fact Sheet* 8 below) for detecting significant pressures, which are exerted by specific water services on the water system within the investigation area that are not (fully) accounted for through existing instruments and measures (i.e. unaccounted for Type I EC).

In those cases, where environmental costs are fully internalised through existing measures or instruments (i.e. the ideal Article 9 implementation), the

Fact Sheet 8: Rough analysis for detecting Type I EC

	Type I EC categories																					
Water service	**Overabstraction**			**Morphological changes (river regulation and flow diversion)**				**Polluted discharges**			**Other**											
	Adverse effects on aquatic species	Loss of biological diversity	Adverse effects on water dependent land ecosystems (wetlands, forests)	Adverse effects on aquatic species	Loss of biological diversity	Adverse effects on water dependent land ecosystems (e.g. wetlands, forests)	Increased flood risk	Water quality degradation (impairment of natural rate of recovery)	Loss of biological diversity	Adverse effects on aquatic species	Land erosion	Degradation of soil quality, loss of soil fertility	Eutrophication	Chemical contamination	Salination	Water logging	Dessication	Reduced natural resiliencee	Other
Water supply			x																
Wastewater management					Filled in for each case study														
Irrigation	x	x	x	x	x	x					x	x			x	x			

Note: "x" indicates the relevance of the Type I EC cost categories, here: illustrative example 'rough analysis' only validates the fact that no environmental costs are left uncovered.[234] In this case, present instruments and mitigating or preventing measures undertaken by the water service provider or e.g. at river basin district level are effective and eliminate all negative consequences of the provision of water services on the environment. Should the *'rough analysis'* outline remaining negative effects of a water service (i.e. salinity caused by irrigation), a more detailed analysis of these effects is undertaken in Step 3.

3.2.4.3 Step 4.3: Taking stock of residual damage – unaccounted for environmental costs

With the help of the *'rough analysis'* (cf. *Fact Sheet* 8), those environmental cost categories were identified, that are relevant to the particular investigation area despite existing internalisation instruments and/or measures. The third step now undertakes a detailed analysis of (potential) residual environmental damages.

As highlighted in Section 2.4.4, a cost-based approach can be followed for assessing (residual) environmental costs. As the WFD implementation process compiles extensive information on both, environmental damages (i.e. expressed as deficits in water status), as well as on the related causing activities (i.e. though an analysis of pressures and impacts as well as through extensive monitoring programmes), using a cost-based approach for assessing residual environmental costs seems straightforward and well-aligned with the overall WFD logic. It supports the WFD objective of proportionality of efforts and exploits information that has already been gathered for other WFD implementation steps.

The cost-based approach links up to the cost-effectiveness analysis and the river basin management plans, which contain the measures needed for reaching good water status and thus for protecting waters from or rectifying existing environmental costs. The required information base for applying the cost-based approach is thus partly collected within the WFD implementation process. However, the programmes of measures address *all* environmental damage costs to the aquatic environment, irrespective of whether they are linked to the provision of water services or not. However, only the latter are relevant to the cost-recovery assessment.

234 However, then the later tasks remain to investigate whether the internalisation mechanism is in line with the polluter-pays principle.

Generally, one can view the distance to the target situation of *'good status'* as a proxy for the magnitude of environmental costs in the investigation area.[235] As outlined in Chapter 1.2, the River Basin Management Plans (RBMP) specify Programmes of Measures (POM) that aim to rectify any environmental damages that trigger impairments in water status. By definition, measures for remedying environmental damages caused by the provision of water services are also included in the POM. While information on the costs of measures required for alleviating environmental costs of water services is included in the POM, it nevertheless requires a clear understanding of pollution pathways as well as cause-effect relationships for singling out those measures and attributing an appropriate share of their costs to a specific water service. Due to summation effects or unclear physical pollution pathways, an unambiguous allocation of costs to specific water services may not always be feasible. Furthermore, measures may have multiple effects, so that their costs cannot be entirely allocated onto a single causing activity, or may only be effective in combination with other measures (although these do not address directly the environmental cost of the water service).[236] Hence, while the relevant information on costs is already compiled within the WFD implementation process, expert knowledge is required for accessing this information appropriately and making use of it for the assessment. Expert knowledge is also required for determining:

- The degree of *certainty* with which relevant measures could be singled out from the overall Programme of Measures and attributed to a particular water service (i.e. in most cases, measures will not *only* be required due to the water service, but rather because of a complex set of interacting causes. Hence an appropriate cost allocation key – which share of the cost is attributed to the water service and on what ground – has to be devised and outlined.)
- The *accuracy* of the cost information (i.e. the costs of measures indicated in the RBMP are projections and may be subject to changes, as measures have not yet been implemented; e.g. construction costs could in practice be higher than originally foreseen).

235 It should however be noted that environmental costs can also be present under a good water status (e.g. new substances that are not being monitored so far, residual damages that do not affect the classification of a water body, etc.).

236 E.g. if *Fact Sheet* 8 indicated that over-abstraction may be a problem under the water service drinking water provision, the costs of measures included in the Programme of Measures for addressing the effects of over-abstraction should be taken into account. If Overabstraction is a problem, it may be linked both to the water service drinking water provision and irrigation. Costs should then be allocated consistently according to a cost allocation key, e.g. the number of m³ extracted.

In concluding, step three aims to extract the costs of measures that address negative impacts of water service provision from RBMP and to indicate related uncertainty of the cost data.

3.2.4.4 *Step 4.4: Assess Type II EC – environmental costs caused by other water uses*

Once Type I EC have been estimated, the second category of environmental costs is assessed, namely environmental costs caused by other water uses that lead to higher costs of water service provision. For example, diffuse pollution from agriculture may lead to higher costs of raw water treatment for the water services drinking water provision and irrigation. *Fact Sheet* 9 gives an overview on potential Type II EC. An indication of relevant water uses that may give rise to Type II EC is provided by the Article 5 river basin characterization reports that identify activities having a significant impact on water status and assess related pressures and impacts. The assessment of Type II EC is taken up again in later phases of the investigation, namely when the adequacy of the contribution to cost-recovery of water uses is studied and the implementation of the polluter-pays principle is assessed (cf. Section 3.4.2).

The main aim of the investigation of Type II EC is to obtain a comprehensive and complete inventory on where environmental costs caused by water uses trigger higher costs of water service provision. Both the analysis of pressures and impacts as well as the monitoring programmes provide information as to the origin of environmental damages and pollution pathways. The following *Fact Sheet* 9 provides a non-exhaustive list of possible interactions between environmental impacts of water uses and the costs of water service provision. One should note that water services may also influence each other. However, considering the environmental costs caused by ineffective wastewater treatment as a Type II EC would result in double-counting, as these effects are already addressed through the measurement of Type I EC (Step 1 to 3 above).

While the detection of Type II EC may not always be straightforward, the related cost information is generally included in the financial accounts of water service providers (i.e. by the definition outlined above, Type II EC refer to *'high-er costs for water service provider'*).[237] At the level of the service entity, the differentiation of relevant cost data should be feasible. It should be noted that Type II EC may be substantial in magnitude. For example, if nitrate pollution

237 An exception to this generalized rule are future costs, such as the depletion of resources or the pollution of a source, which imply that new sources have to be developed or long distance transfers have to be relied upon for future water service provision.

from agriculture renders a local water source unsuitable for drinking water abstraction so that alternative supplies have to be accessed, high capital costs may have to be encountered. More often, however, higher water purification costs will be the result. The *Fact Sheets* on financial costs (cf. Section 3.2.2) include relevant indications on Type II EC in the column *'influencing factors / driving forces for cost category'*. Difficulties for the assessment may be related to the indivisibility of costs or the clear attribution of costs to a specific water use. Therefore, it is important to include information on the certainty and the accuracy of the cost information and its attribution to different water uses (similar to Step 3 above) and to sketch Type II EC qualitatively.

Fact Sheet 9: Identifying Type II EC

Water service	Effect	Related water uses	Affected cost category
Drinking water provision	Degraded raw water quality leading to higher water purification costs	Diffuse pollution of sources through agricultural activities Inadequately treated discharges from industrial users or fisheries Navigation Salination through agriculture or industry	Operation and maintenance costs (e.g. material costs, energy costs, monitoring costs) Capital costs Other direct costs
	Lowering of water tables	Over-abstraction through agriculture (not linked to irrigation) or industry Mining activities	Operation and maintenance costs (e.g. material costs, energy costs) Capital costs (e.g. if alternative long distance supply systems have to be established or existing systems become inoperational)
	Morphological changes with negative impacts on raw water quality or access to water sources	Agriculture Navigation Industry Tourism and recreation	Operation and maintenance costs (e.g. material costs, energy costs) Capital costs

Water service	Effect	Related water uses	Affected cost category
Wastewater management	Reduced self-cleaning capacity of receiving water bodies	Diffuse pollution of sources through agricultural activities Inadequately treated discharges from industrial users or fisheries Navigation (through impact on wetlands)	Operation and maintenance costs (e.g. material costs, energy costs, monitoring costs) Capital costs
	Harmful discharges requiring special treatment (e.g. membrane filtration)	Industry Agriculture Mining activities	Capital costs Operation and maintenance costs
Irrigation	Degraded raw water quality leading to higher water purification costs	(Households) Diffuse pollution of sources through agricultural activities, Inadequately treated discharges from industrial users or fisheries Salination through agriculture or industry	Operation and maintenance costs (e.g. material costs, energy costs, monitoring costs) Capital costs Other direct costs
	Morphological changes with negative impacts on raw water quality, access to water sources	Agriculture Navigation Industry Tourism and recreation	Operation and maintenance costs Capital costs
	Lowering of water tables	Over-abstraction through agriculture (not linked to irrigation) or industry Mining (Households)	Operation and maintenance costs Capital costs (e.g. if alternative long distance supply systems have to be established or existing systems become inoperational)

The assessment of Type II EC provides an important input to later assessment steps, most notably, the implementation of the polluter-pays principle and the analysis on the adequacy of the present contribution of water uses to the cost-recovery of water services.

3.2.5 Assessing resource costs and their recovery

As discussed in Chapter 2, the CIS process produced two diverging definitions of resource costs. This study adheres to the broader definition of resource costs proposed by the working group DG Eco II. In this understanding, resource costs arise if the provision of water services constitutes *"an economically inefficient allocation of water, either in terms of water quantity or water quality, over time and across different water use(r)s"* (emphasis added) (CIS 2004B). For resource costs to be present in the provision of water services, the difference between the economic value in terms of net benefits of water service provision and the economic value in terms of net benefits of the best alternative water use (now or in the future) has to be negative. Resource costs hence exist, if the water service does not constitute the highest value use.

In principle, in order to assess whether resource costs are present in the provision of water services, the use and non-use values of all potential alternative uses have to be investigated. However, one can question, whether resource costs can in effect be encountered *at all* in the provision of the water services *'drinking water supply to households'* and *'sewerage management'*, due to their high societal preference. Most Member States give drinking water supply a priority in water use, often even enforced by legislative regulations.[238] This societal preference may be interpreted as an indication of the underlying non-use value associated with the service's provision. However, an alternative industrial use may be more productive in terms of monetary equivalents generated than the water service *'drinking water provision'*, high non-use values are attached to the latter, which overcompensate lower financial returns. Analogously, the provision of the water service *'sewerage management'* is given high societal preference, signalling, for example, the high non-use values attached to the control of public diseases. It can thus be concluded, that resource costs are seldom present in the case of these two water services, because there are only rarely higher value uses from a societal perspective.

Yet, a number of caveats apply to this reasoning. First, there could be regional alternatives, e.g. water supply could be arranged differently albeit at higher

238 For the German context, the hierarchy of water uses ranges from drinking water provision to the cooling of power plants to agriculture. For Italy, water provision to agriculture is only second to drinking water provision.

costs (and in violation of the water sustainability principle of locality of resource use). Second, in the case of the water service *'drinking water provision'*, the high societal preference is only linked to essential household uses and not e.g. to the use of water for watering gardens, filling swimming pools, etc. Hence, it has to be questioned, whether the reasoning that no resource costs stem from the water service 'water supply' applies to the entire range of water use by households. Nevertheless, on grounds of proportionality, it may be questioned how much effort is warranted for the assessment of resource costs for the water services *'drinking water provision to households'* and *'sewerage management'*, given the high societal preference associated with them. Generally, the investigation of resource costs will be case-specific and the assessment framework should allow for different degrees of in-depth investigations, depending on the relevance attested to resource costs in a specific investigation area.

For the third water service investigated in this study, the provision of *'irrigation water'*, a differentiation is needed. While some Member States attest a high societal preference to the provision of water for subsistence farming, the argumentation above cannot be generalised to the overall provision of this water service. Here, alternative water uses will generally have to be investigated (although agriculture may be given priority in the hierarchy of water uses, also expressing the societal preference sketched above). If instead of agriculture, an alternative industrial production or an ecosystem service could generate higher net benefits, resource costs are present and need to be considered in the analysis.

While information on alternative use values may be more readily available and even be accessible in monetized form (e.g. in the case of alternative productive uses), potential alternative uses may not always be known e.g. due to traditionally determined water use patterns which discourage alternative uses altogether. Thus, in order to enable an area-wide approximation of resource costs for the purpose of the cost-recovery assessment, a pragmatic approach has to be opted for. The central aim of the analysis is to increase transparency on costs and trade-offs between water uses within the river basin (district) / the investigation area. Hence, a pragmatic approach is sufficient for the assessment and can be substantiated during further implementation cycles, e.g. through in-depth investigations of trade-offs between different uses. The focus for this study is on eliciting sufficient information, in order to determine whether the cost-recovery assessment conducted so far has to be adjusted due to uncovered resource costs. To this end, an abbreviated three step approach is proposed for an overview investigation on the relevance of resource costs in the investigation area, as outlined in Figure 17 below.

Figure 17: Step-wise approach for the assessment of resource costs

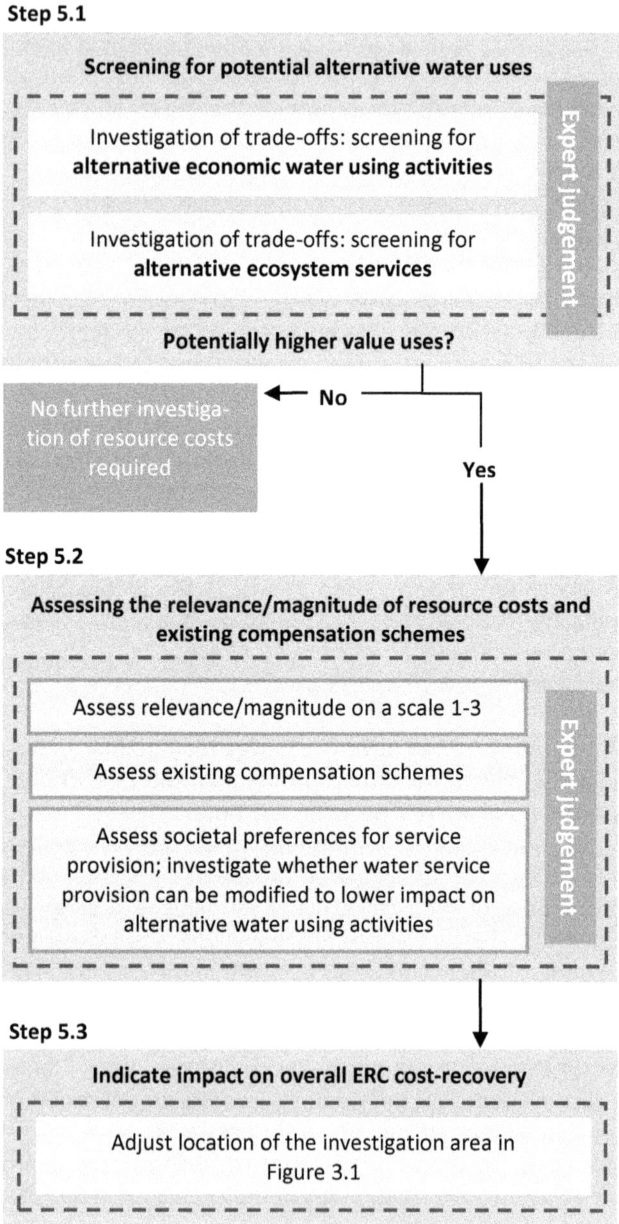

3.2.5.1 *Step 5.1: Screening for potential alternative water uses*

In a first screening step, it has to be investigated whether alternative water using activities could generate a higher (use or non-use) value. Alternative water using activities are inherently site-specific and thus cannot be examined generically (cf. Chapter 4 for illustrations). For example, the assessment may boil down to the investigation of trade-offs between (intensified) touristic development and the provision of irrigation water services. To facilitate a pragmatic evaluation, the first screening for potential resource costs is based on expert judgement, as local knowledge on potential alternatives is crucial for the analysis. The economic analysis of water uses conducted as part of the Article 5 reports by the year 2004 was already intended to outline economic tradeoffs within the river basin (districts). This analysis can be built upon and extended for the assessment of resource costs.

In addition, it needs to be explored whether ecosystem services may constitute economically more efficient water using activities than the provision of the particular water service. *Fact Sheet* 10 provides the framework for conducting a rough analysis whether ecosystem services may constitute relevant alternatives giving rise to resource costs. The assessment draws on the Millennium Ecosystem Service Assessment (MA) as developed by UNEP (2005). The MA distinguishes three categories of ecosystem services, namely

- **Provisioning services**, including food, fiber, genetic resources, biochemicals, natural medicines, pharmaceuticals and fresh water;
- **Regulating services**, including air quality regulation, climate regulation, water regulation, erosion regulation, water purification and treatment, disease regulation, pest regulation, pollination, natural hazard regulation;
- **Cultural services**, including spiritual and religious values, aesthetic values and recreation and ecotourism.

In order to ensure a pragmatic procedure for the assessment of resource costs, a relevant core set of ecosystem services that may be affected by water service provision and thus constitute alternative uses has been selected (i.e. only those ecosystem services that may be affected by water service provision are of relevance to the investigation). The following *Fact Sheet* summarises the relevant ecosystem services.

If no significant alternative economic water using activities are identified and the rough analysis of alternative ecosystem services guided by *Fact Sheet* 10 does not indicate the presence of resource costs, there is no need for further investigations.

Fact Sheet 10: Rough analysis of resource costs stemming from ecosystem services

Ecosystem service (potential alternative uses)	**Definition (UNEP 2005)**	**Affected by water service**			
		Water supply	Sewerage Management	Irrigation	Other
Fresh water provision	People obtain fresh water from ecosystems and thus the supply of fresh water can be considered a provisioning service.		x	x	(x)
Water regulation	The timing and magnitude of runoff, flooding, and aquifer recharge can be strongly influenced by changes in land cover, including, in particular, alterations that change the water storage potential of the system, such as the conversion of wetlands, the replacement of forests with croplands or croplands with urban areas.			xx	(x)
Erosion regulation	Vegetative cover plays an important role in soil retention and the prevention of landslides.	xx		x	
Water purification and treatment	Ecosystems can be a source of impurities (e.g. in fresh water) but also can help filter out and decompose organic wastes introduced into inland waters and coastal and marine ecosystems and can assimilate and detoxify compounds through soil and subsoil processes.		x	xx	
Disease regulation	Changes in ecosystems can directly change the abundance of human pathogens, such as cholera, and can alter the abundance of disease vectors, such as mosquitoes.		x	x	
Recreation and ecotourism	People often choose where to spend their leisure time based in part on the characteristics of the natural or cultivated landscapes in a particular area.	xxx		xxxx	(x)

Filled in for each case study

Source: own compilation drawing on UNEP (2005).
"x" indicates relevance of ecosystem service as alternative water use affected by water service provision (here: illustrative examples).

If the rough analysis indicates that other water uses would be economically more efficient, these have to be investigated more closely in a subsequent analytical step.

3.2.5.2 Step 5.2: Assessing the relevance/magnitude of resource costs and existing compensation schemes

For the purpose of the assessment at hand only an indication of the relevance/magnitude of resource costs is required and not an exact specification in monetary terms. The assessment should allow a judgement on the degree to which uncovered resource costs impact on the overall cost-recovery level.[239] To this end, a three-tiered classification scheme based on expert judgement suffices. The magnitude/relevance of resource costs is assessed on a scale from 1 to 3:

(1) Signalling a low magnitude/relevance of resource costs;
(2) Indicating a medium magnitude/relevance of resource costs, translating into a downward trend in cost-recovery levels in the integrative assessment (cf. Step 3);
(3) Indicating a significant magnitude/relevance of resource costs, necessitating a downgrading of cost-recovery levels in the integrated assessment (cf. Step 3) and the analysis of potential alternatives approaches for the provision of these water services in the future (cf. Section 3.3.3).

The classification should be qualified by presenting supportive evidence as far as possible regarding the relevance of resource costs in the specific investigation case (e.g. information on existing compensation schemes).

Box 7: The role of expert judgement

Expert judgement is taken recourse to at several stages of the methodological assessment. While partially opening the assessment to subjectivity, expert judgement also allows for the incorporation of local expert knowledge into the decision-making process. During the first implementation cycles, iterative improvement to cost and revenue accounting should lessen the influence of qualitative expert judgement in the process. However it can provide a valuable input by judging the quality of information provided as well as by evaluating

239 Inter-linkages to Type I EC need to be considered. There it was assessed, if there are effects on ecosystems – but not whether the service provided by the ecosystem would be more valuable than the water service provision. Hence, Type I EC indications in Fact Sheet 8 may support the assessment of resource costs.

outcomes on the basis of local in-depth knowledge. It allows for outlining in how far e.g. costs could clearly be allocated to the specified cost categories, where there have been problems in reporting and where these problems stem from, on which aspects further activities / research / data collection efforts should focus on, etc.

3.2.5.3 Step 5.3: Indicating the impact of uncovered resource costs on overall environmental and resource cost-recovery rate

Once the relevance and magnitude of resource costs have been determined, their impact on the overall environmental and resource cost-recovery rate has to be indicated. If significant unaccounted for resource costs have been attested to in Step two above, the location of an investigation area within Figure 13 (above) has to be adjusted downward to a lower degree of internal environmental and resource cost-recovery through a move to the left on the 'x'-axis.

For the assessment, it is not required to be able to pinpoint the impact of uncovered resource costs exactly (e.g. whether overall cost-recovery has to be adjusted from 96 per cent to 91 per cent or to 85 per cent). What is required, however, is a better understanding of trade-offs related to the provision of water services and their potential influence on the overall degree of cost-recovery. While a more thorough and refined analysis could be adopted in later implementation cycles (replacing e.g. the expert evaluation of the magnitude/ relevance of resource costs through a refined multi-criteria analysis or in-depth local investigations), the proposed framework provides a pragmatic and proportional perspective on the issue of resource costs and their impact on cost-recovery in terms of Article 9.

3.3 Linking the assessment components: implementation pathways and adaptation needs

Before the strands of the assessments undertaken in Sections 3.2.2 to 3.2.5 can be combined to an integrative assessment of current implementation pathways towards cost-recovery in Section 3.3.3, further background information is required on the investigation area in order to facilitate a meaningful interpretation of results (Section 3.3.1). Furthermore, assessing the nature of divergences from the ideal Article 9 path to cost-recovery (i.e. case 1 in Figure 13) necessitates studying the incidence of Article 9 exemptions more closely (Section 3.3.2).

3.3.1 Complementary information for a comparative evaluation: background information on the investigation area

With the assessment framework developed up until now, an overview of financial costs and associated revenues in the provision of water services as well as an indication of accounted for and unaccounted for environmental and resource costs has been gained. While the basic information required for assessing the first objective of Article 9, i.e. the recovery of financial and environmental and resource costs of water services, has thus been gathered, the current information base is insufficient for interpreting results comparatively across European river basins. For this, further information on the framework conditions under which the provision of water services takes place, such as the study context, is required. Water services are provided in a complex web of interaction framed by natural conditions, legal requirements and social norms (cf. Chapter 2.1). Financing of water services is inherently context-driven; hence, the context of service provision needs to be considered for an integrated assessment of cost-recovery pathways.

Part of the required information base was already compiled through the structured *Fact Sheets* above. Table 9 outlines a core set of complementary indicators for the assessment. Additional indicators may be required at case study level in order to depict local/regional particularities.

Table 9: Indicators on the general characteristics of the investigation area

Indicator	Unit / computation	Scale	water service
Population Size of the service area Population density Number of households Average household size	inhabitants in the assessment area km² inhabitants / km² in total number of inhabitants per household	National, river basin, regional, local, river basin district level	Water supply Sewerage management
Connected inhabitants (if relevant min-max e.g. in the case of strong variations due to tourism / seasonal impacts)	number of inhabitants connected to the water service and % of inhabitants connected (connected inhabitants / population)	National, river basin, regional, local, river basin district level	Water supply Sewerage management
Rural – urban share in investigation area	%	National, river basin, regional, local, river basin district	Water supply Sewerage management Irrigation

Indicator	Unit / computation	Scale	water service
Water use per connected inhabitant and day	m^3 Indicate underlying definition: water abstracted or water supplied to connected inhabitants, leakage subtracted, etc.	National, river basin, regional, local, river basin district level	Water supply Sewerage management
Length of the pipe system (size of the distribution network)	km in relation to population density		Water supply Sewerage management Irrigation
Sewerage treatment capacity Primary treatment Secondary treatment Tertiary treatment Capacity	% of sewerage treated	National, river basin, regional, local, river basin district level	Sewerage management
Number of service providers in the service area Share in service provision	in total % in terms of connected inhabitants and water service use	National, river basin, regional, local, river basin district level	Water supply Sewerage management Irrigation
Role of agriculture in the area: – share in GDP – number of people working in agriculture – number of farms (if possible with indication of average farm size or in groups of farm sizes)	 % in total number and hectare	National, river basin, regional, local, river basin district level	Irrigation
Evidence on role of rain-fed agriculture: Cultivated area Irrigated area	km^2 %	National, river basin, regional, local, river basin district level	Irrigation
Farms connected to the irrigation network	in total %	Service area	Irrigation
Number of private wells for irrigation purposes	in total	Service area	Irrigation

In addition, basic information on the actors involved in water service provision (cf. Section 3.4.2.4 on the implementation of Article 9 objective 4 below) is needed for depicting and understanding the financial flows within the system of water service provision.

3.3.2 Testing the relevance of Article 9 exemptions in justifying divergences from the *'ideal path'* to cost-recovery

As outlined in Chapter 2, Article 9 (4) allows for exemptions from a strict internal cost-recovery approach:

> *"Member States may in so doing have regard to the* ***social, environmental and economic effects*** *of the recovery as well as the* ***geographic and climatic conditions*** *of the region or regions affected"* (emphasis added)

While this paragraph implies that divergences from a strict internal cost-recovery are acceptable in terms of Article 9, the addition *"in doing so"* highlights that it cannot be interpreted as allowing the cost-recovery principle to be neglected in full (UNNERSTALL 2005). Divergences from this principle warrant a clear justification and a demonstration of how the objective will be approached in the further implementation process. However, it allows Member States to take considerations of fairness (e.g. by introducing cost recovering water prices in a staged process or by introducing financial aid for vulnerable groups) or regional development into account. In any case, Article 9 (4) cannot be interpreted as freeing Member States from the general obligation of achieving cost-recovery in accordance with the polluter-pays principle in the long-run (cf. discussion in Chapter 2).

Hence, evoking an exemption from the cost-recovery objective on the basis of Article 9 (4) requires a transparent procedure. The following sections aim to map out the relevant aspects that need to be considered in the analysis. These will then have to be concretised and amended to address the regional specificities of each investigation area (cf. Chapter 4).

3.3.2.1 Justifying divergences due to social effects

According to Article 9 (4), adverse social effects may justify a divergence from a cost-recovery of water services in line with the user/polluter-pays principle. With this provision Article 9 addresses the social good dimension of water (cf. Section 2.1.2) and acknowledges that pricing water in a way that reflects environmental and efficiency concerns can sometimes be controversial due to social considerations (OECD 2003:18). Integrating aspects of social fairness and affordability into water service financing is thus in agreement with the provisions of Article 9.

To address affordability concerns in water service provision and financing[240], two main policy measures can be utilised: first, tariff-related measures (i.e. specific tariffs for vulnerable groups or reduced tariffs for a specified minimum consumption level) and second, income-related measures (i.e. tariff rebates and discounts, income assistance or payment assistance loans) (OECD 2003).[241]

A wide variety of indicators/criteria may be drawn upon for assessing the social effects of water prices and charges (cf. KLAUER ET AL. 2007, OECD 2003). For assessing the relevance of social effects regarding cost-recovery of the three water services under investigation in this study, the following two indicators may be applied:

- Drinking water supply and sewerage services: Average expense for water service in excess of 3–4% of average available per capita income in the service area (KLAUER ET AL. 2007, COURTECUISSE 2005, OECD 2003);
- Irrigation water service: Average percentage of farm income spent on irrigation water service exceeds a predefined (politically determined) threshold value (KLAUER ET AL. 2007).

Furthermore, the policy measures in place for addressing affordability issues of water service financing should be documented, e.g. by outlining:

- How affordability of water services is guaranteed for all customer groups (e.g. whether income or tariff related measures are applied);
- How customers are identified that may be subject to affordability concerns;
- Whether special social groups are addressed through policy measures (e.g. prolific families, elders) and why.

Furthermore, it should be outlined, whether social effects are likely to persist in the future, and whether measures are undertaken to remedy social effects in the longer run. The investigation of recovery mechanisms under Section 3.2.3 already provided part of the required information (cf. *Fact Sheets* 6 and 7).

3.3.2.2 Justifying divergences due to environmental conditions

Unfavourable environmental conditions may lead to considerably higher costs of water service provision, which may induce or heighten affordability concerns

240 Cf. also OECD (2009) for a discussion of the role of affordability in water pricing.

241 In view of the other objectives of Article 9, income support measures are to be preferred, as they do not interfere with the economic and environmental signals sent by the size of the water bill and thus do not impede with the incentive properties of the tarification scheme (*objective 2*).

for certain water service user groups within a river basin district. COURTECUISSE (2005) found for the Artois-Picardie river basin that areas in which cost-intensive programmes of measures where required for reaching good water status tended to coincide with low average disposable household income. Hence, financing the required measures for reaching good water status while simultaneously establishing strong internal cost-recovery of water services may be at conflict in certain areas, i.e. raise social concerns. In this case, a divergence from a strong internal cost-recovery approach may be justified under Article 9 (4).

Additional arguments connect to the social and economic justifications for exemptions from Article 9. For example, if the raw water quality for water supply or irrigation is very low, comparably higher costs will be encountered for water service provision under an economic cost approach. These may induce socially undesirable effects for low income groups, may hinder the development of the agricultural sector (irrigation), or may affect regional economic development (e.g. if water-intensive industry refrains from locating in the area due to higher production costs induced by water prices). Alternatively, local environmental conditions (e.g. topography, carrying capacity of receiving waters) could require one to establish cost-intensive collection and/or treatment techniques (e.g. specialised decentralised solutions for mountain areas). This could lead to social hardship for certain segments of society, thus justifying on social grounds, or under the policy principle of regional equality, a divergence from a strict internal cost-recovery approach.

3.3.2.3 Justifying divergences by economic effects of cost-recovery

Establishing a strong internal cost-recovery of water services may have adverse effects on economic sectors, e.g. water-intensive production processes or irrigated agriculture. In this case, Article 9 (4) provides the possibility to justify a divergence from the internal cost-recovery approach. Irrigated agriculture may, for example, be an important economic sector in an investigation area, indispensable for regional development and employment. Establishing strong internal cost-recovery may lead to the sector's extinction, with considerably adverse effects at local or regional level.

In order to assess the economic effects of a strong internal cost-recovery approach, information on the following aspects has to be provided:

- The number of people employed by the economic sector;
- The contribution to GDP by the economic sector;

- The average share of profit of sector undertakings spent on the investigated water service;
- The projected sector development (cf. Baseline Scenario);
- The expected impact of higher expenses for water services on the sector's development.

In addition, information should be provided on:

- The sector's financial contribution to programmes of measures (financial involvement in WFD implementation);
- The sector's contribution to Type II EC and related internalisation mechanisms of the investigated service.

As with the other justification mechanisms under Article 9 (4), economic effects may only serve as a justification for postponing but not for entirely neglecting internal cost-recovery. It may, however, justify a staged-approach to its implementation. Accordingly, it should be outlined how the adverse economic effects may be alleviated in the future.

3.3.2.4 Justifying divergences by geographic and climatic conditions

According to Article 9 (4), geographic and climatic conditions may also serve as a justification for diverging from a strict internal cost-recovery approach. Section 2.4.3 discussed the influence of changing climatic conditions on the costs of water service provision due to required adaptations of infrastructural systems to manage extreme events (i.e. heavy irregular rainfalls and/or periods of prolonged droughts) or negative effects on the availability or quality of water resources. While water pricing may act as a support tool in implementing an adaptation strategy to climate change (EUROPEAN COMMISSION 2009), such an (incentivizing) pricing scheme is likely to contradict a strict internal cost-recovery approach (cf. Section 2.2.1.2). Another example of climatic effects which may lead to a divergence from cost-recovery can be observed during periods of prolonged droughts. In this type of scenario environmental costs of irrigation water services may be significantly higher, but are often not recovered from irrigators in order to prevent social mishap. The same argumentation may apply to water supply.

Unfavourable geographic conditions may considerably influence the costs of water service provision (HOLLÄNDER ET AL. 2008). With the aim of establishing comparable living standards on an area-wide basis, these areas are often cross-subsidized by other parts of the service area or issued direct subsidies e.g. for establishing the infrastructure in remote rural areas. For the assessment, it is necessary to point out, what impact climatic and/or geographic conditions have on water service provision and why they necessitate a divergence from an internal financing approach. The background information provided in Section 3.3.1 may aid the required evaluation.

3.3.3 Mapping out present approaches to cost-recovery and adaptation needs

Combining the different strands of the assessment, one can analyse the extent to which the ideal assessment framework could be applied in the investigation case at hand (i.e. drawing on the *'evaluation'* column in the *Fact Sheets*); the reasons for divergences from the accounting framework (e.g. cost attribution difficulties, etc., cf. *Fact Sheets*); and the likely impact of these divergences on the overall assessment. Second, an integrative perspective on how costs of water service provision are currently recovered can be obtained from combining the different strands of the assessment conducted so far. Third, divergences from the ideal cost-recovery approach can be assessed along with the relevance of Article 9 exemptions.

Furthermore, first insights on the additional objectives or Article 9 have been gathered through the analysis to this point. These have to be expanded on and studied in greater detail in the next sections in order to make relevant trade-offs between objectives explicit (cf. Chapter 1.4) and allow assessing the overall compliance with WFD Article 9.

3.4 Assessing the implementation of the complementary objectives of WFD Article 9

In order to complete the analysis of the implementation of WFD Article 9, the Article's additional objectives must be studied more closely. While the analysis of costs and revenues provided initial insight on the degree of implementation of the additional objectives of Article 9, further information is required for investigating their compliance in greater detail. To this end, one needs to evaluate whether present financing schemes for water services (cf. Chapter 1.4):

- Set incentives for efficient water resource use (objective 2);
- Make the polluter pay and ensure adequate contributions by sectors (objective 3);
- Establish transparency of financial flows in the European water sector and guarantee the sustainability of infrastructural systems (objective 4).

Part of the required knowledge base for assessing compliance with *objective 2* (setting incentives for efficient resource use) has already been provided through the evaluation of resource costs. This indicated the efficiency of the present allocation of water among different uses. However, an investigation of the incentive mechanisms linked to the present financing scheme is still missing, even though part of the required information base has been established through the revenue accounting matrix.

To assess the degree to which *objective 3* (making the polluter pay and ensuring adequate contributions by water uses) is met, it is necessary to investigate how water uses participate in the generation of costs and revenues of water service provision respectively. A framework for assessing the degree to which an adequate contribution of environmental costs is made by different water uses has already been provided in Section 3.2.4.. However, further information is required on the extent to which the polluter-pays principle, which forms the basis for assessing the adequacy of a contribution of water uses, has been implemented overall (i.e. are revenues generated through a certain water use in line with its contribution to the water service's provision costs and more particular with its contribution to environmental costs). The *Fact Sheets* of Section 3.2.2 and Section 3.2.3 collected part of this information by outlining, whether there are specificities in cost and revenue generation with regard to particular water service users.

In order to study the compliance with Article 9 *objective 4* (establishing transparency of financial flows in the European water sector and guaranteeing the sustainability of infrastructural systems), it is first necessary to combine the information reported in the *Fact Sheets*, both for cost accounting and for revenue accounting, on the quality of data and on divergences from the ideal Article 9 approach to cost-recovery.[242] Furthermore, through complementary indicators, the sustainability gap can be further substantiated and made transparent.

Table 10 summarises the information requirements for assessing compliance with the additional objectives or Article 9 and highlights which part of the information base has already been compiled through the analysis conducted so far. The following sections develop pragmatic procedures for gathering the residual information required for an assessment of the objective's degree of

242 This also entails questions of data collection and allocation (i.e. if revenues / costs cannot be clearly allocated as their present documentation does not allow for it along with an assessment of the related uncertainty. It has to be evaluated whether present difficulties justify a change in data collection, or whether the additional certainty in information would not be proportional to the improvements in decision making that could be derived from a better data collection system or another information aggregation scheme).

Table 10: Remaining information requirements on additional Article 9 objectives

Article 9 objective	Information provided through previous analysis	Additional information requirements

Objective 2
Set incentives for efficient resource use
	– Analysis of resource costs: indication of efficiency of present allocation of water among different uses
	– Properties of endogenous revenue generation mechanisms (e.g. payment basis, collection mechanisms, structure of charges, exemptions, etc.)
	– Incidence of exogenous revenue sources
	– Incidence on other revenue components
		– Information on elasticity of water demand
		– Evaluation of incentive properties of tarification scheme, e.g.:
		  – Variable tariff component
		  – Degree of metering (signal effect)
		  – Scarcity costs addressed through tarification scheme
		– Assessment of rewards for sensible water use behaviour[243] within tarification scheme
		– Assessment of incentive schemes linked to subsidies (do they promote efficient resource use, e.g. by furthering a change to locally adapted crops or water saving technical appliances).

Objective 3 Make the polluter pay and ensure adequate contributions by sectors	– Evidence on Type II EC – Analysis of outliers in contribution to costs and/or revenues (→ *Fact Sheets*) – Analysis of internalization mechanisms for environmental costs (polluter-pays)	– Closer investigation of the implementation of the polluter-pays principle – Information on pollution-pathways – Information on contribution to costs and revenues of water service provision by sectors
Objective 4 Establish transparency of financial flows in the European water sector and guarantee the sustainability of infrastructural systems	– Information on costs of water service provision – Information on financing mechanisms of water services provision – Information on investments – Information on approach adopted to account for capital costs	– Link information on costs and revenues with actors: graphical representation of financial flows between actors – Assess sustainability of infrastructural system through appropriate indicators

implementation. The relevant analytical aspects are outlined along with possible indicators for their assessment. While a set of indicators/interrogatory aspects, which may support the analysis, has been selected for each objective, it

243 With regard to the notion of setting '*adequate incentive for efficient resource use*' it has to be noted that efficient does not automatically have to translate into '*water saving*'. Incentive schemes need to be adapted to local circumstances and reflect the respective pressing water policy issues.

is by no means exhaustive or inadaptable. For each objective, a large variety of alternative indicators could be drawn upon. Depending on the available information base (e.g. from benchmarking activities, WFD monitoring programmes, the IMPRESS analysis, etc.) different indicators may be less time consuming to compile, while being equally suited for the assessment. Furthermore, all indicators are linked to specific drawbacks and thus need to be interpreted with great care.[244]

An integrative assessment of the implementation of Article 9 requires reporting on the additional objectives in a transparent manner that furthers an understanding of the underlying issues and water management problems addressed by the objectives. For reasons of consistency and in order to avoid duplication, information that has already been compiled within the overall WFD implementation process, most notably within the economic analysis of water use and the analysis of pressures and impacts as well as during the monitoring exercises, should be drawn upon as far as possible.

3.4.1 Assessing the adequacy of incentives for efficient water resource use

The analysis conducted in Chapter 3.2 and 3.3 already provided part of the required information base for assessing whether the present financing scheme provides adequate incentives for efficient resource use. First, the analysis of resource costs gave an indication of the economic efficiency of current water use patterns. Second, the *Fact Sheets* on endogenous revenue generation mechanisms indicated, the degree to which the present charging scheme accommodates individual consumer behaviour through variable tariff components and discourages excessive water use (e.g. through special tariffs applicable during seasonal water scarcity or through increasing charges for higher levels of consumption). In addition, information on exemptions for certain water users can be extracted from the *Fact Sheets*. Third, the information gathered in the *Fact Sheets* on exogenous revenue sources allows for assessing whether subsidisation schemes are targeted at encouraging efficient water resource use (e.g. by fostering the cultivation of less water-consuming crops, by

244 The selection of appropriate indicators should be guided by the identified deficits detected in the analysis so far, as the indicators aim to further clarify the origin and nature of these deficits and set the basis for designing appropriate adaptation strategies for future implementation cycles. As a guiding principle it should be noted that a concise set of indicators should be selected to promote the overall aim of the additional indicators of increasing the transparency of the analysis; gathering information on too many indicators would not necessarily achieve this aim.

promoting investments in less water consuming irrigation techniques, or by advancing the installation of water saving appliances in households). Finally, the evidence on other revenue components gave insights into the overall impact they have combined with exogenous revenue sources on the overall financing scheme. The incentive properties of the endogenous revenue sources will only develop an impact, if they make up for a considerable share in overall revenue generation.

While a considerable information basis has thus already been established, there are nevertheless additional aspects that warrant further investigation. These include the exact composition of variable and seasonal tariff components (i.e. to which degree is consumption metered, in how far are seasonal tariffs actually applied, etc.), and the elasticity of water demand (i.e. to which degree is consumption responsive to price changes (cf. Chapter 2.3)). As the discussion in Chapter 2.3 revealed, only scattered evidence on the elasticity of water demand is presently available for most European Member States. Hence, during the first implementation cycles, investigations of water demand elasticity will often draw exclusively on pilot investigations. Furthermore, the issue of scarcity costs needs to be studied more closely (i.e. the degree to which scarcity costs are addressed through the present tarification scheme). Finally, water pricing is understood in the relevant CIS guidance documents as a complementary measure, which may contribute actively to reaching the overall objective of good water status (EUROPEAN COMMISSION 2007, 2000, CIS 2003A). Hence, the potential of water pricing to contribute to this end in the particular investigation area must be assessed.[245]

It has often been argued in the ongoing implementation process that the adequacy of incentives for efficient resource use only needs to be investigated if the present status of environmental resources demands curbing down water demand via prices (or other instruments). However, pricing schemes always set incentives for certain behaviour (e.g. a flat rate for water encourages higher water use). The proposed investigation only makes the incentives inherent in the present tarification scheme transparent and should thus be conducted irrespective of the resources' status and scarcity.

245 This aspect is closely linked to the debate on the elasticity of water demand: if the effect of water pricing on water demand is not known, the (exact) potential of water pricing in curbing down water demand and thereby alleviating e.g. water quantity related problems cannot be determined.

3.4.2 Assessing the implementation of the polluter-pays principle and the adequacy of contributions from other water uses

As outlined in Chapter 3.1, the Directive aims at a high degree of internal financial and environmental and resource costs recovery, with a use-/pollution-dependent allocation of costs onto different user groups and no costs being transferred to the next generation

As outlined in Chapter 1, the third objective of Article 9 asks for an investigation of the adequacy of contribution of water uses to the cost-recovery of water services. In judging whether the contribution of a certain water use is adequate, the polluter-pays principle serves as the relevant yardstick. Accordingly, if a certain water use contributes over-proportionately to the costs of water services provision, e.g. through Type II EC, its contribution to the revenue side of the equation should be equally over-proportionate. For example, if an industrial sector requires additional investment in sewerage treatment facilities, which would otherwise not be needed (to this extent), the associated costs should be attributed to the investment initiating sector. Similarly, if nitrate pollution from agriculture causes Type II EC for the drinking water service provider, these costs should be borne by the agricultural sector. Hence, the adequacy of a sector's contribution to the cost-recovery of a certain water service is determined by its influence on the costs of the service's provision.

The analysis in the preceding sections already provided part of the information base required for the assessment. The *Fact Sheets* on financial costs (cf. Section 3.2.2) analysed whether certain users or activities contribute over-proportionately to certain cost categories (e.g. mining may lead to higher pumping costs; industry may necessitate specific investments).[246] Complementary to this analysis, the *Fact Sheets* on financing mechanisms documented the set-up of revenue generation mechanisms (i.e. who contributes according to which rationale to the financing of a water service's provision) and outlined whether differentiations between user groups are undertaken (e.g. special charges or exemptions for certain water uses or customer groups). The in-depth investigation of the financing scheme's incentive properties as outlined in Section 3.4.1 above provided additional insights on the degree to which the polluter-pays principle figures in the present financing scheme.

246 E.g. the *Fact Sheets* indicate whether the magnitude of a cost category is significantly influenced by other water uses (e.g. the degree to which material costs for raw water treatment are linked to water pollution caused by agriculture), and if yes, which cost components are most likely to be affected for the different services.

This information must now be combined in order to gain an overall picture of the polluter-pays principle's implementation and the adequacy of contribution by water uses. To this end, it needs to be concluded whether:

- The polluter-pays principle serves as a guiding principle in revenue generation (e.g. through charges / fees that vary with the contribution to pollution/use);
- Revenue generation mechanisms reflect cost structures and the degree of contribution to costs by service users[247];
- Type II EC are borne by their causing activities; and
- The costs of the provision of side services (e.g. fire protection) is borne by their specific user groups.

As discussed in Chapter 1, the implementation of the second and the third objective of Article 9 may be contradictory. The assessment therefore aims to gain an understanding, why a certain implementation way is selected, e.g. why the polluter-pays principle is not applied to certain activities, or why the cost structure is not reflected in revenue generating mechanisms (e.g. to better employ the potential incentive setting properties of pricing schemes). The analysis may also necessitate the investigation of the adequacy of inter-generational subsidies between user groups (e.g. between households and agricultural water users) or between regions (e.g. rural and urban areas).

Box 8: Earmarking of revenues (internalisation instruments)

Earmarking can be implemented in the following ways:

- The simplest form of earmarking involves allocating resulting revenues to the group that paid for the original taxes. Usually, different formulae are used to raise and disburse revenues. Such simple earmarking usually has little environmental effect and transaction costs can be high. However, such systems can be useful, for example, in mutual insurance schemes.
- More complex is the French model of raising "*redevance*" which is largely recycled back to those who contributed to the revenue of the *Agence*. In fact, those who contributed have a moral claim on their contribution and can expect a subsidy when it is their turn to make pollution control investments. Such system can be useful to spread the burden of heavy investments, especially in the context of comprehensive investment programs

247 I interpreted broadly. E.g. high fixed costs – if only variable pricing scheme then consumer groups which use very low amounts of water may not be contributing adequately to their share in infrastructure costs.

implemented over a long period of time. Depending on the degree of solidarity among the water users within a river basin area, such schemes can either work for the basin as a whole or be segmented according to sectors or regions.
- Beyond these possibilities, earmarking can be relaxed so that money does not go back to those who paid, but is used instead to finance typical governmental functions, such as water and groundwater monitoring, modelling, research and technical development, or information disbursement. In this case, earmarking produces a double dividend. This is because of the incentive function on the revenue side, and also by financing activities that are beneficial to the environment.

Earmarking has economic advantages and disadvantages and can be politically useful.

- Earmarking can, if badly designed, favour capital-intensive investments over others because of the availability of capital. Similar to the Averch-Johnson effect, it can thus lead to over-investment.
- Earmarking should not be used to subsidize activities that have significant negative externalities. Instead they should only be used to compensate for positive externalities (external benefits) that cannot be captured through the market or regulated prices or charges.
- Experience has shown that earmarking is useful to raise political acceptance for the introduction of economic instruments.
- Over time, earmarking tends to be relaxed, so that general water management functions are financed first, and later the revenue from originally earmarked charges is treated as general revenue and becomes part of general taxation.

Contrary to the principles of public finance, earmarking has proven to be a useful tool in environmental policy.

Source: Kraemer et al. (2003).

3.4.3 Assessing the sustainability of infrastructural systems and the transparency of financial flows

In order to study the compliance with Article 9 *objective 4* (establishing transparency of financial flows in the European water sector and guaranteeing the sustainability of infrastructural systems), it is first necessary to combine the information reported in the *Fact Sheets*, both for cost accounting and for revenue accounting, on the quality of data and on divergences from the ideal Article 9 approach to cost-recovery.[248] Furthermore, through complementary indicators, the sustainability gap can be further substantiated and made transparent.

The sustainability of infrastructure can be explored through a large variety of indicators. The selection of indicators is not trivial, as all indicators only address certain aspects of the sustainability of a water service's provision. Furthermore, underlying indicator definitions tend to vary greatly. Even if the seemingly same indicator is reported on in national statistics, there may be great variations in the indicator's exact meaning (e.g. the classification of investments may differ in terms of what is considered as renewal or as rehabilitation).

Generally, information on the following aspects will be relevant for assessing the sustainability of water service provision:

- Does the approach to depreciation allow for accumulating sufficient financial means for reinvesting after the infrastructure's economic lifetime has expired?
- Are reserves available at the time of reinvestment?
- What is the status of the infrastructural system and of service provision at present (e.g. security of supply, leakage rate, share of population with access to the service, compliance with relevant EU Directives (e.g. UWWTD, Nitrate Directive, Drinking Water Directive), customer satisfaction with service provision, etc.)? Is it considered sustainable?
- What is the role of exogenous revenue sources? Will they pose a problem to long-term refinancing of infrastructure (e.g. if they cannot be relied upon in the future)?
- How are future investment requirements allocated between consumers, the water service provider's shareholders and taxpayers?

For investigating the transparency of financial flows in the provision of water services, the financial interaction between actors can be graphically represented (cf. Chapter 4), with at least differentiating between the river basin (district) level (e.g. water boards), state actors at national / regional level, municipalities, water service providers, intermediate levels such as irrigation districts or water user associations, water service users, and other (relevant) water uses (i.e.

248 This also entails questions of data collection and allocation (i.e. if revenues / costs cannot be clearly allocated as their present documentation does not allow for it. Should include an assessment of the uncertainty caused by this – would it justify a change in data collection, or is the additional certainty in information not proportional to the improvements in decision making that could be derived from a better data collection system / another information aggregation scheme).

those water uses identified as significant during the environmental and resource costs analysis).

Ultimately, the accounting framework aims to identify *"who pays what to whom for which service"* in order to establish transparency of financial flows within the European water sector. Thus, investigating the general applicability of the accounting framework ultimately boils down to investigating the degree to which the transparency objective could be achieved. An investigation of the following aspects may help to improve transparency in subsequent implementation cycles:

- Are severe divergences encountered in the application of the cost accounting framework? What are the underlying reasons for these divergences (outlined under 'evaluation' in the Fact Sheets)?
- Is the quality of available data acceptable?
- Do confidentiality concerns critically affect the analysis of cost-recovery?
- Which uncertainties are related to the available data?
- Are adaptations to data collections necessary for subsequent implementation cycles or is the additional certainty in information not proportional to the improvements in decision making that could be derived from a better data collection system / another information aggregation scheme?

3.5 Plan of procedure for assessing cost-recovery and overall attainment of Article 9 objectives

Chapter 3 proposed an integrative framework for evaluating the implementation of the multiple objectives of WFD Article 9. The following paragraphs provide a concise summary of the different steps of the methodology. Special attention is given to *'linking steps'* (i.e. Step 3, Step 6 and Step 10) that combine intermediate results of the analysis. They have not figured in the assessment framework above as they do not require a further substantiation in terms of guiding rules, indicators or *Fact Sheets*.

The overall assessment framework allows for investigating which costs are encountered in water service provision and how these are recovered. By reference to the ideal implementation path for meeting the objectives of Article 9 derived from the analysis in Chapters 1 and 2, the methodology allows for detecting divergences along with the related explanatory factors, which may justify an exemption in terms of Article 9. In conclusion, the integrative framework allows one to identify in how far the present approach to cost-recovery is in line with the requirements of WFD Article 9, whether the additional objectives of Article 9 are met, and how extensively adaptations are required in subsequent implementation cycles.

Step 1: Accounting for financial costs of water service provision

The first step of the analysis investigates the financial costs of service provision on the basis of structured *Fact Sheets* for each cost category. The *Fact Sheets* (1 to 5) detail which information has to be provided for each cost category, while also providing Article 9 accounting guidelines. As a result, they document difficulties encountered in the assessment along with divergences from the accounting guidelines. On the basis of the information provided in the *Fact Sheets*, the influence of these divergences on the overall assessment of cost-recovery can be determined in later steps of the analysis. In addition, the *Fact Sheets* document *'influencing factors / driving forces for the cost category'*, which provides an important contribution to the determination of additional appropriate (remedying) activities for improving the implementation of Article 9 in future implementation cycles and delivers initial information for the analysis of the implementation of the polluter-pays principle (cf. Step 9). An important input to subsequent analytical steps is also given by the category *'evaluation,'* which contains expert judgement on the degree to which full cost accounting and attribution to the cost categories could be attained.

After Step 1

- ✓ The analysis provides information on the **degree of divergence** from the accounting rules regarding financial costs (documented by the *Fact Sheets* and supported by expert opinion); (i.e. divergence factor 1).
- ✓ Information is documented on potential **influencing factors and drivers** of financial costs, which have to be taken into consideration in adaptation strategies.
- ✓ Initial information for the analysis of the implementation of the **polluter-pays principle** can be derived from the analysis.

Step 2: Accounting for revenues

The second analytical step mirrors the analysis of financial costs by investigating the nature and composition of associated revenue streams. The assessment framework provides guiding *Fact Sheets* for analysing endogenous revenue sources (*Fact Sheet 6*), exogenous revenue sources (*Fact Sheet 7*) along with other, non-water service related revenue sources. Next to documenting the different revenue generating mechanisms, by indicating from whom revenue is collected and according to which structure the payment scheme is set up (e.g. variable and fixed components, variation factors, exemption regulations and differentiations between user groups), the *Fact Sheets* allow for mapping out divergences, e.g. if revenues cannot be clearly allocated to a specific water service. Furthermore, the analysis provides information on the degree to which different cost categories factor into revenue generation mechanisms.

After Step 2

- ✓ An overview of the **composition, set-up and relative magnitude** of different revenue generating mechanisms is provided.
- ✓ Initial evidence for Article 9 **objective 2**, i.e. on assessing the properties of endogenous revenue generation mechanisms (e.g. payment basis, collection mechanisms, structure of charges, exemptions, etc.), allowing for an analysis of the degree to which incentive pricing results from the interlinked revenue structure at later stages of the assessment. Furthermore, the incidence of exogenous revenue sources is documented along with the relevance of other revenue components, which also impact the incentivising nature of water prices.
- ✓ Initial evidence for Article 9 **objective 3**, i.e. on assessing the degree to which the polluter-pays principle is applied and an adequate contribution of water uses is present can be derived from the interlinked revenue structure for later stages of the assessment.

Step 3: Mapping out Type I cost-recovery and identify distance to WFD target (ideal path)

Once financial costs and existing revenue generating mechanisms have been accounted for through *Fact Sheets*, a first explorative view on the status of financial cost-recovery can be obtained by combining the two strands of the investigation. To this end, the documented information needs to be brought together and combined. However, in order to ensure a context-specific interpretation, it is necessary to include additional information on the investigation area and local surrounding factors. Otherwise, a misinterpretation of results is possible.

Step 3 provides an intermediate result for the implementation of the WFD Article 9 cost-recovery objective. It pinpoints major divergences in terms of cost accounting, outlines the importance of exogenous revenue sources and gives a first indication of the overall distance to the ideal WFD cost-recovery approach (cf. Figure 13).

Furthermore, first information on other Article 9 objectives was already gathered through the analysis conducted so far. However, one very important component is still missing from the assessment and has so far only marginally been touched upon (i.e. if strict cost separations were not feasible), namely environmental and resource costs.

<u>*After Step 3*</u>

- ✓ An indication of the status of financial cost-recovery and the overall distance to the ideal WFD cost-recovery approach for financial costs.
- ✓ An identification of the **cost-recovery model** followed along the continuum outlined in Section 2.2.1.2 ranging from a strong internal financial cost-recovery to no internal financial cost-recovery.
- ✓ On the basis of the gathered information, a **graphic representation** of the present cost-recovery status with regard to financial costs is possible in Figure 13.

Step 4: Assessing environmental costs and their recovery

Step 4 presents a pragmatic approach for assessing environmental costs. To this end, first those environmental costs are estimated that are directly linked to the water service's provision. These so-called Type I EC are approximated in a three step procedure, namely by taking stock of and investigating present internalisation mechanisms (cf. Box 6, Step 4.1); undertaking a rough analysis for identifying potential residual Type I EC (cf. *Fact Sheet 8*, Step 4.2); and investigating residual damage, i.e. unaccounted for environmental costs, through a detailed analysis (Step 4.3). Next, the second component of environmental costs caused by other water uses and leading to higher financial costs of water service provision, so-called Type II EC, are estimated through a guided analysis based on *Fact Sheet 9* (Step 4.4).

<u>*After step 4*</u>

- ✓ The analysis provides information on the **degree of internalisation** of environmental costs, partly amended through expert judgement.
- ✓ It delivers information on the degree to which **environmental costs caused by other water uses** lead to higher costs of water service provision, which constitutes an input to the assessment of WFD Article 9 **objective 3**.

Step 5: Assessing resource costs and their recovery

The assessment of resource costs in Step 5 focuses on eliciting sufficient information for determining whether the cost-recovery assessment conducted so far has to be adjusted due to uncovered resource costs. It increases transparency on costs and trade-offs between water uses within the investigation area. To this end, an abbreviated three step approach is applied that allows for an overview investigation on the relevance of resource costs in the investigation area.

Step 5.1: Screening for potential alternative water uses

The first screening step investigates whether alternative water using activities could generate a higher (use or non-use) value. Both are inherently site-specific. To facilitate a pragmatic evaluation, the first screening for potential resource costs is based on expert judgement, as local knowledge on potential alternatives is crucial for the analysis. In addition, the analysis explores whether ecosystem services may constitute economically more efficient water using activities than the provision of the particular water service. *Fact Sheet* 10 provides the framework for conducting a rough analysis of whether ecosystem services may constitute relevant alternatives giving rise to resource costs. If no significant alternative economic water using activities are identified and the rough analysis of alternative ecosystem services guided by *Fact Sheet* 10 does not indicate the presence of resource costs, there is no need for further investigations. Otherwise, further assessments are undertaken in Step 5.2.

<u>*After Step 5.1*</u>

✓ An **indication of potential alternative higher value water uses**, which may generate a higher use or non-use value than the present provision of the water service under investigation, necessitating further investigations.

Step 5.2: Assessing the relevance and magnitude of resource costs and existing compensation schemes

The assessment at hand only requires an indication of the relevance/magnitude of resource costs and not an exact specification in monetary terms. The analysis hence allows for judging the degree to which uncovered resource costs may impact on the overall cost-recovery level. To this end, a three-tiered classification scheme based on expert judgement suffices. The magnitude/relevance of resource costs is assessed on a scale from 1 (low magnitude/relevance) to 3 (significant magnitude/relevance of resource costs necessitating a downgrading of cost-recovery levels in the integrated assessment and the analysis of potential alternative approaches for the provision of these water services). The classification should be qualified by presenting supportive evidence as far as possible regarding the relevance of resource costs in the specific investigation case. Once relevant resource costs have been identified, existing compensation mechanisms are documented and assessed.

<u>*After Step 5.2*</u>

- ✓ An indication of the **relevance and magnitude of resource costs** based on a three-tired classification scheme.
- ✓ An analysis of **existing compensation mechanisms**.

Step 5.3: Indicating the impact of uncovered resource costs on overall environmental and resource cost-recovery rate

Once the relevance and magnitude of resource costs have been determined, their impact on the overall environmental and resource cost-recovery rate has to be indicated. If significant unaccounted resource costs have been attested to in Step 2 above, the location of an investigation area within Figure 13 has to be adjusted downward to a lower degree of internal environmental and resource cost-recovery through a move to the left on the 'x'-axis.

<u>*After Step 5.3*</u>

- ✓ A better understanding of **trade-offs** related to the provision of water services and their potential influence on the overall degree of cost-recovery.
- ✓ An identification of the **required adjustments** to the identified environmental and resource costs recovery degree.

Step 6: Mapping out Type II cost-recovery and identifying distance to WFD target (ideal path)

Step 6 amends the results of Step 3, which assessed financial cost-recovery, and evaluates the degree to which environmental and resource cost-recovery is achieved in line with WFD Article 9. Based on the assessment steps conducted so far, the cost-recovery approach presently applied can be depicted along the continuum of cost-recovery approaches outlined in Section 2.2.1.2. On these grounds, the investigation area can be displayed in Figure 13. It makes transparent, to which degree there is a divergence from the ideal cost-recovery path in line with WFD Article 9 and maps out the maximum scope for Article 9 exemptions. In any case, any acceptable cost-recovery quadrant will become smaller as implementation progresses, i.e. the distance to the target, which cannot be justified, has to be within the sketched limits in order to be in line with implementation requirements.

After Step 6

- ✓ A depiction of the combined cost-recovery level (i.e. financial and environmental and resource costs) in **Figure 13**.
- ✓ Evidence on the **maximum distance to the target** of WFD Article 9 implementation for which the relevance of exemptions has to be tested.

Step 7: Testing for the relevance of Article 9 exemptions

Article 9 (4) allows for exemptions from a strict internal cost-recovery approach due to the social, environmental and economic effects of the recovery as well as the geographic and climatic conditions of the region or regions affected. Step 7 proposes guidelines for testing the relevance of each exemption justification individually, which may provide a reason for a divergence from the ideal Article 9 approach to cost-recovery. Hence, Step 7 identifies in how far divergences identified in Step 6 are covered through the exemption regulation of Article 9. Evoking an exemption from the cost-recovery objective on the basis of Article 9 (4) requires a transparent procedure. Step 7 maps out the relevant aspects that need to be considered in the analysis on each exemption justification. The procedure allows for further regional specificities to be integrated through a concretisation of the proposed evaluation guidelines.

After Step 7

- ✓ Information on the role of the social, environmental, economic effects as well as the local geographic and climatic conditions in **explaining the degree to which the present cost-recovery approach diverges** from the ideal Article 9 pathway to an internal cost-recovery approach.
- ✓ Knowledge on the degree to which **exemptions** justify the divergence from a strict internal cost-recovery approach identified under Step 6.

Step 8: Outlining adaptation needs

Once the present approach to cost-recovery and its divergence from the ideal Article 9 pathway is known, and the possibility to justify identified divergences through the exemption regulations have been evaluated and documented, the adaptation needs for subsequent implementation cycles can be outlined.

Combining the different strands of the assessment, it can first be analysed the extent to which the ideal assessment framework could be applied in the investigation case at hand (i.e. drawing on the '*evaluation*' column in the *Fact Sheets*); the reasons for divergences from the accounting framework (e.g. cost attribution difficulties, etc., cf. *Fact Sheets*); and how these are likely to impact the overall assessment. Second, an integrative perspective on how costs of water service provision are currently recovered can be obtained from combining the different strands of the assessment conducted so far. Third, divergences from the ideal cost-recovery approach can be assessed along with the relevance of Article 9 exemptions.

<u>After Step 8</u>

✓ Overview on **specific adaptation needs** for subsequent implementation cycles.

Step 9: Assessing the attainment of the additional Article 9 objectives

Step 9 addresses the additional Article 9 objectives and determines the degree to which they are implemented through the present implementation framework and water policy approach. Part of the required information base has already been compiled through the analytical steps to this point. These have to be amended through additional investigations to make relevant trade-offs between objectives explicit (cf. Chapter 1.4) and allow for assessing the overall compliance with WFD Article 9.

Step 9.1: Assessing the adequacy of incentives for efficient water resource use

While considerable evidence has already been gathered on the adequacy of incentives for efficient resource use in preceding steps of the analysis, most notably Step 2 and Step 5, further aspects have to be studied to complete the analysis. Step 9.1 therefore provides guidelines for assessing the incentive mechanisms inherent in the composition of the tarification scheme, the elasticity of water demand and the role of scarcity costs. As a side-effect, the analysis also delivers additional information on the degree to which the present pricing scheme promotes the polluter-pays principle. In addition, the potential of water pricing in facilitation the implementation of the overall WFD objectives in a particular investigation area is assessed as part of this investigative step.

<u>After Step 9.1</u>

- ✓ An evaluation of the **incentive properties** of the present tarification scheme along with an assessment of rewards for sensible water using behaviour within the present tarification scheme.
- ✓ Information on the **elasticity of water demand**.
- ✓ An investigation of the **incentive schemes** linked to the provision of **exogenous revenue sources** (e.g. do they promote efficient resource use, by furthering a change to locally adapted crops or water saving technical appliances).
- ✓ Additional information on the degree to which the present pricing scheme promotes the **polluter-pays principle**.

Step 9.2: Assessing the implementation of the polluter-pays principle and the adequacy of contribution from other water uses

Preceding assessment steps already provided considerable input for the assessment of the implementation of objective 3, i.e. the implementation of the polluter-pays principle and the adequacy of contribution by different water uses. The *Fact Sheets* compiled in Step 1 of the analysis document, whether certain users or activities contribute over-proportionately to certain cost categories, while Step 2 provided information on whether differentiations between user groups are undertaken in the financing scheme, which address differences in cost contribution. Finally, the investigation of the incentive properties of the financing scheme in Step 9.1 provided additional insights on the degree to which the polluter-pays principle figures in the present financing scheme. On this basis, Step 9.2 investigates whether the polluter-pays principle serves as a guiding principle in revenue generation; whether revenue generation mechanisms reflect cost structures and the degree of contribution to costs by service users; and whether Type II EC are borne by their causing activities.

<u>After Step 9.2</u>

- ✓ Investigation of the implementation of the polluter-pays principle.
- ✓ Information on the (relative) contribution to costs and revenues of water service provision by relevant sectors.

Step 9.3: Assessing the sustainability of infrastructural systems and the transparency of financial flows

An analysis of the degree of implementation of *objective 4* of Article 9 requires combining the information reported in the *Fact Sheets*, both for cost accounting (Step 1) and for revenue accounting (Step 2), on the quality of data and on divergences from the ideal Article 9 approach to cost-recovery (Step 3, Step 6 and Step 7). To complement the investigation base, Step 9.3 proposes a procedure, which delivers complementary indicators and guiding questions that allow for further substantiating the sustainability gap and making it transparent. The analysis of the transparency of financial flows in the provision of water services is completed by a graphic representation of the financial interaction between actors. Finally, an analysis of the importance of divergences encountered within the application of the cost and revenue accounting framework; the overall quality of data; the role of confidentiality concerns and uncertainties of available data supplements the investigation of the transparency of financial flows.

After Step 9.3

- ✓ An assessment of the sustainability of infrastructural system through appropriate indicators.
- ✓ An investigation and graphic representation of financial flows between actors.

Step 10: Linking the assessment components – an integrated evaluation

Once all the assessment steps have been completed, the degree of overall attainment of the WFD Article 9 objectives can be evaluated through an integrated analysis, which likewise allows for identifying potential trade-offs between different objectives in the specific implementation case at hand. On this basis, future implementation steps can be prioritised, i.e. by determining which divergences should first be addressed by adequate policy responses. In addition, Step 10 allows for documenting implementation trends (e.g. expected effects of ongoing activities and reforms) and interactions between problems or divergences which may reinforce or alleviate each other. Finally, it summarises the degree to which Article 9 has been implemented in line with the requirements on each objective and in how far exemptions justify identified divergences.

4 Illustrative cases: testing the methodological framework

"Although the right information may not be available today, it is important to start the analysis and develop it in iterations. Thus, as important as the results of the analysis for the different steps is the assessment of the most significant information gaps and the development of activities aimed at filling these gaps."
CIS (2003a)

The following chapter contains illustrative cases, which put key steps of the methodological framework presented in Chapter 3 to an empirical test. The illustration cases allow for detecting difficulties in applying the methodology and provide first *'lessons learned'*. The main objective of this chapter is to investigate whether the proposed methodology is pragmatic and applicable to different national and river basin (district) settings. Moreover, it sets the basis for the evaluative and concluding Chapter 5.

4.1 Aim and scope of the illustration cases

In the following chapter, selected and well-framed illustration cases are conducted for testing key steps of the methodological approach. They focus on those aspects of the methodological framework that are deemed most problematic or where insights are considered most valuable for assessing and improving the applicability of the methodology. While the empirical illustrations form an integral part of this study, it should be noted that the study focuses on delivering new insights by developing an appropriate methodological framework and not on conducting an analysis of cost-recovery for selected river basin districts. Hence, the illustration cases are only considered as a means toward the end of an improved and verified methodological framework and not as an end in themselves.

Testing is concentrated on the household sector in Germany and the agricultural sector in Italy.[249] The German cases illustrate steps of the methodology on the provision of drinking water services to private households in Berlin and the provision of wastewater services in the Emscher river basin. For the Italian

249 Please note that the case study selection has been explained and justified in Chapter 1.6.

case, the provision of irrigation water services in the region Emilia-Romagna, and more specifically, in the *Consorzio di Bonifica della Romagna Occidentale* has been selected for the illustration.

Chapter 4.2 provides the required background information on the selected illustration cases (cf. Section 3.3.1) and outlines the framework conditions under which the provision of water services takes place. Subsequently, key steps of the methodological framework are put to an empirical test. Section 4.3.1 applies the accounting framework for financial costs to the provision of drinking water services to private households in Berlin and highlights specificities of an application to the agricultural sector for the case of Emilia-Romagna. The revenue accounting matrix is likewise applied in Section 4.3.2 to the case of Berlin, complemented again by illustrative evidence on the methodology's application to the agricultural sector in the Emilia-Romagna region. Step 4 of the methodological framework, i.e. accounting for environmental costs and their recovery, is studied for the Emscher river basin in Section 4.3.3 and mirrored in the assessment of resource costs in Section 4.3.4 with complementary information being provided on the agricultural sector for the illustration case Emilia-Romagna. The relevance of exemptions from Article 9 is tested on the illustration case Emscher (Section 4.3.5). Finally, the three complementary objectives of Article 9 are applied to the illustration cases Emscher and Berlin (Section 4.3.6).

This chapter aims to deliver an integrated view on the applicability and the practicability of the assessment framework proposed in Chapter 3 and to draw initial conclusions. Hence, it sets the basis for the methodology's evaluation in Chapter 5.

4.2 Background information on the selected illustration cases

Section 3.3.1 outlined that a basic set of information is required for interpreting results comparatively across illustration cases. While it is not the aim of this chapter to conduct complete case studies, it is necessary to gain a basic understanding of the study context, i.e. the framework conditions under which the provision of water services is taking place, as the provision and financing of water services is inherently context-driven. Hence, information is not only required on the illustration cases themselves, but also on the wider national context. Next to the proposed set of general indicators from Section 3.3.1, basic information on the main actors involved in water services provision is provided in order to be able to depict the financial flows within the system at later stages of the investigation.[250]

4.2.1 The German water sector and the illustration cases of Emscher and Berlin

The methodological framework on the provision of water services to households is illustrated on two German cases, namely on the provision of wastewater services in the Emscher river basin and the provision of drinking water in Berlin, the capital of Germany.

4.2.1.1 The German water sector

The Federal Republic of Germany is a country rich in water, despite a number of regions encountering seasonal water scarcity (such as the Hessian Riet, or northern parts of Bavaria). Germany's total annual renewable water resources amount to about 188 billion m³ (BMU 2006). Only 18 percent of these resources are currently employed by various water users. Public water supply uses approximately 5.4 billion m³ per year, i.e. 2.9 percent of available water resources (ATT ET AL. 2008).[251] Groundwater constitutes the most important water resource for public water supply, with a share of about 65 percent, rendering the protection of groundwater bodies an important task.[252] 26 percent of the resources used for public water supply are surface water, including impounding reservoirs, bank filtrate, enriched ground water, as well as direct extractions from rivers and lakes (BMU 2006).[253] An important characteristic of the German water management system is the preference to use local water resources for drinking water supply, which is established by water legislation.

250 The case studies Emscher and Berlin draw in part on a report conducted within the project "Evaluation of Sustainability of Sewerage Systems in Metropolitan Areas – Case Studies Berlin and Emscher", cf. HERBKE ET AL. (2006).

251 With about 22.5 billion m³ (12 percent of available water resources) the bulk of water abstractions was attributable to thermal power stations for cooling purposes. Mining and manufacturing industries accounted for 7.7 billion m³ of water abstraction. Agriculture makes up for a mere 0.1 percent of water abstractions. (BMU 2006)

252 These national figures hide variations in the exact composition of groundwater and surface water sources for public water supply at regional level (e.g. some regions in Eastern Germany rely primarily on surface water sources for public water supply).

253 Industry covers 96 percent of its water demand through self-supply; only 4 percent of the demand is provided by public water supply (ATT ET AL. 2005).

The territory of the Federal Republic of Germany has a surface of 357.030 km^2. With 82.2 million inhabitants (status as of 31.12.2007) the country has an average population density of 230 inhabitants per km^2 (FEDERAL STATISTICAL OFFICE 2007), which differs greatly at regional level.[254] The degree of connection to the public water supply system is above 99 percent and about 96 percent of the German population is connected to the public sewerage network.[255] 90 percent of the population is connected to sewage treatment plants with nutrient elimination (i.e. the third purification stage corresponding to the UWWTD) (ATT ET AL. 2008). Households which are not connected to central sewerage management systems are connected to decentralized sewage purification plants. Hence, the degree of connection to wastewater treatment plants equals almost 100 percent (99.5 percent in 2004 according to FEDERAL STATISTICAL OFFICE 2006).

Since the beginning of the 1990s, the average per capita use of water dropped by 15 percent and has stabilized at a low level of 125 litres per inhabitant and day (ATT ET AL. 2008). The average per capita water consumption in the eastern German *Länder* (States) is significantly lower than in the western *Länder* (ranging from 85.3 litres in Saxony to 143 litres in Schleswig Holstein) (FEDERAL STATISTICAL OFFICE 2008). From a techno-operational perspective, further reductions in water consumption would pose significant difficulties to the existing infrastructure system, as adequate capacities need to be made available for peak demand, especially in dry seasons (LEIST 2007).

In 2007, the average price for one cubic meter of drinking water amounted to 1.85 Euro (ATT ET AL. 2008).[256] Given a water consumption of 125 litres per inhabitant and day, the average citizen paid 7 Euro per month or 84 Euro per year for the provision of drinking water. For wastewater services, the average charge amounted to 10.75 Euro per month or 129 Euro (2005) per year.[257] Water prices are subject to an interwoven net of controls in Germany, of which

254 The population density differs greatly between the various regions of Germany: Berlin constitutes the area with the highest population density (3.846 inhabitants per km^2), while only 73 inhabitants per km^2 are settled in Mecklenburg-Western Pomerania (FEDERAL STATISTICAL OFFICE 2006).

255 The connection rate to wastewater services increased considerably during the 1990s from 90 percent to 96 percent due to the implementation of the UWWTD (BMU 2006).

256 This average price per cubic meter includes VAT of 7 percent and is calculated on the basis of the volumetric tariff component and the basic fixed tariff component. Connection fees and other tariff components have not been included (ATT ET AL. 2008).

257 These averages hide important regional variations caused by the marked differences in the conditions for water service provision (HOLLÄNDER ET AL. 2008).

Table 11: General characteristics of the German water sector

Indicator	**Water supply**	**Wastewater management**
Population Surface Water covered surface Population density Number of households Average household size	82.2 million (as of: 31.12.2007) 357,030 km^2 8,279 km^2 231 inhabitants / km^2 39.722 million 2.07 (Federal statistical office 2007)	
Connected inhabitants	99%	96% to centralised sewerage management systems 99.5% including decentralised solutions
Available water resources Share of available water resources employed for water service provision	188 billion m^3 per year 2.9%	(not applicable)
Water use per connected inhabitant and day	125 l/d Ranging from 83 l/d in Saxony to 143 l/d in Schleswig-Holstein.	(not applicable)
Length of the pipe system (size of the distribution network)	500,000 km (without house connections) 6.06 m per inhabitant	515,000 km (combined and separate sewer systems) 6.24 m per inhabitant
Role of central vs. decentralised water service provision	Large predominance of centralised systems	Mainly centralised systems; regional variations[258].
Sewerage treatment capacity Primary treatment Secondary treatment Tertiary treatment	(not applicable)	99.6% of sewerage is treated, thereof: 100 % primary treatment 100% secondary treatment 97% biological treatment with nutrient elimination
Number of service providers in the service area	6,400	6,900

Sources: Federal Statistical Office 2007, ATT et al. 2008, BMU 2005.

258 Decentralized solutions to wastewater collection and treatment may however play a more prominent role on a regional basis, most notably in the Eastern States (i.e. in Thuringia).

the exact structure and the related administrative responsibilities are determined by the legal arrangement chosen by the service provider for delivering water services to its customers (AMMERMÜLLER ET AL. 2009).[259]

The following Table 11 summarises a number of key data on the German water sector (cf. Section 3.3.1).

As a federal State, the competences for water management are distributed in Germany between the federal level (*Bund*) and the States (*Länder*).[260] The principles of federalism also underlie the municipal autonomy (*kommunale Selbstverwaltung*) of water management.[261] The constitution of the Federal Republic as well as the constitutions of the States guarantee municipal control over local affairs, including the right to manage the provision of public water supply and sanitation services at the local level (KRAEMER AND JÄGER 1998). Municipalities may choose the institutional arrangement of service provision and may opt for an organisational structure under public or under private law. The possible institutional arrangements (in order of decreasing integration into municipal administration) which can be selected include the following[262]:

---

259 The tarification schemes of municipal utilities are generally controlled by the municipal or local council at municipal level as well as subject to supervision by the local supervisory authority. If water service providers charge their services directly to customers, i.e. have chosen an arrangement for delivering their services under private law and thus apply charges under private law, these utilities (which can be both, public or private in their organisational arrangement) are subject to the additional supervision of the cartel authorities. If fees under public law are instead charged by the water service provider, which is only allowed for organisational arrangements under public law, a review of these fees can be initiated by customers through control by a civil or administrative court. In addition, they are reviewed by municipal control authorities (*Kommunalaufsichtsbehörde*) and controlled by the municipal or local council. (AMMERMÜLLER ET AL. 2009, ATT ET AL. 2008)

260 With the revised Federal Water Act passed on 10 July 2009, the Federal State has received more competence in water management. Before, the federal level was only allowed to pass framework legislation, which then had to be transposed to legislation at the level of the 16 German *Länder*. For implementing the WFD into national legislation, 33 administrative acts were required, with one revision of the Federal Water Act, the revision of 16 *Länder* water acts and 16 accompanying regulations. With the revised Federal Water Act the Federal State may now directly pass binding legislative requirements on water management, which do not require a transposition at *Länder* level. Nevertheless, the *Länder* retain important competences so that differences in water management across the German *Länder* are likely to persist.

261 A variety of actors at the different administrative levels plays a part in water management in Germany as Figure A4.2 in the Annex depicts (cf. PIELEN ET AL. 2004 for an in-depth introduction to the role and competences of the different actors).

262 Cf. BMU (2002), CRONAUGE (2001), KRAEMER AND JÄGER (1998) for an in-depth description of the different institutional arrangements.

- Organisational arrangements under public law include: direct labour (*Regiebetrieb*); semi-autonomous municipal agency (*Eigenbetrieb*); inter-municipal association (*interkommunale Kooperation*); public-law corporation (*Anstalt öffentlichen Rechts*); water and soil management association (*Wasser- und Bodenverband*).
- Organisational arrangements under private law include: municipal enterprise (*Kommunale Eigengesellschaft*); public enterprise (*öffentliche Gesellschaft*); delegation to a private enterprise (*Beauftragung privater Unternehmen*); operator Model (*Betreibermodelle*); concession agreement (*Konzessionsmodell*).

For the provision of water supply services, the last decades saw a trend towards institutional arrangements under private law (from 12.7 percent in 1986 to 30.2 percent in 2003) and a sharp increase in public-private holdings (from 3.3 percent in 1986 to 20.8 percent in 2003) (ATT ET AL. 2005).[263] While the percentage of fully privatised water supply companies has gradually increased from 1 percent in 1986 to 3.5 percent in 2003, the overall percentage is still low compared to other organisational forms. The majority of the 6,400 water supply companies are comprised of small undertakings serving a relatively small number of inhabitants in rural areas. In contrast to this, a small number of undertakings supply a large number of inhabitants in the metropolitan regions, i.e. half of the water delivered is supplied by only 1.5 percent of the undertakings (ATT ET AL. 2008).

As the provision of wastewater services is classified as a sovereign obligation of municipalities, the largest share of service providers is held by municipal utilities or by special-purpose and water associations (ATT ET AL. 2008). Similar to the organisation of water supply, the provision of wastewater services is ensured by 6,900 utilities. There has been a trend away from organising wastewater services as part of the municipal administration structure, such as direct labour (*Regiebetrieb*), towards the introduction of autonomous public organisational forms, such as public-law corporations (*Anstalt öffentlichen Rechts*). Undertakings with an organisational form under private law still play a minor role in the provision of wastewater services (about 10 percent for wastewater discharge and 12 percent for wastewater treatment in relation to

263 In interpreting these figures it should be noted that they relate to a water statistic compiled by the industry association bdew in 2005, which covers about 1,300 undertakings. These undertakings represent approximately 79 percent of total water provision in Germany. Detailed data on the more than 5,000 undertakings not covered in this statistic, providing approximately 21 percent of total water delivery, are not collected. However, it is to be assumed that these undertakings are predominantly small state-run and municipal utilities.

the inhabitants connected (ATT ET AL. 2008)), but the private sector is involved through sub-tasks (e.g. planning, building and operating).

Figure A4.1 in the Annex provides an overview on the actors involved in water management in Germany.

4.2.1.2 Emscher river basin: wastewater services provision

The Emscher river basin[264] forms part of the Rhine river basin and covers around 865 km^2 (EMSCHERGENOSSENSCHAFT 2008). The Emscher has its source east of Dortmund and flows into the river Rhine near Dinslaken around 83 km to the west.[265] It is located in-between the catchments of the rivers Lippe and Ruhr (cf. Figure 18).[266] The Emscher region forms part of the Ruhr area, which once constituted the industrial centre of Germany. The region is characterised by coal mining, steel and chemical industries. In the second half of the 19th century, due to conveniently situated coal deposits, the industrialization of the region between the river Ruhr and the river Emscher began. Within a few decades, a predominantly agricultural area with a relatively low population density developed into the centre of Germany's heavy industry.

In the middle of the 20th century the industrial boom stopped and industrial plants started to close down. Since then, the region has undergone deep structural and social transformation processes, but is still characterised by a high unemployment rate of about 13.5 percent, which is significantly above the average value for the State North Rhine-Westphalia (9.0 percent) and that of Western Germany (6.9 percent) (as of June 2009).[267] The Emscher river basin belongs to the most densely populated areas in Germany with about 2.3 million

---

264 According to the WFD Article 5 report for the river Rhine basin the Emscher river basin is defined as a '*working area*' (*Arbeitsgebiet*), corresponding to the C-level of WFD implementation. The related B-level is the sub-river basin (*Bearbeitungsgebiet*) Lower Rhine. However, for simplification reasons, the term '*Emscher river basin*' is used throughout this illustration case. The Emscher river basin is located within the administrative boundaries of three local government districts, namely Münster, Düsseldorf and Arnsberg.

265 The Rhine-Herne Channel, an important navigation route, is situated in parallel to the Emscher from Duisburg to Castrop-Rauxel. From its source in Holzwickede to its estuary in the Rhine, the Emscher used to surmount 120 m of altitude over 109 km of river course. Through the reconstruction of the Emscher as an open sewer system, the river course has been shortened to 83 km (described in detail below).

266 Six different river classifications can be differentiated within the Emscher river basin. The Emscher itself is classified as a sand and clay dominated lowland river.

267 See http://www.ihk-nordwestfalen.de/volkswirtschaft_statistik/arbeitsmarkt.cfm (visited 6 August 2009).

inhabitants and a population density of 2,642 inhabitants per km^2 (EMSCHERGENOSSENSCHAFT 2009A).

Figure 18: The Emscher river basin (embedded in the area commonly known as 'Ruhr Area')

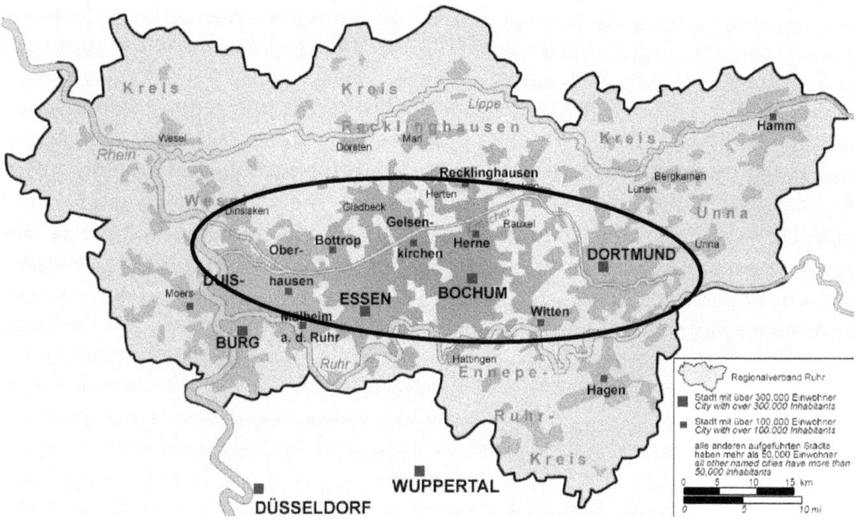

Source: http://www.emscher.nrw.de/offen/start.htm.

The period of industrialization and intensive mining activity deeply affected the Emscher river basin, both in terms of water quality and river hydromorphology. The rapid economic growth at the beginning of the 20th century was accompanied by an overexploitation of water resources and severe water pollution.[268] The rising accumulation of wastewater, induced by an increasing domestic and industrial consumption of water, was discharged into the rivers untreated. In addition, there was a considerable water inflow into the Emscher River due to the draining of mines. Within twenty years (1890–1910), the deposition

268 From the beginning of the 1840s, the mining sector extended into the Ruhr area from the South, including the regions of Oberhausen, Altenessen, Gelsenkirchen, Herne, Castrop and Dortmund of the Emscher river basin. Today, mining is only actively pursued in the Emscher river basin upstream of the river Boye close to Bottrop-Grafenwald. The Emscher itself has not been influenced by mining activities since the mid-1990s. Nevertheless, in order to maintain underground mining, the mining sector has to continually drain subsurface flowing water both from actively used as well as from formerly used mining areas. To this end, the mining sector operates draining systems for mine water at seven locations within the Emscher river basin (Status as of May 2008), which were reduced from 20 in 1990. (EMSCHERGENOSSENSCHAFT 2009A)

of coal increased 20 times, leading to an increased number of subsidences caused by mining and thus a decreased water flow from the Emscher and its tributaries (HERBKE ET AL. 2006). The large amounts of wastewater posed severe difficulties to the region at the turn of the century, especially in times of low water flows. Even small floods often led to large and persistent inundations with untreated sewage. These facilitated the transmission of diseases and epidemics of cholera and typhoid fever.[269]

In response to these challenges, special laws were enacted in the following years that allowed for establishing independent water associations within the region, whose operating areas corresponded to the hydrological boundaries of the rivers Ruhr, Lippe and Emscher respectively. The water association for the Emscher river basin, the *Emschergenossenschaft,* was founded in 1899 as an independent public law corporation (*Körperschaften des öffentlichen Rechts*) and was granted self-administration rights.[270] It was established by representatives from municipalities, industry and the mining sector in the Emscher region to address the occurring wastewater problems and related diseases at supra-regional level.[271]

From its beginning, the *Emschergenossenschaft* carried out several tasks. Specifically, it sought to establish a viable system for wastewater services provision given the intensive mining activities at the time, to avert floods and swamping of land due to subsidence caused by mining, to ensure the drainage of mine waters, to provide flood protection measures, especially behind dams and in polder areas, and to guarantee compliance with legal requirements for wastewater treatment (TRESELER AND GÜTLING 2006). According to § 2 of the EmscherGG, the *Emschergenossenschaft* is today also responsible for

- Managing surface water bodies along with the related infrastructural assets;
- Reconstructing surface water bodies in a near-natural condition;
- Managing groundwater levels;

---

269 Every second worker in the Emscher region was infected by dysentery, typhoid fever, diphtheria, cholera or pneumonia. Every 10th worker contracted malaria (MUNLV 2008A).

270 The basis for the Emscher Association Law of 1904 (*Gesetz über die Emschergenossenschaft – Emschergenossenschaftsgesetz, EmscherGG,* last amended on 11 December 2007) was the so-called '*Middeldorf'sche Plan*', which considered the provision of water flow, flood protection and wastewater treatment within the Emscher river basin in unison. For the German context, it constituted the first management plan that addressed an entire river basin. The *Middledorf'sche Plan* constituted the basis for the creation of the open sewer system in the Emscher region. (EMSCHERGENOSSENSCHAFT 2009A).

271 From the beginning, the Emschergenossenschaft co-operated with the *Lippeverband*; both work together under one umbrella named *Emschergenossenschaft / Lippeverband* (EGLV).

- Preventing, mitigating, rectifying and balancing (1) expected or existing negative ecological impacts on the water system related in particular to the impact of the mining sector on groundwater levels; and (2) expected or existing negative impacts on surface water bodies induced by the provision of wastewater transportation and other aspects of wastewater services provision; and finally for
- Monitoring water status.

Today, the Emschergenossenschaft is responsible for 343 km of watercourses, of which 290 km still function as an open sewer system. The Emschergenossenschaft operates four wastewater treatment plants with about 4.8 million population equivalents and about 69 infrastructural assets for rainwater and combined water management (EMSCHERGENOSSENSCHAFT 2009A). The capacity of the wastewater treatment plants ranges from 500,000 population equivalents for the wastewater treatment plant Duisburg-Old Emscher to 2,400,000 population equivalents for the treatment plant Emscher Estuary (see Annex, Table A4.1 on the capacity of the treatment plants operated by the *Emschergenossenschaft*).[272]

The following Table 12 provides an overview on general characteristics of the Emscher river basin illustration case.

Background to the provision of wastewater services

Despite hygienic concerns, the sewerage management system in the Emscher area was not constructed as an underground closed pipe system, but rather as an open surface canal.[273] Cost considerations, on one hand, and technical

272 The wastewater treatment plants Bottrop, Emscher-Estuary and in part Dortmund-Deusen (until the end of 2009) act as river treatment plants, which may take up the entire Emscher for biological treatment, depending on water flows. The biological wastewater treatment plant Emscher-Estuary became fully operational in 1977 and had a major impact on water quality of the river Rhine. At the end of the 1970s planning for the wastewater treatment plant Duisburg-Old Emscher started, which began its operations in 1988 and was able to eliminate nitrogen and phosphorous compounds. In line with state-of-the-art wastewater treatment approaches the treatment plant Dortmund-Deusen followed in 1984 and Bottrop in 1997. Once the reconstruction of the Emscher river basin has been completed, the remaining river treatment plants will be refitted to regular wastewater treatment plants to which wastewater flows through an underground closed pipe network.

273 Due to the expected continuous subsidence and in line with consolidated findings at the time, to transport and discharge wastewater into the nearest larger river, it was decided to construct a surface system of technically and artificially constructed (wastewater) watercourses. At the time, about 3.7 percent of the area had lost its natural drainage capacity to the river Rhine due to subsidences.

Table 12: General characteristics of the Emscher river basin

Indicator	Wastewater management
Population Size of the service area Population density Number of households Average household size	2,285,813 inhabitants (as of 31.12.2007) 865 km² 2468 inhabitant per km² (ranging from 1,201 inhabitants/km² in Bottrop to 3,395 inhabitants/km² in Herne) 1,827,754 2.01 (Emscher region), 2.12 (average value for Northrhine-Westfalia)
Connected inhabitants to wastewater services provision	99.4%
Water use per connected inhabitant and day	141 l/d (sub-river basin Lower Rhine) Between 120 l/d and 140 l/d in the municipalities of the Emscher region.[274]
Length of the sewerage pipe system (size of the distribution network)	199 km sewerage pipe system 293 km of river used as open sewer canal (Note: these figures only reflect the present status of reconstruction of the Emscher river basin (cf. discussion below, Section 4.3.3) and are subject to change as the construction of the closed sewer system progresses.)
Sewerage treatment capacity Primary treatment Secondary treatment Tertiary treatment Capacity	99.4% of the population in the basin is connected to sewerage collection and treatment. 100% primary treatment 100% secondary treatment 100% tertiary treatment[275] 4 wastewater treatment plants with 4.8 million population equivalents; cf. Table A4.1 (Annex)
Number of service providers in the service area Share in service provision	1 (sewerage treatment) (note: sewerage collection is partly conducted by the cities / municipalities) 99.4% in terms of connected inhabitants

Sources: Emschergenossenschaft (2008, 2009), MUNLV (2005), MUNLV (2008), IT.NRW (2006).

problems and the fear that structural damages caused by coal-mining would result in subsidence, on the other hand, led to this decision. Accordingly, the Emscher and its tributaries (today a total of around 343 km) were gradually straightened, deepened, diked and got concrete formwork. At the same time,

274 Divergences from these averages occur for selected municipalities (EMSCHERGENOSSENSCHAFT 2009A).

275 While all four sewerage treatment plants operated within the Emscher river basin adhere to the standards of tertiary treatment and eliminate nitrogen and phosphorous compounds, their treatment efficiency is presently still reduced by the dilution of wastewater encountered in the river treatment plants (cf. footnote 24 above).

several pumping stations (two at the beginning, 102 today (EMSCHERGENOSSENSCHAFT 2009A)) delivered surface and ground water from the lower-lying surrounding hollows to the diked water courses (TRESELER AND GÜTLING 2006, EMSCHERGENOSSENSCHAFT 2008). To ensure the unhindered runoff into the Rhine and reduce the back pressures of Rhine floods, the Emscher course was altered twice in a northward direction.[276] In order to avoid the unpleasant odour caused by the decomposition of sludge in the hollows, the *Emschergenossenschaft* established from 1910 onwards mechanical treatment plants in the densely populated areas. Later, biological treatment plants were added to the system in order to further protect the quality of the Rhine.

Due to the raising environmental awareness of the public since the beginning of the 1970s, the disadvantages of the open sewer system came to the fore. It was considered non-compliant with legal requirements and the demands on a modern urban infrastructure. The unpleasant odour was regarded as unacceptable and the technical modifications of the river were perceived as too visibly artificial. At the same time, the mining era drew to a close, with an increasing number of coal-mines closing down. Due to the reduction in subsidences caused by mining, an open sewer system was no longer needed. In response, an extensive restructuring process of the Emscher river basin towards a sewage-technical reconstruction was discussed.[277]

In 1991, the specific objectives of the reconstruction of the Emscher river basin were decided upon, namely: (i) to remove the open discharge of effluent; (ii) to construct state-of-the-art decentralised wastewater treatment plants; and (iii) to reconstruct the river's hydromorphology to its full potential[278] (EMSCHERGENOSSENSCHAFT 2009A, TRESELER AND GÜTLING 2006). The implementation timeframe was expected to extend over more than one generation. In the long-run, the Emscher system is to be rebuilt in such a way as to reach good water status or good potential in all water bodies, to ensure flood protection and to

276 Today, about 37.8 percent of the Emscher river basin is drained artificially through pumping stations (EMSCHERGENOSSENSCHAFT 2008). As a result, the original estuaries '*Old Emscher*' and '*Small Emscher*' remained disconnected from today's main course of the Emscher. Today's river beds diverge to a large part from their original course and have been dislocated to different geographical heights: partly they are guided over dykes over the territory, but often they are deeply recessed, in order to be able to drain deep subsidences in the backland. As a side-effect, these deep recesses fulfill an important groundwater regulation function. (MUNLV 2008A:37)

277 Cf. EMSCHERGENOSSENSCHAFT (2009A), SEMRAU ET AL. (2009).

278 The river's original wetland cannot be fully reconstructed due to the overall structural changes within the region.

enable recreational water uses (TRESELER AND GÜTLING 2006).[279] The Emscher river basin is to be restored as far as possible to its natural hydromorphological and structural conditions, considering its present anthropogenically influenced status, while integrating the Emscher as far as possible into today's surroundings (SEMRAU ET AL. 2009).[280] Furthermore, the reconstruction of the Emscher system aims to increase the infiltration of rainwater in order to (re-)close the natural water circle, which was heavily disturbed over a long period through coal-mining and its consequences. While the *Emschergenossenschaft* is responsible for managing and redeveloping water bodies influenced by mining activities and fulfils a central planning and coordinating function, the municipalities and cities in the Emscher river basin are also involved in the river basin's reconstruction (EMSCHERGENOSSENSCHAFT 2009A) (cf. Section 4.3.3 below).

The costs for the overall rehabilitation project of the Emscher, which is expected to be finished by the year 2020, are estimated to amount to about 4.4 billion Euro (SEMRAU ET AL. 2009). However, the river basin's hydromorphology is expected to take approximately another ten years for fully recovering and accommodating the effect of the measures. The finalisation of the technical construction of the closed sewer system is envisaged for 2017 (EMSCHERGENOSSENSCHAFT 2009A). By 2009, nearly 50 percent of the budget had already been invested in the construction and modernisation of wastewater treatment plants, the construction of part of the closed sewerage network and the re-modification of watercourses, which were partly financed by the State North Rhine-Westphalia (VAN BÖMMEL 2009). For the upcoming years, an annual budget of 200 million Euro has been foreseen for the reconstruction of the Emscher river basin (EMSCHERGENOSSENSCHAFT 2008).[281]

Actors in the provision of wastewater services in the Emscher river basin

The *Emschergenossenschaft* is responsible for the provision and management of wastewater services in the Emscher river basin. It is a not-for-profit public-law corporation and has self-administration rights. Its activities are regulated through the statutes of the *Emschergenossenschaft* and financed through its

279 With a few exceptions in upstream rivers with near-natural conditions (such as the Hörder Bach, Schondelle, Deininghauser Bach, Ostbach, Dorneburger Mühlenbach and Handbach, the entire Emscher river basin has been designated as heavily modified.

280 Due to the variant intensity of uses within the Emscher river basin, a diversified concept has been developed, which formulates individual development objectives for different parts of the river basin subject to existing structures and constraints (SEMRAU ET AL. 2009).

281 A detailed description of the financing schemes and the contributions by different actor groups is provided in Section 4.3.3.

members' contributions. Municipalities, industry and the mining sector are members in the *Emschergenossenschaft*. The association's management and the control of its activities are subject to democratic principles. The organisational structure of the *Emschergenossenschaft* is based on the following three distinct bodies (EMSCHERGENOSSENSCHAFT 2008):

- The **Association Assembly** (*Genossenschaftsversammlung*) is the primary organ of the Association (like a parliament). Members delegate representatives to the assembly for a period of five years. Voting rights are based on membership fees and regulated through the statute (Article 11 and 12, Emscher Association Law).
- The **Association Council** (*Genossenschaftsrat*) consists of fifteen representative elected by the Association Assembly for a period of five years (Article 15, Emscher Association Law).[282] As a supervisory organ, it oversees all activities of the managing board and acts on behalf of all members.
- The **Board** (*Vorstand*) is elected by the Association Council for a period of five years and manages the association.

Furthermore, the following two bodies fulfil supplementary organisational functions:

- The **Investment Committee** (*Investitionsausschuss*) consists of four members, which are delegated by the Association Council for a five-year term. It evaluates all investments proposed by the Board with regard to their necessity, urgency, costs and financial viability and proposes a decision to the Association Council.
- The **Appeal Committee** (*Widerspruchsausschuss*) consists of three civil servants (delegated by the supervisory authority, see below) and six representatives elected by the Association Assembly. According to the Emscher Association Law (Article 28 and 29), members may appeal against their membership fees (*Beitragsbescheid*). If the board does not accept the appeal, the appeal committee has to review its legitimacy. Due to a new law enacted on 1 November 2007 the appeal committee is no longer required and members have to appeal directly at the administrative courts. Nevertheless, the appeal committee remains responsible for appeals against membership fees which were raised prior to the enactment of the law.

282 The exact composition of the representatives (i.e. number of representatives from municipalities, towns, mining corporations, traffic facilities, etc.) is regulated through Article 16(1) of the Emscher Association Law (1995).

Finally, the *Land* North Rhine-Westphalia acts as a supervisory authority with respect to legal matters (*Rechtsaufsicht*).[283] Operational decision-making is unaffected by this supervision. The following Figure 19 summarises the organisational structure of the *Emschergenossenschaft*.

Figure 19: Organisational structure of the Emschergenossenschaft

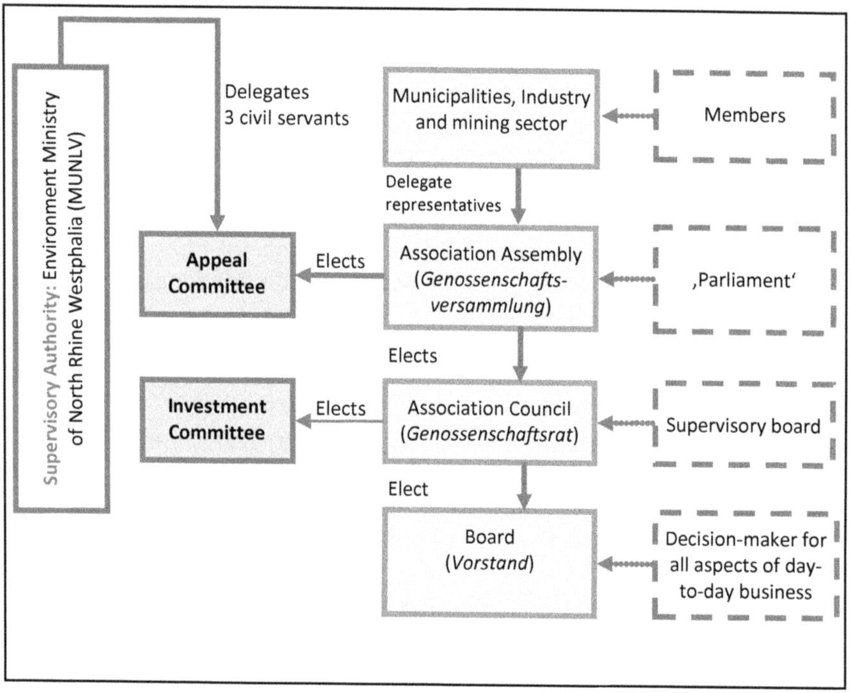

Source: based on Emschergenossenschaft (2008).

4.2.1.3 Berlin: drinking water provision to private households

Berlin is Germany's capital and with 3.39 million inhabitants also its largest city.[284] Berlin is spread over a comparatively large surface of 892 km², leading to

283 According to Article 33 of the Emscher Association Law.

284 With 3.4 million inhabitants in its service area Berlin, the *Berliner Wasserbetriebe* serve more customers than the cities of Munich, Hannover, Frankfurt and Stuttgart combined (www.bwb.de).

a relatively low population density of 3.797 inhabitants per km² for a capital.[285] Berlin is located within the Elbe River Basin. Nine additional German *States*[286], as well as part of the Czech Republic and, to a minimal extent, Poland and Austria form part of the international Elbe River Basin. The Spree River crosses the city, flowing from south-east to north-west where it meets the Havel River, one of the largest Elbe tributaries. Berlin also profits from an extensive web of canals and lakes.

Table 13: General characteristics of drinking water service provision in Berlin

Indicator	**Drinking water provision in Berlin**
Population Size of the service area Population density Number of household connections Number of households Average household size	3.4 million inhabitants in the assessment area 892 km² 3.797 inhabitants / km² 262,000 1,930,300 households 1,76 persons per household
Connected inhabitants	99.8% of inhabitants connected to water service
Rural – urban share in investigation area	100% urban area
Water use per connected inhabitant and day	111 l/day per person
Domestic consumption	138.4 million m³ in 2007 (compared to 174.5 million m³ in 1992)
Total consumption	192.6 Mio m³ in 2007 (compared to 270.9 million m³ in 1992)
Number of service providers in the service area Share in service provision	1 100% in terms of connected inhabitants and water service use (own water supply not considered).

Sources: BWB (2009a), www.bwb.de, Statistical Office of Berlin.

The provision of water supply services in Berlin has undergone considerable transformations since the late 19th century, triggered by important organisational and political changes. The borders of the town of Berlin itself changed

285 Data derived from www.statistik-berlin.de and www.berlin.de. In comparison, Paris and London have a population density of 20,500.9 inhabitants per km² and 4,717.3 inhabitants per km² respectively (EUROSTAT Website, visited on 18 June 2009).

286 The German part of the Elbe river basin district extends over (parts of) the territory of Schleswig-Holstein, Hamburg, Lower Saxony, Mecklenburg-West Pomerania, Saxony-Anhalt, Thuringia, Saxony, Brandenburg, Bavaria and Berlin.

repeatedly, with considerable impact on the organisation of services provision and the corresponding technical facilities.[287]

The following Table 13 and Table 14 provide an overview of the general characteristics and basic technical facts regarding the provision of drinking water services in Berlin.

Table 14: Basic facts on the technical system (Berlin)

Indicator	**Drinking water provision in Berlin**
Number of water works Total capacity of water works Number of wells	9 1,140,000 m³/day 952 of which 687 are currently used for drinking water production
Water abstraction	201,000,000 m³/year
Water provision	193,000,000 m³/year
Length of the pipe system Water abstraction per km of pipe system	7.888 km 25,481 m³/km
Water sources used for water supply	Natural groundwater recharge: 32% Artificial groundwater enrichment: 14% Bank filtrate: 54%

Source: BWB (2009a), www.bwb.de.

Background to the provision of drinking water services to households in Berlin

Economic difficulties after the political unification in 1990 together with an increasing use of water efficient appliances led to an unexpected decrease in water demand in Berlin of nearly 30 percent between 1992 and 2007 (cf. Table 15), which posed severe difficulties to the management of infrastructural systems.[288]

287 For an in-depth investigation of the historic evolution of water supply services in Berlin cf. BÄRTHEL (1997).

288 After the political change the city was projected to grow to about 6 million inhabitants. This figure served for several years as the basis for infrastructural planning.

Table 15: Evolution of water demand in Berlin, 1992–2007

	Unit	2008	2007	2004	2000	1996	1992
Inhabitants of Berlin	million	3.39	3.39	3.39	3.33	3.46	3.47
Total consumption	million m^3	192.7	192.6	201.7	212.6	229.0	270.9
Consumption by private households	million m^3	140.7	138.4	144.8	152.9	161.0	174.5
Consumption by industry and trade	million m^3	29.7	27.4	26.6	23.0	26.5	40.0
Consumption by other consumers	million m^3	18.8	23.5	27.2	33.5	38.7	54.0
Consumption in surrounding areas	million m^3	3.5	3.3	3.1	3.2	2.8	2.4
Private households: consumption per person and day	L/d per person	113	111	117	126	128	138
Total consumption per person and day	L/d per person	155	156	160	172	181	214

Source: BWB (2008b).

The region of Berlin is well-endowed with water resources. Available groundwater resources amount to 350 million m^3 per year, exceeding water demand by far. In 2007, 201.1 million m^3 of water were extracted for the provision of public water supply services (BWB 2009), about 0.75 percent more than in the previous year (199.6 Mio m^3).[289] In order to guarantee a long term sustainable water supply, a '*Water Supply Concept 2040*' has been developed by the *Berliner Wasserbetriebe* in cooperation with the Berlin Senate, which analyses different development scenarios for the metropolitan region of Berlin (MÖLLER AND BURGSCHWEIGER 2008), e.g. in terms of climatic change. It details on which areas integrated water resources management should focus in order to maintain safe water supply conditions.[290]

The main source for drinking water provision is Berlin is groundwater (partly as bank filtrate). In the 1960s and 1970s high water consumption led to sinking groundwater tables in parts of Berlin: an increase in artificial groundwater

289 The average daily rate of water abstraction amounts to 549.570 m^3. Higher rates of water abstraction in 2007 can be explained by exceptionally dry summer months (BWB 2009A).

290 Among others, the concept formulates minimum flow requirements for the rivers Spree and Havel to guarantee groundwater abstraction from bank filtrate; it outlines quality enhancements on the river Spree; and proposes the development of a new water work, which would allow to spread out water abstraction for water provision in Berlin from 2014 onwards more evenly, thereby guaranteeing long term security of water supply provision and enabling ecologically sustainable groundwater management.

enrichment became necessary and regulation to that end was enacted. The water used for this enrichment consists of bank filtrated surface water, so that the quality of surface water also plays an important role for drinking water provision (BWB 2004). Groundwater management in Berlin is not unproblematic and the needs vary greatly between rather scarcely populated green areas and forests on the one hand, and more densely populated urban areas on the other. Whereas residential areas can be plagued with damp cellars and parks in these areas can turn into swamps due to a rising groundwater table, forests, lakes and wetlands on the outskirts of Berlin (where most of the groundwater abstraction takes place) may suffer from lowering groundwater tables. Under present conditions, groundwater enrichment thus remains necessary for maintaining these areas (ABGEORDNETENHAUS Berlin 2005).

The Berlin Water Law mandates that all of the water needed to cover Berlin's public water supply has to be taken from Berlin territory (§37a (4) of the Berlin Water Law, *Berliner Wassergesetz*), unless an exception is granted. In order to ensure the quality of drinking water, groundwater abstraction takes place in water protection zones. These zones are established on the basis of §19 of the Federal water legislation (*Wasserhaushaltsgesetz*, WHG). Water protection zones are determined by the Berlin Senate. There are presently thirteen Provisions Establishing Water Protection Areas (*Wasserschutzgebietsverordnungen*).[291] These provisions prohibit or restrict damaging actions such as wastewater discharge and infiltration (except in the case of unpolluted rainwater), industrial or intensive agricultural activities.

The pipe system for water supply has a length of about 7,888 km and an annual rate of renewal of 0.8 percent. A low percentage of pipe breaks of 0.1 percent testifies to a high security of supply. With 5.18 percent of abstracted water resources (OELMANN ET AL. 2009), Berlin's leakage rate is significantly below the national average of about 7 percent (ATT ET AL. 2008).

Actors in the provision of drinking water services to households in Berlin

Drinking water services are provided in Berlin by the *Berliner Wasserbetriebe*. The formerly 100 percent municipally owned *Eigenbetrieb* was converted into a public law corporation in 1994 in order to allow the undertaking to pursue commercial activities. In 1998, the government of Berlin took the political deci-

291 Overall, about 212 km² of the area of Berlin are designated as water protection zone. Hence, in relation to Berlin's total surface, about one quarter of the city is covered by the thirteen Provisions Establishing Water Protection Areas (cf. http://www.berlin.de/sen/umwelt/wasser/wasserrecht/index.shtml).

sion to privatise its water utilities and entered (in 1999) into a public-private partnership with a consortium consisting of Vivendi (now Veolia), the German multi-utility company RWE and the insurance company Allianz. To this end, the *Berliner Wasserbetriebe* were integrated in a new holding, *Berlinwasser Holding plc.*, while legally keeping the status of a public law corporation. The consortium paid the city of Berlin 1.7 billion Euro to acquire 49.9 percent of both entities (LANDESPRESSEDIENST VON BERLIN 1999). The 5 percent of the shares originally held by the insurance company Allianz were sold to Veolia and RWE in 2002, both of whom now own 24.95 percent of the holding each.

In order to ensure the influence of the *Land* Berlin on key management aspects, certain agreements were made in a non-public consortium agreement. These agreements aimed to keep water prices and sewerage charges constant until 2003 and to foster their long-term decrease, as well as at guaranteeing the economic independence and international positioning of the *Berliner Wasserbetriebe*. Furthermore, investments of about 2.5 billion Euro in technical infrastructure over the following ten years were agreed upon, together with the promotion of new technologies. In addition to these accords, the consortium agreement prohibited operational lay-offs for the next 15 years and envisaged the creation of 2,160 new jobs. Besides these conditions settled upon in the consortium agreement, the *Land* Berlin retains important responsibilities for technical issues and staff-related decisions through its seats in the governing board of the *Berliner Wasserbetriebe*. (LANDESPRESSEDIENST VON BERLIN 1999)

The organisational and management structure of the *Berliner Wasserbetriebe* is centred on the following three bodies (BWB N.Y.):

- The **Executive Board of Directors** (*Vorstand*) consists of four members, namely the chairman of the board; the executive financial director; the executive director of personnel and social affairs; and the executive technical director. They are appointed by the supervisory board (*Aufsichtsrat*) for a maximum period of five years.
- The **Supervisory Board** (*Aufsichtsrat*) advises and controls the Executive Board of Directors (§9 BerlBG, §5 Corporate Governance Codex). It is chaired by a member of the Berlin Senate and consists of seven members proposed by the Berlin Senate and appointed by the Assembly of Guarantors, as well as eight members appointed by the employee committee. Members may contribute to the Advisory Board for a maximum period of five years.[292]

292 A prolongation of contracts beyond the quinquennial period is not encouraged (see § 5.1.2 Corporate Governance Codex of the *Berliner Wasserbetriebe*, www.bwb.de).

- The **Assembly of Guarantors** (*Gewährträgerversammlung*) is appointed by the Berlin Senate and consists of up to five members. Among its responsibilities is the appointment and dismissal of the members of the Supervisory Board. Furthermore, it decides upon the allocation of unappropriated profits. The Assembly of Guarantors meets at least once a year.

In addition, an **Audit Committee** (*Prüfungsausschuss*) is appointed by the Supervisory Board. It covers accounting and risk management related issues, and is responsible for ensuring the independence of the auditor of the annual accounts. Figure 20 graphically summarises the organisational structure of the *Berliner Wasserbetriebe*.

Figure 20: Organisational structure of the Berliner Wasserbetriebe

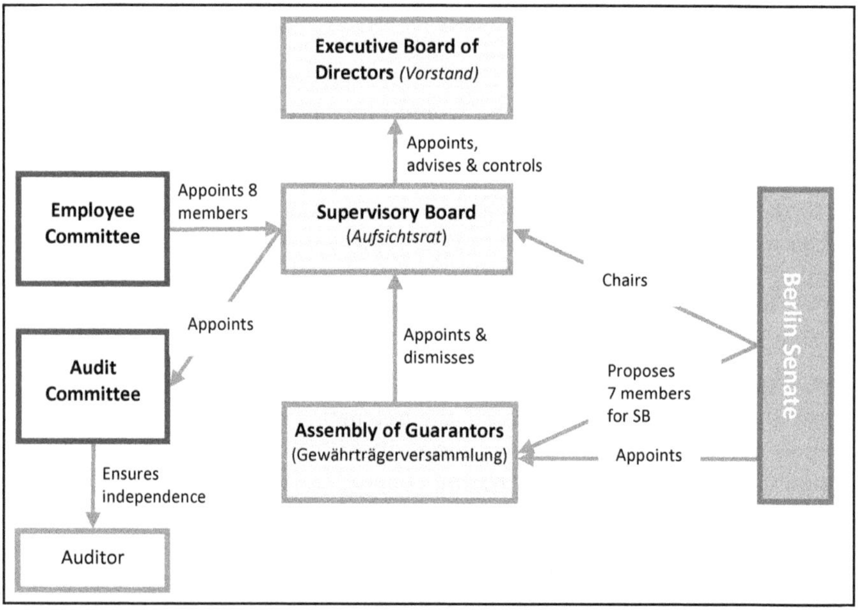

Source: BWB (2009a), OELMANN ET AL. (2009).

4.2.2 The Italian water sector and the illustration case of Emilia-Romagna

The following sections briefly introduce the Italian water sector for agriculture along with the illustration case *'Consorzio di Bonifica della Romagna Occidentale'* in the region Emilia-Romagna.

4.2.2.1 The Italian water sector for agriculture

Italy is a country with an uneven distribution of water resources. While overall water resources are relatively abundant with a yearly net average rainfall per head of around 5.200 m^3 (MASSARUTTO 1999), one must account for a great regional and seasonal variability. Given irregular outflow paths as well as technical and natural constraints, de facto availability of water resources is far lower, i.e. about 2000 m^3 if all potentially usable resources are considered and only about 928 m^3 if only the amount of water is considered that can be used, given available infrastructure and storage capacities (BAZZANI ET AL. 2005, MASSARUTTO 1999).

Most water reservoirs are located in the north (due to the Alps and the natural storage capacity provided by glaciers and lakes), while the centre and the south are less endowed and prone to recurrent droughts and high seasonal variability of run-offs (BAZZANI ET AL. 2005, GORIA AND LUGARESI 2002, MASSARUTTO 1999). A large part of the south suffers from consecutive periods without rain of 100–150 days (MASSARUTTO 1999). Due to the mountainous nature of large parts of Italy, the scope and technical feasibility of internal water transfers are reduced, thereby forcing many regions to rely exclusively on their own resources, or to incur very high costs for long(er) distance water transfers (MASSARUTTO 1999). Water scarcity phenomena are common in several Italian regions with about 12 percent of the total population, mainly in the South and the islands, affected by discontinuity of supply (BAZZANI ET AL. 2005). In southern Italy water uses are highly interdependent, since they often rely on the same large water storage and transfer schemes, which are operated by state-owned organisations and/or irrigation boards (MASSARUTTO 1999).

Historically, municipalities were responsible for the provision of water supply and wastewater services, but the Galli law in 1994 introduced a new organisation through the creation of optimal territorial areas (ATO – *Ambito Territoriale Ottimale*) with new authorities (AATO – *Autorità di ATO*) under the responsibility of the regions (BREUIL ET AL. 2004). 91 ATOs were defined by the regions through regional legislation. Although the law mandates ATO boundaries to reflect hydrological features and homogeneity of geographical conditions, they were rather delineated in line with administrative boundaries, e.g. the district level (MASSARUTTO AND PACCAGNAN 2007). Through this process municipal management is gradually being abandoned in favour of bigger territorial units, as ATO should be served by one operator.

Italy is a parliamentary republic, which is divided into 20 regions, 100 provinces and about 8,000 municipalities with an average of approximately 6,000 inhabitants (MASSARUTTO 1999). The territory of Italy has a surface of 301.338 km^2 (EUROSTAT 2007) with a population of about 60 million inhabitants (EUROSTAT 2009) resulting in a population density of 199 inhabitants per km^2. The country is divided into the urbanised northern regions and the mostly rural south. The more prosperous North contains most of Italy's larger cities and about two-thirds of the country's population. The primarily agricultural south has a smaller population base and a more limited economy (BREUIL ET AL. 2004). Generally, there is a relatively low concentration of population in great metropolitan areas and a great number of small and medium towns, with a tradition of autonomy and self-government (MASSARUTTO 1999).

Using about 50 percent of available water resources, agriculture constitutes by far the largest water use in Italy, due mainly to irrigation (BARTOLINI ET AL. 2007, INEA 2005). This figure is even higher in the south of Italy (56.3 percent) and on the islands (63.6 percent) (GORIA AND LUGARESI 2002). Households and industry each account for about 20 percent of total water resources use and energy constitutes about 11 percent (MASSARUTTO 1999).[293] Average per capita water consumption amounts to 197 litres per day, if only household uses are considered and 301 litres per day if all uses supplied by public networks are taken into account.[294] If total abstractions for public supply are considered, average per capita consumption amounts to 390 litres per day. (MASSARUTTO 1999)

The largest fraction of water used by agriculture (around 66 percent) is surface water abstracted from rivers (BAZZANI ET AL 2005). Only about 10–15 percent of irrigation water stems from groundwater sources, mainly in mountain and hilly areas, where irrigation is often practiced during the winter in order to prevent damage from hard frost. In the south and on the islands on the contrary, groundwater is used intensively during the summer, either in mainland areas or along the coastal plains. Overall, due to the strong reliance on surface water for irrigation, abstraction of groundwater by the agricultural sector constitutes

293 Updated figures could not be accessed.

294 As MASSARUTTO (1999) outlines a high level of leakage should be accounted for given the bad maintenance of pipelines and/or bad monitoring of effective consumption.

only in selected regions a significant pressure on water status giving rise to unsustainable patterns of groundwater use (MASSARUTTO 1999).

Irrigation water service provision is generally managed collectively, through farmers associations with public status (Reclamation and Irrigation Boards – RIB). They control the management and distribution of water resources for agricultural purposes over a certain area and often manage land drainage, reclamation and flood defence services as well (BAZZANI ET AL. 2002, BARTOLINI ET AL. 2007). In a few minor cases, they also contribute to water delivery for non-agricultural purposes (BARTOLINI ET AL. 2007). Water use regulation is based on a complex system of locally specific property rights. RIB have undergone a process of concentration, and there are now nearly 200 nationwide, supplying roughly 90 percent of irrigation water demand (MASSARUTTO 1999).

In 2000, irrigated agriculture in Italy covered about 29 percent of usable farmland and 40 percent of farms (BARTOLINI ET AL. 2007). The regions that have the highest share of irrigable land on the agricultural land are: Lombardia (54 percent), Veneto (40 percent), Emilia Romagna (1.581.285 ha or 32 percent), Piemonte (1.539.078 ha or 29 percent) and Puglia (20 percent) (WADI 2006). While irrigation is most common in the north, due to higher water availability, it tends to be a crucial issue in the south, where traditionally important productions may be cultivated only where irrigation is possible (e.g. citrus, vegetables) (BARTOLINI ET AL. 2007). The share of irrigated area in total agricultural area differs greatly between regions, ranging from 8.2 percent of Marche to 66.2 percent of Lombardia (ISTAT 2006). For some crops, (e.g. orchards, vegetables, flowers) the irrigated area is virtually 100 percent of the total cultivated area. 60 percent of Italian agricultural export is made up of irrigated crops (LAMOGLIE 2001 in BAZZANI ET AL. 2002).

While the total irrigation area has increased over time, the use of water in agriculture decreased in relative and absolute terms during the 1990s, due to the concurrent effects of improved irrigation technology and the abandonment of demanding irrigation practices (e.g. marshland), shortages of water during this period (with frequent droughts during the summer, often leading to a halt in winter distribution), the move towards more extensive crop patterns fostered by the Common Agricultural Policy (CAP) and the economic difficulties of irrigated agriculture (BARTOLINI ET AL. 2007, MASSARUTTO 1999).

According to BARTOLINI ET AL. (2007), water pricing for agriculture usually applies area-based tariffs aimed to recover the operational costs of RIB. Investment costs are generally borne by the state and not considered in tarification schemes, along with environmental and resource costs. Hence, the provision of water services is generally characterised by a large gap between actual tariffs and the cost of water services provision (BAZZANI ET AL. 2005). In some regions, where pressure pipe distribution systems allow for metering water demand, volumetric pricing of irrigation water is practised (BARTOLINI ET AL. 2007, BAZZANI ET AL. 2002). Water saving is usually not promoted as a water policy objective through pricing schemes. As BARTOLINI ET AL. (2007) note, water policy has traditionally been a key promoter of irrigated agriculture, through the provision of subsidised water due to state funding of water storage and transportation infrastructure (cf. Section 4.3.2 below).

In Italy, irrigation impacts the environment in a myriads of ways (MASSARUTTO 1999). It affects river outflow patterns and the management of lakes and reservoirs leading to environmental problems linked to low water flows. Groundwater resources are affected through diffuse pollution arising from intensive agriculture and disposal of livestock waste, imposing additional costs on drinking water supply (i.e. Type II EC for water supply services). Many areas in Italy are vulnerable to nutrient pollution, which also stems from inadequate sewerage treatment, affecting large lakes and the upper Adriatic Sea in particular (MASSARUTTO 1999).[295] Furthermore, soil contamination due to landfills and abandoned industrial sites as well as rainwater contamination constitute important sources of diffuse pollution.

4.2.2.2 Emilia-Romagna (Consorzio di Bonifica della Romagna Occidentale): irrigation water service provision

The region Emilia-Romagna, which is located in northern Italy, is delineated in the north by the river Po, in the east by the Adriatic Sea and in the south by the Apennin and the Republic San Marino. It has a surface of 22.123 km² and about 4.17 million inhabitants. The region consists of the provinces Parma, Piacenza, Modena, Reggio nell'Emilia, Bologna, Ferrara, Ravenna, Forlì-Cesena and Rimini. Emilia-Romagna is one of the wealthiest regions in Italy. About 19 percent of the national value of vegetables is cultivated in the region and an important share of Italian fruit tree (27.5 percent) and sugar beet (21.6 percent) cultivation takes place in Emilia-Romagna (ISTAT 2005).

The *Consorzio di Bonifica della Romagna Occidentale* (CBRO) is located in the eastern part of Emilia-Romagna in Lugo[296], in the province of Ravenna. It manages an area of about 195,000 hectare, encompassing Sillaro in the west,

---

295 While important efforts have been undertaken from the 1976 onwards to provide adequate sewerage treatment, these activities are still lacking; a significant percentage of wastewater in Italy is still discharged untreated.

296 Lugo has about 31,939 inhabitants (as of 31 January 2006) and is located in the northwest of the productive and fertile lowland between the rivers Santerno and Senia.

Lamone in the east, Reno in the north and the water divide in the south. The CBRO extends over parts of five provinces, namely Ravenna (largest part), Bologna, Forlì-Cesena, Ferrara, Firenze and 35 municipalities (cf. Figure 21 below). The region is part of a river basin with very low water availability and thus diverts water from a large canal, the *Canale Emiliano Romagnolo*, originating in the Po River. Cropping patterns consist of annual crops, such as cereals, sugar beet, seed crops, vegetables, and fruits such as peach, nectarine, kiwi, apricot, pears and wine grapes (VIAGGI 2003, ISTAT 2006).

Figure 21: The case study area of the Consortio di Bonifica della Romagna Occidentale

Source: CBRO (2009).

The CBRO is divided into a lowland area and a hillside and mountain area. In the lowland area, mainly available water resources from the *Canale Emiliano Romagnolo* (C.E.R.) are used for irrigation purposes. The C.E.R. constitutes one of the most important Italian water engineering structures, which derives 68 m^3/s of water from the river Po, has a length of 150 km and covers an area of

3000 km^2.[297] It is used for transporting surface water into a region, which is characterized by a water intensive agricultural land use and dispersed settlements and industries. Through the C.E.R. an overall additional surface of 389,000 hectare is irrigated in a very productive part of the Po valley, where climate conditions are particularly favourable due to the proximity to the sea (BAZZANI 2005). The provision of water through the C.E.R. aims to mitigate sinking groundwater tables by distributing surface water to areas, where it would not be available otherwise as a substitute for groundwater usage.

The mountainous regions of the CBRO account for about 120,000 hectare. In these areas, the redirection of water for irrigation is based on the natural hydrografic network. The region is characterized by extensive fruit cultivation, which constitutes an important economic production sector, and low precipitation during summer months. Specific measures are required for maintaining soil stability and for protecting hillsides from erosion and landslides. To maintain an optimal water flow, the natural flow velocity is generally influenced through locally adapted construction measures.

The rural water engineering structures which are supported by numerous springs, serve about 2000 scattered farms. These are located in areas where the irrigation network is normally not extended due to prohibitively high costs. The CBRO undertakes the preservation of infrastructure in cooperation with local administrations. Flood irrigation is most efficient in these areas and in line with the demands of agricultural water use in this region (CBRO 2009).

All *Bonifica* are organized as public entities, with an association structure representing the relevant sectoral actors. They act according to the principles of organizational autonomy (i.e. local regulations and statues are decided on independently) and financial self-autonomy (i.e. actors pay for their membership (cf. Section 4.3.2 below) and the *Consorti di Bonifica* decides autonomously on the level of charges and membership fees). A distinct feature of the *Consorti di Bonifica* is the delineation of their territories in line with hydro-geographical instead of administrative boundaries. It allows the *Consorti di Bonifica* an integrated management within its territory, facilitates an adapted water management approach and is well in line with WFD provisions.

As with all *Bonifica*, the CBRO is organized as a public law corporation (CBRO 2009). The legal background for the CBRO and its underlying regulations is pro-

297 The C.E.R. is fuelled by water from the Po River via the *Cavo Napoleonico*. The *Cavo Napoleonico* was erected in the 19th century and fulfils a double-function: next to providing the C.E.R. with water from the Po River, it aims to regulate water flows and to manage floods of the Reno River.

vided by the national law RD 13/02/1933 n. 215 and, at regional level, the LR Emilia-Romagna 2/08/1984 n. 42.

The CBRO is responsible for the planning, organisation and maintenance of construction work (including e.g. the construction and maintenance of water works and canals; the lowering of existing canals; safeguarding and maintaining aqueducts in the mountain areas) and resources management (including e.g. water diversions and flow regulations; the protection of river/canal sides; the consolidation of river beds; the control of landslides) as well as for the operation of the *Bonifica* itself (e.g. setting tariffs, deciding upon membership, setting rules and regulations). These functions are laid down in the statutes of the CBRO. Overall, *Bonifica* are allowed to set tariffs, to pass regulations for their territories as well as to penalize internally the neglect of its regulations (disciplinary actions).[298] Contributions to the CBRO are raised from all its members. The level of the contribution is determined by the degree to which a member benefits from the *Bonfica*, which is classified through technical and economic indices (cf. Section 4.3.2).

In the lowland areas of the CBRO, the work of the Bonifica is usually commissioned and financed by the State, while in the mountainous areas, with their multitude of small aqueducts and constructions, maintenance work is often directly implemented by the personnel of the *Bonifica* (unless this is considered economically inefficient) (CBRO 2009). The irrigation period within the CBRO normally starts in March and extends through October or November. The lowland parts of the CBRO are largely served with irrigation water from the C.E.R. The distribution of irrigation water from the C.E.R. is conducted through two different systems, either though open communal canals owned and operated by the CBRO or through pressure pipes.

Irrigation through pressure pipes relies on pumping works and a network of pipelines for the extraction and distribution of water, including hydrants for connecting individual farms to the system. Lowland areas, which are located above the C.E.R. at a higher altitude, can only be served at a higher water pressure. Hence, this irrigation technique is specifically well suited for this region. Nevertheless, two irrigation areas located below the C.E.R. are also served through pressure pipes.

In the territory of the CBRO, irrigation through open canals in particular draws on water from the C.E.R., which is made possible by gravity (at times pumping stations are needed). In order to guarantee agricultural water abstrac-

298 The competence to levy fees and to set tariffs is regulated at national level through Articles 10, 11 and 59 of RD 13/02/1933 n. 215, and at regional level through Article 13 of LR Emilia-Romagna 2/08/1984 n. 42 (CBRO 2009).

tions from the canal at other locations along the canal, minimum water flow regulations apply. Those canals which are fuelled by the C.E.R. often fulfil a double function of drainage and irrigation. There is a combined pipe system of over 400 km. Drainage is mainly conducted as a flood prevention measure. The canal system is not entirely owned by the CBRO, but is partly in the hands of the municipalities. Nevertheless, the management of irrigation and the maintenance and management of the network are under the obligation of the CBRO. Likewise, the management of the rivers *Senio* and *Lamone*, which are fuelled with water from the C.E.R. is not conducted by the CBRO, but by the region. They are also used for irrigation purposes, particularly during dry summer months.

Table 16: General characteristics of irrigation water service provision in Emilia-Romagna

Indicator	
Size of the service area Population Population density	Emilia Romagna: 22,123 km² CBRO: 195,000 ha 4.17 Mio inhabitants (Emilia Romagna) 188.49 inhabitants / km² (Emilia Romagna)
Rural – urban share in investigation area	Lowland – mountain division more relevant in this study case: 38.5% lowland and 61.5% mountain areas.
Length of the pipe system (size of the distribution network)	canal network: 900 km thereof combined canals (irrigation and drainage): 400 km
Number of service providers in the service area	1 (only CBRO considered)

Sources: CBRO (2009), INEA (2005).

In order to gain access to the open irrigation canals, a permit has to be issued by the CBRO. For this permit, information on the irrigation area (e.g. surface, location of abstraction, canal to be abstracted from) has to be provided. One prerequisite for granting a permit is that the irrigation water use by the applicant may not harm other agricultural water users, who are located behind the point of abstraction. If structural changes are required for enabling the desired abstraction, a permit is not sufficient and more extensive procedures apply. Once the permit for water abstraction has been issued, the applicant is required to pay a yearly fee (cf. Section 4.3.2 below). The right for water abstraction is generally prolonged on a yearly basis.[299]

299 Different procedures apply to farms with an irrigation surface above 15 hectare, to which an irrigation permit is only granted on an annual basis.

In the CBRO, the price of water is based on a volumetric basis if irrigation water is delivered in pressure pipes, and charged on an area basis, if it is provided through open canals. A detailed description of the financing scheme and the contribution by users is provided in Section 4.3.2. The following Table 16 provides information on the general characteristics of the case study area Emilia-Romagna.

4.3 Testing the framework: illustrative examples on key steps

Chapter 4.2 introduced three investigation areas to illustrate the methodological assessment framework developed in Chapter 3. In the following sections, key steps of the methodology are exemplified on one illustration case respectively, while insights from other cases on key aspects are added in complementary information boxes. For each step, data and information sources are outlined along with difficulties encountered in the assessment as an input to Chapter 4.4, which summarises *'lessons learned'* from the illustration cases.

4.3.1 Testing Step 1: accounting for financial costs in water service provision

The following paragraphs apply the cost accounting framework of Chapter 3, i.e. Step 1 – accounting for financial costs –to the provision of drinking water services to households in Berlin. The required information for constructing this illustration case was mainly derived from publicly available information sources.[300] Complementary information was gathered through interviews to resolve open issues in applying the methodological framework and to verify interpretations by expert judgement.[301]

---

300 In particular, the following sources were used: the annual reports of the *Berliner Wasserbetriebe* for the years 2008 and 2007; the tariff calculations for 2008 and 2009; the publication *"Grundlagen der Tarifkalkulation – Dokumentation"* (BWB 09/2007), *"Vertragsbestimmungen für die Wasserversorgung von Berlin"* (last adapted on 21 May 2007, operational as of 1 July 2007 (ABl. No. 23 published on 1 June 2007, p.1454)). All of these documents are available free of charge on the website of the *Berliner Wasserbetriebe* (www.bwb.de).

301 Cf. Annex. Complementary information has been compiled through expert interviews with additional data provided by email by the *Berliner Wasserbetriebe* (18 November 2009).

Article 5 of the *Berliner Wassertarifordnung* (the Water Tarification Ordinance for Berlin – WTO), details which costs qualify as financial costs of drinking water service provision, i.e. as eligible costs, and may thus form the basis for tariff calculations (cf. Step 2 of the analysis in Section 4.3.2). It also sets the applicable rules for the calculation of capital costs and the division of costs among different accounting periods.

The annual reports of the *Berliner Wasserbetriebe* and accompanying publications together with the annual tariff calculation provide information on the magnitude of different cost categories along with their mode of calculation. However, data is generally provided for the entire water supply segment and not separately for private households in Berlin (thus excluding other water user groups as industry, agriculture and forestry and the provision of water to surrounding areas of Berlin). However, as no special costs are linked to different customer segments in Berlin, the share of household water consumption in total water consumption can be used for calculating the appropriate cost share for private households for each category.[302] In 2008, the consumption of private households accounted for 72.9 percent of total water consumption (BWB 2009B, cf. Table 15), which is hence used as the adjustment factor.

Operation and maintenance costs

Personnel costs make up for the largest share of operation and maintenance costs.[303] The cost category '*personnel costs*' also includes staff involved in routine preventive maintenance, repairs, and construction for minor rehabilitation. In 2008, total personnel costs for water supply entered the tariff calculation with 122,088 T Euro (BWB 2008B); hence a share of 72.9 percent or 89,002 T Euro can approximately be assumed for the provision of water supply to private households.

Material costs consist of raw materials and supplies for maintenance as well as energy costs, which contribute the largest share (60–70 percent) of this cost category (BWB 2007). In 2008, the cost category '*material costs*' was included in

302 This simplified procedure has been discussed and approved in interviews with experts from the *Berliner Wasserbetriebe*.

303 The tarification contract for provisioning services utilities (*Tarifvertrag für Versorgungsbetriebe (TV-V)*) as of 5 Octobre 2000 in the revised form enacted on 1 February 2005 of the third amended tarification contract of 1 June 2005 forms the basis for personnel costs.

tariff calculation with 25,173 T Euro (BWB 2008B), with 18,351 T Euro (i.e. 72.9 percent) linked to water supply to private households.

Hired and contracted services mainly include costs for contracted out maintenance or rehabilitation services on water works and the pipe network, vehicle maintenance as well as administrative tasks (i.e. there is an overlap with the cost category administrative costs, cf. below). For the tariff calculation in 2008, *'hired and contracted services'* were included in the tariff calculation with 31,409 T Euro for the entire segment of water supply (BWB 2008B), of which 72.9 percent or 22,897 T Euro can be allocated to private households.

Maintenance costs are partly provided by internal personnel, or contracted out to external providers. Hence, the cost category maintenance cost is accounted for in *'personnel costs'*, *'material costs'* and in *'hired and contracted services'*. While no separate data is publicly available on this cost component, the costs (as defined in Section 3.3.2) are included in the overall cost category *'operation and maintenance costs'* and thus accounted for fully.

Similarly, information on the magnitude of the cost component *'monitoring costs'* in *'operation and maintenance costs'* cannot be extracted on the basis of published information. For the provision of water supply services, an important component of monitoring costs as defined in Section 3.2.2 is given by laboratory services. The *Berliner Wasserbetriebe* operate their own laboratory, which performs services for the water supply division as well as for wastewater management. While the costs for each division are accounted for separately, these are not reported as *'monitoring costs'*, but rather integrated into the cost categories *'personnel costs'*, *'material costs'*, *'hired and contracted services'* and *'other direct costs'* as well as in the capital cost categories *'depreciation'* and *'opportunity costs of capital'*. On the basis of additional information provided by the BWB (SCHMITZ 2009, HÖHNDORF 2009), it can be derived from cost centre accounting that laboratory services for the water supply division account for about one percent of total costs of service provision.

Depreciation

The calculation of depreciation for tariff calculation by the *Berliner Wasserbetriebe* is based on the replacement value of existing assets[304], which is determined on a yearly basis for all infrastructure assets. To this end, an appropriate basket of goods is constructed with 21 standardised indices, and is then used

---

304 This was changed in the revised version of the law on Part Privatisation of 14 December 2003. Beforehand, depreciation was based on historic (i.e. acquisition or production) costs.

for transforming historic costs to their replacement costs in the future (BWB 2007).

Depreciation is calculated on the basis of the *economic lifetime of assets.*[305] Quality improvements or technical change are not taken into consideration on a yearly basis, but through updates of the composition of the basket of goods. Depreciation is calculated on a linear basis and takes inflation into account.

Investment subsidies and allowances as well as investment contributions by users (cf. Section 4.3.2) are not subtracted from the depreciation base. As discussed in Section 3.2.2.2, this procedure allows for accumulating sufficient financial means for reinvestment. The contributions are booked as a liability and linearly released in line with the economic lifetime of the respective assets.

For the 2008 tariff calculation, depreciation was projected with 101,585 T Euro for the water supply company division (BWB 2008B), of which 72.9 percent or about 74,056 T Euro can be accounted for the provision of water supply services to private households.

Opportunity costs of capital (imputed interest)

Opportunity costs of capital are calculated for the necessary operating capital for service provision (according to § 1 (3) and the Annex to § 2 of the Water Tarification Ordinance for Berlin – WTO). The necessary operating capital is composed of operating assets reduced by retained earnings. Retained earnings are composed of liabilities, which have been provided to the company without the calculation of interest (including e.g. investment contributions by users). The appropriate balance sheet items are to be considered on the basis of average bounded assets for the current accounting year. The resulting relevant asset base forms the basis for calculating the opportunity costs of capital. For 2008, an interest rate of 7.3 percent was applied. The applicable interest rate is determined on a yearly basis by the *Land* Berlin (*Verordnungszinssatz*) and applied both to equity and borrowed capital.[306] According to the tariff calculation for 2008, opportunity cost of capital sum up to 84,014 T Euro (BWB 2008B), of which a share of 61,246 T Euro (72.9 percent) can be attributed to service provision to private households.

305 The usual time of operation applied to assets is as follows: buildings (25–50 years), pipe system (30–40 years), household connections (25 years), water meters (15 years), vehicles (6–10 years), IT-Hardware (3–10 years) (BWB 2007).

306 The minimum level for the applicable interest rate, the so-called '*Verordnungszinssatz*' is calculated on the basis of the average returns on ten-year Federal Government Bonds over a period of 20 years preceding the calculation period.

Administrative costs

On the basis of published information, administrative costs cannot be directly determined in line with the methodological framework. As indicated above, there is an overlap with the cost categories *'operation and maintenance costs'* and *'other direct costs'*. This is due to the fact that administrative costs are linked to nearly all cost categories, starting from personnel costs to the depreciation of computers, other direct costs in the form of consultancy services or travel costs of administrative units, rent for administrative office space, etc. As indicated by expert interviews, the delineation of administrative costs also may not be straightforward in every case. For example, technical personnel may also perform administrative tasks. Accordingly, expert judgment is needed for assessing the magnitude of administrative costs and their relevance.

While administrative costs cannot be strictly separated, they are nevertheless included in other financial cost components and thus do not qualify as unaccounted costs. Therefore, overall *'financial costs'* are complete. Furthermore, all of the (overlapping) cost categories form part of financial costs; hence there is no effect on the combined assessments in Step 3 and Step 6.

According to the additional data provided by the BWB through expert interviews, administrative costs amount to approximately 9 percent of total costs for water supply services.

In addition to the administrative costs encountered in service provision at the *Berliner Wasserbetriebe*, the definition of this cost category (cf. Sections 2.2.2.1 and 3.2.2.3) also extends to the administrative costs at the level of the *Land* Berlin, e.g. related to its control over price setting. These costs are however partly already accounted for in administrative costs of the *Berliner Wasserbetriebe* outlined above, as e.g. compensation is paid to representatives from the administration of Berlin for attending meetings of the advisory board or the assembly of guarantors. Nevertheless, additional costs by the administration of Berlin are not included in the assessment and thus not accounted for. As a share of overall costs, they can however be expected to be negligible.

Other direct costs

The category *'other direct costs'* includes rent, insurance payments, office supplies, telecommunication, marketing activities, travel costs and consultancy services. Additionally, it comprises concession fees to the municipality and payments made to the *Land* Berlin in return for using public roads to lay and maintain the pipe network (*Sondernutzungsentgelt*). Since 2004, this has to be paid to the *Land* Berlin for about 6,809 km of the total water supply network

(BWB 2007). The annual fee amounts to 1.02 Euro/m of pipe network. Furthermore, an additional fee has to be paid to the *Land* Berlin for using public roads for construction sites, i.e. when building work is undertaken on the pipe network (*Straßennutzungsentgelt*). This fee equals 0.51 Euro/m² and depends on the size of the construction area and the duration of construction work (BWB 2007). In 2008, other direct costs entered the tariff calculation for water supply services with 41,048 T Euro (BWB 2008B), of which a share of 29,554 T Euro can be allocated to the household sector.

Other aspects relevant for the analysis / difficulties in the assessment

The following aspects may warrant further attention in an integrated case study analysis over all of the assessment steps, and particularly the analysis of environmental and resource costs in Step 4 and Step 5 respectively:

- According to expert judgement, no resource protection costs are included in the cost categories above that may qualify as internalized Type I EC (SCHMITZ 2009) that would have to be extracted from financial costs for the analysis.
- While costs are encountered for the operation of wells for preventing contaminations on one water work, the costs for their construction and operation is directly charged to the *Land* Berlin. This is because the environmental costs are not caused by the provision of the water service itself, but rather relate to historic environmental damages.
- Equally, the costs of groundwater management services provided to the *Land* Berlin are accounted for separately and charged directly, thereby avoiding an overlap with the cost categories above (SIMON 2009).
- The same applies to the treatment of surface water, with the aim of improving water quality in the case of the water work Tegel. The lakes *Tegeler See* and *Schlachtensee* were successfully cleaned up by means of drastic reductions in the external phosphorus levels (BWB 2007B).
- As an internalisation instrument of Type I EC, the *Berliner Wasserbetriebe* are liable to pay a water abstraction charge of 0.31 Euro/m³ for each cubic meter of water abstracted for the purpose of water supply (i.e. groundwater management abstractions are excluded, cf. discussions in Section 4.3.6.1). This abstraction charge is also not included in the cost categories above, but rather listed separately and paid to the *Land* Berlin.
- According to expert judgement, Type II EC, i.e. environmental costs caused by other water users leading to higher costs of service provision, are presently not encountered in the provision of water supply services to private households in Berlin (NOBIS 2009, SCHMITZ 2009). Specific raw water contaminations, e.g. from agriculture or industry, are not present. In the future, sul-

phate stemming from mining activities in the region of Lusatia may, however, lead to additional costs in drinking water provision (HORLACHER 2009), unless preventive measures are undertaken.

- The operational costs of the side service fire protection are not included in the cost categories above. They are accounted for separately and recovered directly from the *Land* Berlin as well as from fire protection services (cf. Section 4.3.2 below).
- The *Berliner Wasserbetriebe* operate about 230 km² of water protection zones for the provision of drinking water for Berlin (cf. Figure 22 below). One of its waterworks is located in the *Land* Brandenburg, but equally used for supplying water to the city of Berlin (cf. Section 4.3.6.1). According to SCHMITZ (2009), no specific investments are undertaken in these water protection zones, which could qualify as avoided purification costs.

Figure 22: Water protection zones operated by the Berliner Wasserbetriebe

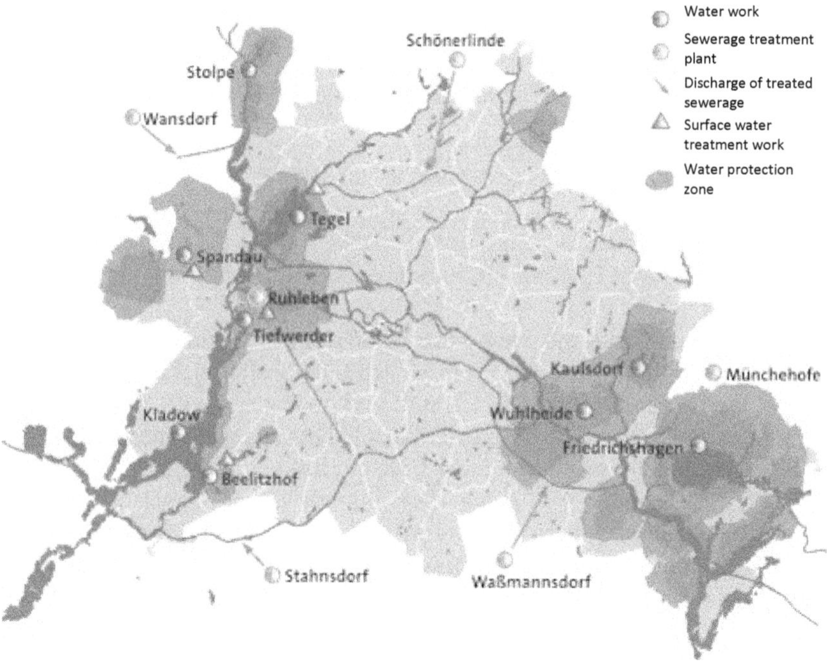

Source: www.bwb.de (visited: 18 June 2009).

Table 17 provides an overview on the magnitude of the individual cost categories of financial costs of drinking water service provision in Berlin.

Table 17: Overview on cost categories for drinking water service provision in Berlin

Cost category	Cost allocation to private households (in T Euro)	%	Total costs for water supply section (in T Euro)	%
Operation and maintenance costs	130,250	44.08	178,670	44.08
Personnel costs	89,002	30.12	122,088	30.12
Material costs	18,351	6.21	25,173	6.21
Hired and contracted services	22,897	7.75	31,409	7.75
Maintenance costs	(not separated, cf. *Fact Sheet* below)		(not separated, cf. *Fact Sheet* below)	
Monitoring costs	(not separated, cf. *Fact Sheet* below)		(not separated, cf. Fact Sheet below)	
Capital costs	135,302	45.80	185,599	45.80
Depreciation	74,056	25.06	101,585	25.06
Opportunity costs of capital	61,246	20.74	84,014	20.74
Administrative costs	(not separated, cf. *Fact Sheet* below)		(not separated, cf. *Fact Sheet* below)	
Other direct costs	29,924	10.12	41,048	10.12
Total	295,476	100.00	405,317	100.00

Note: The percentage calculation is based on the reported cost categories. The tariff calculation of the *Berliner Wasserbetriebe* in addition includes income taxes, other taxes, the groundwater abstraction charge, as well as imputed risks and intercompany sales.

It should be noted that the water abstraction charge for Berlin (included with 52,780 T Euro in the tariff calculation for 2008, cf. Section 4.3.6.1) as well as taxes (i.e. income taxes with 15,521 T Euro and other taxes with 1,483 T Euro) have been excluded from the assessment in line with the guidelines of Chapter 3. Likewise, imputed risks (4,627 T Euro) have not been considered.[307]

The following *Fact Sheet* 11 summarises the analysis of financial costs, and outlines difficulties encountered in the assessment along with their relevance for the overall interpretation of results and the applicability of the accounting framework.

307 The *Berliner Wasserbetriebe* assess the costs of imputed risk (in line with § 1 (3) WTO). These include the risk of assets (based on actual losses), the risk of reduction in material value and losses and the risk of non-payment by customers, which is calculated on the basis of information for previous years.

Fact Sheet 11: Accounting for financial costs (Berlin)

	Operation and maintenance costs	**Depreciation**	**Opportunity costs of capital**	**Administrative costs**	**Other direct costs**
Data requirements	*Personnel costs*: 87,903 T Euro *Material costs*, including energy costs: 18,124 T Euro Hired and contracted services: 22,614 T Euro *Maintenance costs*: included in personnel costs, material costs, hired and contracted services *Monitoring costs*: included in other components of this cost category; account for approximately 1% of total costs of water service provision (expert information)	73,141 T Euro Replacement value of existing assets Linear depreciation basis Economic lifetime of individual assets considered Inflation considered; Money available at the time of reinvestment There are no direct contributions from the public budget (no subtraction from depreciation base required) Contributions from users are linearly dissolved in line with usage of infrastructural assets	60,490 T Euro Necessary asset base Share of equity: ~30 % Rate of return applied: *'Verordnungszinssatz'* determined by the *Land* Berlin and applied to the entire necessary asset base; in 2008: 7.3%	Account for about 9% of total costs of water service provision (expert information) Administrative costs at the level of the *Land* Berlin not included.	29,554 T Euro Includes concession fees to the *Land* Berlin.
Fixed or variable (if both: approximate % division)	Estimated: fixed 80%, variable 20%.	Fixed.	Fixed.	Partly fixed (e.g. personnel), partly variable (travel costs, consultancy services, etc.).	Partly fixed, partly variable.

	Operation and maintenance costs	Depreciation	Opportunity costs of capital	Administrative costs	Other direct costs
Quality of data	Good, updated annually. Expert judgement required for extracting information on potential overlapping cost categories. A1 – A2	Good, updated annually. A1	Good, updated annually. A1	Publications: not accounted for separately. Expert interviews required for assessing dimension of administrative cost. A2	Sufficient. Accounted for separately, but mixture of different elements.
Scale of data availability and periodicity of update	Service provision area, entire water supply service, i.e. not differentiated for private households.	Service provision area, entire water supply service, i.e. not differentiated for private households.	Service provision area, entire water supply service, i.e. not differentiated for private households.	Service provision area, entire water supply service, i.e. not differentiated for private households.	Service provision area, entire water supply service, i.e. not differentiated for private households.
Confidentiality	Data mostly available from published reports, no confidentiality concerns.	Partly problematic, not for aggregated data.	Partly problematic, not for aggregated data.	Not relevant.	Not relevant.

	Operation and maintenance costs	**Depreciation**	**Opportunity costs of capital**	**Administrative costs**	**Other direct costs**
Applicability of proposed accounting guidelines in line with Article 9 economic cost approach	Good. No overlap with other side-services. Type II EC are not encountered, therefore not hidden in this account. Attribution of relevant cost share to the household sector approximated by household share in total water delivered (procedure approved by expert judgement). Difficulties related to the separation of '*administrative costs*'.	Accounting guidelines fully applied. Attribution of relevant cost share to the household sector approximated by household share in total water delivered (procedure approved by expert judgement.)	Accounting Guidelines fully applied. Entire necessary asset base addressed to ensure preservation of substance. Attribution of relevant cost share to the household sector approximated by household share in total water delivered (procedure approved by expert judgement.)	Could not be extracted from existing accounts; expert judgement needed. Administrative costs at the level of the *Land* Berlin not included.	Attribution of relevant cost share to the household sector approximated by household share in total water delivered (procedure approved by expert judgement.)
Influence of divergence from accounting rule [depends upon magnitude of cost category in total costs]	Divergence not significant as it mainly affects inter-cost category allocations between cost factors. Exception: administrative costs, minor divergence.	No divergence.	No divergence.	Low due to minor share in overall costs.	Low due to minor share in overall costs.
Influencing factors / driving forces for cost category	Development in water demand.	Demographic change and reductions in water demand may lead to stranded assets in future.	Development in '*Verordnungszinssatz*'.	(not relevant)	(not relevant)

	Operation and maintenance costs	**Depreciation**	**Opportunity costs of capital**	**Administrative costs**	**Other direct costs**
Difficulties encountered in the assessment	Maintenance costs included in other cost components of this category – no mix-up between cost categories. Monitoring costs cannot be determined from published accounts (expert judgment needed), but included in this cost category – no confusion between cost categories. Administrative costs partly included in this category; separated and accounted for by expert judgment.	(not relevant)	(not relevant)	Overlap with '*operation and maintenance costs*' and '*other direct costs*'. Information on administrative costs cannot be extracted from published material. Expert judgement required. Administrative costs at the level of the *Land* Berlin not included.	Overlap with '*administrative cost*' category. Regulatory costs of the *Land* Berlin could only in part be accounted for.
Evaluation Full cost accounting and attribution to cost categories – *Expert judgement* –	(1a) All costs have been accounted for (2b) Minor divergences and mixture of cost components (only related to financial cost categories, not extending to Type I /II EC or resource costs).	(1a) All costs have been accounted for. (2a) full cost attribution achieved on the basis of existing data.	(1a) All costs have been accounted for. (2a) full cost attribution achieved on the basis of existing data.	(1b) not all costs could be accounted for (administrative costs of the *Land* Berlin excluded) (2c) mix-up of cost components and categories, but all within financial costs; no impact on Type I CR assessment	(unclear due to conglomerate nature of the cost category)

Extending the information base: Emilia-Romagna – irrigation water services

In the case of irrigation water provision in the region Emilia-Romagna and more specifically in the CBRO, service provisioning costs differ for the source of water used, i.e. whether irrigation water is abstracted from open canals or from the C.E.R., or if it is provided to farmers through pressure pipes. Most of the infrastructural investment has been commissioned and financed through public budgets (CBRO 2009). Hence, cost accounts do not contain the entire cost information for retaining and financing the provision of irrigation water services in the long-run. Nevertheless, as the analysis in Section 4.3.2 documents, a differentiated financing scheme is operated, taking account of the type of irrigation service provided (pressure pipes or open canals) and the irrigated (or cadastre) area.

Particular difficulties in applying the cost accounting framework relate to the fact that the CBRO (and Reclamation and Irrigation Boards in general) do not only provide irrigation water services, but also manage land drainage and flood defence services (CBRO 2009, BARTOLINI ET AL. 2007). Hence, the costs for these services have to be subtracted for the assessment, e.g. aided by expert judgment on their respective magnitude and contribution to total costs.

4.3.2 Testing Step 2: accounting for revenues in water service provision

In order to complement the structured analysis of financial costs in *Step 1*, an analysis of revenue generating mechanisms (*Step 2*) is undertaken in the following sections. Together, *Step 1* and *Step 2* then allow for assessing Type I cost-recovery and the financing structure of water service provision. As a reminder, Article 9 aims at a strong internal cost-recovery approach, i.e. a financing of water services through user/polluter contributions. Hence, the investigation of revenue streams provides the necessary complementary information for assessing divergences from the strict internal cost-recovery approach intended by Article 9. The revenue accounting framework focuses on ensuring transparency with regard to different revenue streams by documenting who contributes to which revenue stream; and by which mechanism the level of contribution is determined. Furthermore, the revenue assessment sets part of the required information base for evaluating the adequacy of contribution of water uses and the implementation of the polluter-pays principle (*objective 3*). In addition, basic information on the incentive properties of revenue generation mechanism is compiled. Thus, the following analysis contributes to the later assessment of

Article 9 *objective 2* (i.e. the adequacy of incentives for efficient resource use), which is also conducted for the illustration case Berlin (cf. Section 4.3.6.1).

The following sections provide an application of the revenue accounting framework from Section 3.2.3 to the illustration case of drinking water supply to private households in Berlin. In addition, the information base is extended by an illustrative application on the region Emilia-Romagna.

The case of Berlin – drinking water provision to private households

General background on water tarification scheme

The tarification scheme in Berlin has undergone marked changes since the part-privatisation in 1999. The last important change was implemented in July 2007, with the introduction of a two-part tarification scheme.[308] Before, the tarification scheme consisted of a variable component depending on water use and a connection fee. But as reductions in water demand had to be compensated with higher variable prices on a yearly basis in order to achieve cost-recovery (cf. Chapter 2.2), this pricing scheme was under severe public criticism and impeded the objective of a stable pricing scheme. Hence, a fixed price component was introduced in 2007, which lead to a stabilisation of water prices.

Since the change in legislation in 2000 the *Berliner Betriebegesetz* (BerlBG) and the Water Tarification Ordinance for Berlin (*Wassertarifordnung – WTO*) provide distinct guidelines for the calculation of water prices, including a definition of the costs that are to be included and recovered from users through endogenous revenue sources, i.e. prices and charges. Tariff setting procedures involve several actors and external auditors. First, the tariff calculation presented by the *Berliner Wasserbetriebe* for the upcoming year, which draws on cost estimates and expected water sales, is subject to an investigation by a certified accountant appointed by the supervisory board (and approved by the Berlin Senate). Once the accountant has testified that tariffs are in line with the equivalence principle and adhere to the equal treatment requirement, the advisory board validates the tariffs. Subsequently, the responsible control authority (*Genehmigungsbehörde*) has to approve the proposed tarification scheme.

308 The *Berliner Wasserbetriebe* had tried before to introduce a two-part pricing scheme, consisting of a fixed and a variable tariff component in 2005. But the proposal was rejected by the Berlin Senate. The proposal was rejected, as it was considered that the advantages of a change in pricing system would primarily accrue to enterprises with high water consumption, while the disadvantage would hit owners of one-family houses. The completely variable tarification scheme had major drawbacks in terms of achieving cost-recovery objectives, as outlined in Chapter 2 on a general basis.

In July 2006 the BerlBG was amended with detailed regulations regarding cost over or under-compensation. Tariffs are calculated on the basis of projections of costs and water sales. According to § 16 (6) of the revised BerlBG (as of 28 July 2006), divergences from these projections, leading to a failure in cost-recovery – either through prices that exceeded actual costs thus leading to a cost overcompensation, or that underestimated costs leading to a failure to reach cost-recovery – have to be rectified in the upcoming two calendar years.

Contributions from users: endogenous revenue components

The following four tariff components are levied on households for the connection to and the provision of drinking water:

- A variable tariff component depending on water demand;
- A fixed tariff component covering a small share of the fixed costs of service provision;
- A connection fee, accounting for the costs of connecting a household to the public water supply network; and
- Investment contributions for the exchange of house connection pipes.

These tariff components are now described in greater detail.

Variable tariff component

The variable tariff component is levied on each cubic meter of water consumed per connection. It is the same for all customer groups. In 2008, the variable water price amounted to 2.071 Euro/m³ (excluding 7 percent VAT). However, it already contained the water abstraction charge payable by the *Berliner Wasserbetriebe* for each cubic meter of water abstracted for public water supply purposes (cf. Section 4.3.6.1), which amounts to 0.31 Euro/m³ and qualifies in this accounting framework as an internalisation instrument for Type I EC.[309] Hence, the variable tariff component to be considered for the analysis only amounted to 1.761 Euro/m³.[310]

For determining the appropriate level of the variable tariff component, a projection of water sales is made on an annual basis, drawing on present popula-

309 Its usefulness as an internalization instrument would be assessed in Step 4 and Step 9 of the analysis.

310 As outlined in greater detail in Section 4.3.6.1, the *Berliner Wasserbetriebe* abstract 20.7 million m³ of water in the Land Brandenburg, in which a lower water abstraction charge of 0.102 Euro/m³ is operational. For reasons of simplification, these are not considered here. Hence the actual variable tariff component is slightly higher.

tion figures and present per capita water use. The water demand by industry from the preceding year is added. This basic water demand projection is then coupled with elasticity considerations for Berlin to get the final projection of water demand for tariff setting.[311]

Detailed information on the revenue provided by each tariff component is not published. Furthermore, no differentiation for households, industry and surrounding areas can be derived directly from the information that is publicly available. Therefore, additional information was requested through expert interviews.

In 2008, about 140.7 million m³ were provided to private households (cf. Table 15). According to data provided by the BWB, total revenue of 301.9 million Euro was generated from private households through water sales (excluding investment contributions from users, which make up for a comparatively small share of overall revenues).[312] Households provided a share of 73.9 percent of total revenue from water sales in 2008.

To further explore the share of the variable component in total revenues generated by water sales to households the following estimation can be applied: about 140.7 million m³ of water were provided to private households in 2008. Considering a variable tariff of 1.761 Euro/m³, about 247.7 million Euro were generated by the variable tariff component from private households.

Fixed tariff component

The fixed tariff component is intended to cover part of the fixed costs linked to water service provision, i.e. the costs that are incurred by the service provider independent of the actual water demand by its customers. It is calculated per connection and day and is dependent upon the size of the house meter (QN) and the amount of water provided on a yearly basis to this connection. The smallest meter size is QN 2.5, which applies to about 70 percent of the customers of the *Berliner Wasserbetriebe*. To accommodate the very different customer profiles, ranging from a single household to buildings with multiple households and small businesses, and allow for an adequate cost attribution, the fixed tariff component for the meter size QN 2.5 is again divided into five levels, determined by water consumption. Overall, the fixed tariff component increases linearly with meter size. For the next meter size QN 6 a differentiation into two different water consumption levels is made (see Table 18 below).

---

311 According to SCHMITZ (2009), a price elasticity of 5 percent is assumed.

312 Please note that this piece of information has been provided by the consulted experts (HÖHNDORF 2009, SCHMITZ 2009) for the validation of the methodological framework. Due to confidentiality concerns, a clearance of these figures for future publication is still pending.

Table 18: Fixed tariff component (Berlin)

Meter size (QN)	Yearly water demand	Fixed price per day (incl. 7% VAT)
QN 2.5	0 – 100 m³	0.032 €
	101 – 200 m³	0.043 €
	201 – 400 m³	0.071 €
	401 – 1.000 m³	0.141 €
	> 1000 m³	0.214 €
QN 6	0 – 400 m³	0.342 €
	> 400 m³	0.514 €
QN 10	–	0.856 €
QN 15	–	1.284 €
QN 40	–	3.424 €
QN 60	–	5.136 €
QN 150	–	12.840 €

Source: Berliner Wasserbetriebe, www.bwb.de.

In 2008, the fixed tariff component accounted for about 3.5 percent of total revenues for water supply (data provided by the *Berliner Wasserbetriebe* through expert interviews). Individual data on the contribution by households to the fixed tariff component is not available. Nevertheless, households will mainly be concentrated in the smaller meter sizes thus contributing less per connection to this revenue component.

Connection fee

In Berlin, each estate has its own connection to the water supply network and a house meter. Overall, there are 262,000 house connections in Berlin. Furthermore, many apartment houses have separate meters installed for each apartment.[313] The house connection pipe, i.e. the part of the pipe network that is

313 These are installed by request of the house owner and used for distributing the costs of water supply among different flats within the apartment house. However, as tenants are not in a direct contractual arrangement with the *Berliner Wasserbetriebe* these water meters are not registered there. Hence, no figures on the number of water meters in apartments are available from the *Berliner Wasserbetriebe*. There is no legal obligation in Berlin for house owners to distribute the costs of water supply in line with actual water use and thus to meter consumption per apartment. Many house owners sum up all additional housing costs and divide these by some allocation mechanism, e.g. the size of the apartment. Furthermore, the intervals in which additional housing costs are recovered from tenants may extend over

located on the estate itself, is owned by the house owner. Accordingly, he has to bear the costs for its construction, operation and maintenance, as well as for its disconnection. Connections to the public water supply system may only be constructed by certified companies. The metering devise and the connection from the main pipe to the estate border, instead, are owned by the *Berliner Wasserbetriebe*.

The exact regulations on the construction, maintenance and financing of new connections are laid down in the Contractual Regulations for Water Provision in Berlin (*Vertragsbestimmungen für die Wasserversorgung von Berlin* – VBW). The connection fee consists of three components, a basic fee covering planning costs and costs of the construction site, a meter-based fee calculated per front street meter of the estate to be connected, and a construction contribution.

Table 19: Connection fee (in Euro) as of 28 January 2009 (excluding 7 percent VAT) (Berlin)

	Single connection		**Shared service connection**		**Single connection to a shared service connection**
	DN 32–50	DN 80	DN 32–50	DN 80	DN 32–50
Basic fee	1,395.00	4,465.00	1,330.00	4,215.00	945.00
Meter-based fee					
Without personal contribution of house owner to building activity	160.00	285.00	160.00	285.00	160.00
With personal contribution of house owner to building activity[314]	25.00	--	25.00	--	25.00
(subtracted per meter if water <u>and</u> sewerage connection pipe constructed)	55.00		55.00		55.00
Construction contribution per street meter (estate)	40.00		--		40.00

Source: BWB (2009b).

No individual data on the share of the connection fee in total revenue from water sales to households could be retrieved from available data. Expert

several years. Hence, the water tarification scheme may provide only limited incentives to certain water users.

314 The personal contribution by the house owner includes the construction of the trench, the laying of the protective pipe and the closing of the excavation on the house owner's property.

judgement confirmed, however, that a separation would in principle be possible based on existing data collection, but has so far not been required for other reporting needs. Table 19 provides an overview on the structure of this revenue scheme. Connection fees obtained from water service users are recorded as liabilities and liquidated with 5 percent annually (or 2.5 percent for those constructed within the ongoing business year).

Investment contributions

Next to investment contributions paid by house owners for the connections to the public water supply network, investment contributions are levied on the exchange of house connection pipes. Table 20 provides an overview on the payment scheme.

Table 20: Conditions for investment contributions on the replacement of connections (Berlin)

	Exchange (in Euro)		**Exchange conducted in unison with pipe construction work on the main network conducted by the *Berliner Wasserbetriebe* (in Euro)**	
	DN 32–50	DN 80	DN 32–50	DN 80
Basic fee				
Connection provided by BWB, incl. wall works	580.00	1,015.00	460.00	825.00
Connection provided by BWB, wall works undertaken by client	410.00	890.00	310.00	720.00
Meter-based fee				
Without personal contribution	150.00	280.00	125.00	235.00
With personal contribution [315]	25.00	--	20.00	--
With cobbles on property	200.00	330.00	170.00	280.00
Additional costs for changing the part of the pipe system on the property owned by the client	332.00	460.00	281.00	389.00
(subtracted per meter if two water connection pipes within one layout)	55.00	55.00	55.00	55.00

Source: BWB (2009b).

Similar to connection fees, investment contributions are recorded as liabilities and liquidated with 5 percent annually (or 2.5 percent for those constructed within the ongoing business year).

315 The personal contribution by the house owner includes the construction of the trench, the laying of the protective pipe and the closing of the excavation on the house owner's property.

Fact Sheet 12: Accounting for endogenous revenue components (Berlin)

	Description of the structure of the tarification scheme / element						Data			Evaluation
Revenue source	**Payment basis** (e.g. per household, per connected inhabitant, per hectare, per crop type, per meter size, pipe size, etc.; obligatory or on demand only)	**Structure of the charge** (e.g. number of blocks, linear, increasing / decreasing, etc.)	**Collection mechanism and payment sched-ule** (e.g. one-off payment, weekly, monthly, yearly)	**Variation factors** (e.g. season, time-of-day, off-peak)	**Exemptions and eligibility criteria** (e.g. for prolific families, small farms)	**Link to cost categories** (e.g. capital costs of new investments, administrative costs, etc.)	**Rates** (e.g. Euro/m³, Euro/ha, etc.)	**Revenue collection** (Euro/year and% contribution to overall revenue)	**Data quality** (reliability and accuracy of reported data, rated A1 – B3)	**Problems in reporting**
Connection fee and investment contribution	per house/estate	Three components: (1) basic fee covering planning costs and construc-tion site costs; (2) meter-based fee ; (3) construction contribution	One-off payment.	Size of the house meter to be installed/ connected; Individual or shared connection. Construction work can partly be contributed by house/ estate owner.	(only applicable to house/estate owners)	Capital costs of new invest-ments	(cf. Table 19 and Table 20)	Cannot be dissected	--	Data cannot be directly extracted from published material Data available on expert level, specific data provision required.
Variable charge	Per m³ of water delivered	Constant, no variation with different demand levels	Yearly payment for direct contrac-tors of BWB.	No – determined and fixed on a yearly basis, depending on water demand development	No exemptions	No direct link	1.761 Euro/m³	247.7 Mio Euro (imputed)	A2	Detailed information on individual revenue component of the tarification scheme is not published. Imputed on the basis of household water demand.

Revenue source	Description of the structure of the tarification scheme / element						Data			Evaluation
	Payment basis (e.g. per household, per connected inhabitant, per hectare, per crop type, per meter size, pipe size, etc.; obligatory or on demand only)	**Structure of the charge** (e.g. number of blocks, linear, increasing / decreasing, etc.)	**Collection mechanism and payment schedule** (e.g. one-off payment, weekly, monthly, yearly)	**Variation factors** (e.g. season, time-of-day, off-peak)	**Exemptions and eligibility criteria** (e.g. for prolific families, small farms)	**Link to cost categories** (e.g. capital costs of new investments, administrative costs, etc.)	**Rates** (e.g. Euro/m³, Euro/ha, etc.)	**Revenue collection** (Euro/year and% contribution to overall revenue)	**Data quality** (reliability and accuracy of reported data, rated A1 – B3)	**Problems in reporting**
Fixed charge	Determined by meter size and annual water use	Increasing with meter size: QN 2.5: 5 blocks QN 6: 2 blocks	Yearly	Size of meter	(not relevant)	Fixed cost components.	Cf. Table 18	About 3.54 % in total revenues from water sales	B1	Detailed information is not published. Expert consultation provided share in overall revenues of this tariff component, not differentiated for households.
Other fees / charges	Water abstraction charge – Type I EC; per m³ of water abstracted	Constant charge per m³.	Yearly payment with quarterly instalments. Passed on to customers by variable tariff component[316]	No variation factor applied	Abstractions for groundwater management	Type I EC	0.31 Euro/m³	39.89 million Euro (approx. share of house-holds)	A 2	Imputed data on revenue collection.

316 The water abstraction charge has been subtracted from the variable tariff component to separate the share of Type I EC for the assessment.

The following *Fact Sheet* 12 provides a summary overview on the revenue schemes operated by the *Berliner Wasserbetriebe* for the provision of water supply to private households.

Exogenous revenue components

Section 3.2.3.2 outlined a number of exogenous revenue components, which may be present in the provision of water services. For the case of water provision in Berlin to private households, none of these components plays a decisive role (BWB 2007, SCHMITZ 2009):

- There are no transfers from earmarked budgets. While the *Berliner Wasserbetriebe* pay a water abstraction charge and the revenues of this charge are in part earmarked for water management activities (cf. Section 4.3.6.1), these measures are implemented directly by the *Land* Berlin and not used for financing activities of the *Berliner Wasserbetriebe*.
- Investment subsidies and allowances were granted after the political change in the beginning of the 1990s to cope with the massive investments needed at the time. Their ongoing affect on the costs of today's water provision and thus on the need for revenue generation to achieve cost-recovery is considered to be negligible (SCHMITZ 2009). Today, investment subsidies and allowances are no longer granted for the provision of water supply services in Berlin.
- As in the whole of Germany, no direct subsidies are provided to operation and maintenance costs of water service provision in Berlin.
- The *Berliner Wasserbetriebe* do not obtain transfers from river basin level or benefit from European transfer schemes.

The annual report of the *Berliner Wasserbetriebe* does however account for '*allowances*'. These may for example relate to the transfer of ownership of third party investments in the pipe system (e.g. if a new estate area is prepared for construction by an investor, who finances the connection to the existing pipe network of the *Berliner Wasserbetriebe*). Furthermore, if the routing and configuration of streets is changed (necessitating a relocation of the pipe network on request of the public works service, which would otherwise not have been necessary in terms of the economic lifetime of the existing network) part of these costs are reimbursed to the *Berliner Wasserbetriebe* by the public works service (i.e. the City of Berlin). Hence, these recorded allowances rather qualify as polluter-/user pays cost allocations. They are released on a linear basis in line with the economic lifetime of the respective infrastructural assets.

As exogenous revenue sources do not play a role in the provision of water supply services to private households in Berlin, *Fact Sheet* 3 from Section 3.2.3.2 can be ignored.

Other revenue components (non-water service related)

Next to water supply services, the *Berliner Wasserbetriebe* also provide wastewater services to the inhabitants of Berlin. While part of its personnel works for both company segments, the costs of service provision are recorded for each service separately and noted as intercompany sales. Therefore, each water service accounts for its own costs of service provision.

Cross-subsidisation is also not practiced between these two water services (SCHMITZ 2009) and is prohibited by the Berlin Water Law and the WTO, which mandate the internal cost-recovery principle (of financial costs and the water abstraction charge or the effluent charge respectively) for the provision of water services.

Besides, the costs of side-services such as fire-protection or groundwater management are not recovered from all service users, but rather accounted for separately and also recovered directly from beneficiaries, i.e. the *Land* Berlin (cf. Section 4.3.1). In the case of fire protection, houses which require a bigger dimensioning of their connection to the pipe network to fulfil fire protection standards are charged with these extra costs. Furthermore, houses requiring additional peak demand for fire protection purposes are charged a higher basic fee due to the larger water meter required (cf. above '*endogenous revenue components*'). It should however be noted, that the costs of the overall bigger dimensioning of the network, which is in part linked to fire protection provisions, are not recovered from the *Land* Berlin, but from all users through the general tarification scheme.[317] Nevertheless, this procedure may be considered in line with the polluter-pays principle, as all inhabitants of Berlin are potential beneficiaries of these provisions.

317 Furthermore, as outlined in Section 4.2.1.3, the (over-)dimensioning of the pipe network of Berlin is mostly related to the fact that the forecasts of population developments after reunification in 1990 were heavily biased upwards. For several years, investments in the pipe system were undertaken under the assumption that the population of Berlin would increase to about six million inhabitants.

In the CBRO, the tarification scheme applied to the provision of irrigation water depends on the infrastructure system in place. It is either charged on an area basis, if irrigation water is provided through open canals, or based on a two-part tarification scheme, if it is delivered through pressure pipes and metered. The tarification scheme for irrigation water of the CBRO is reviewed and published on an annual basis (CBRO 2009).

Area based tarification scheme – irrigation water provision through open canals

The area based tarification scheme applied to water abstracted for irrigation purposes from open canals varies with cadastre area and/or irrigable surface. The per hectare price to be paid by famers depends on the source from which irrigation water is provided, i.e. whether irrigation water is abstracted directly from the C.E.R., from canals owned and maintained by the CBRO or from rivers and canals not directly managed by the CBRO (e.g. the river Senio or the canal Naviglio-Zanelli).

If irrigation water is provided from the C.E.R, farmers pay between 9.46 Euro and 13.24 Euro per hectare, depending on the '*benefit index*' (CBRO 2009B).[318] This correction mechanism aims to differentiate between the total cadastre surface and the actual area to be irrigated. The pricing scheme seeks to recover the costs of operating and maintaining the shared infrastructure system for water abstraction and distribution from the C.E.R, including related administrative tasks at CBRO. (Re-)Construction costs are, however, covered through public budgets and are hence not included.

For irrigation water provided from canals owned by the CBRO, the amount payable per hectare decreases from 74 Euro per hectare for the first 0 to 1.5 hectare irrigable surface to 28 Euro per hectare for more than 10 hectares (cf. Table 21).[319] The decreasing cost by land size is motivated by the fact that the share of actual irrigation surface decreases with increasing total surface. Generally, the surface considered for charging is the total accounted for at the cadastre, as this information is easily accessible and constant over time, which allows for minimal control costs. The irrigation board considers that a more differentiated tarification scheme (e.g. according to crop type or irrigation technology) would imply higher costs, which could not be compensated by the

318 The CBRO is allowed to divert 12,9 m^3/s from the C.E.R., which has a total flow of 68 m^3/s.

319 For example, for a cadastre surface of 12 hectare, a farmer pays 1.5 x 74.00 + (5–1.5) x 56.00 + (10–5) x 38.00 + (12–10) x 28.00 = 553 Euro.

likely savings (CBRO 2009B, VIAGGI 2003). Nevertheless, farms with more than 15 hectares may choose to pay by actual irrigated area or by cadastre area.

Table 21: Tarification scheme for irrigation water provision from open canals (CBRO)

Area size	**Amount payable in Euro per hectare**
0 to 1,5 ha	74,00 €
1,5 to 5 ha	56,00 €
5 to 10 ha	38,00 €
more than 10 ha	28,00 €

Source: CBRO (2009b).

The level of the charge aims to recover the costs of operating and maintaining the shared infrastructure system. In addition, the costs of measures for preventing erosion or landslides are explicitly accounted for (CBRO 2009B).

If irrigation water is provided through rivers and canals that are not directly owned but managed by the CBRO, the charge paid by farmers varies depending on the source of water provision and ranges from 24.76 Euro per hectare for water provided from the river/canal Pierleone to 84.41 Euro per hectare for water derived from the river Senio (CBRO 2009B).

Furthermore, if water is provided through pressure pipes, but not metered, area based charges are also applied on the cadastre area. For example, farmers connected to the provision work Selice pay on average an amount of 142.26 Euro per hectare of cadastre area in 2009 (CBRO 2009B). The charges aim to recover the costs of operation and maintenance of the different provision works as well as the related administrative costs of the CBRO.

Volumetric charging scheme – irrigation water provision through pressure pipes

If water is provided through pressure pipes, farmers pay a fixed amount per hectare which is based on the cadastre area, and a volumetric charge per measured unit of water consumed (up to 0.15 Euro/m³)[320], which is comparatively high for the Italian context (BAZZANI ET AL. 2005, VIAGGI 2003). The fixed

320 The volumetric charge varies between different provisioning infrastructures. For the provisioning work Tarabina a fixed charge of 44.81 Euro/hectare and a variable charge of 0.15 Euro/m³ is charged in 2009. For the provisioning works Santerno-Senio and Selice-Santerno, a fixed charge of 37.50 Euro/hectare and a variable charge of 0.131 Euro/m³ are applied.

charge aims to recover maintenance costs and administrative costs, while variable costs mainly relate to pumping. The level of charges varies between the infrastructural systems, depending on whether the water work is located downstream or upstream of the C.E.R.

Incentive properties of the tarification scheme

In particular, the volumetric pricing scheme provides incentives for rational water use. VIAGGI (2003) reports that interviews with farmers in the region of the CBRO showed that the price effect discourages the irrigation of water extensive annual crops (e.g. sugar beet), which were traditionally irrigated when water was taken from open canals with a fixed price (so that the only variable cost of water was due to distribution costs, such as fuel and labour). These results confirm studies which show that, at present price levels, irrigation is still highly profitable for fruits and vegetables, but not for other annual crops (BAZZANI ET AL. 2005). In those cases, in which irrigation is still practiced for annual crops, this may be based on the fact that the required irrigation technology is already in place rendering a change in cultivation practices less profitable.

4.3.3 Testing Step 4: assessing environmental costs and their recovery

Step 4 of the methodology, the assessment of environmental costs and their recovery, is illustrated by the case of the Emscher river basin (cf. Section 4.2.1.1). As outlined above, the case study focuses on the *Emschergenossenschaft*, which manages wastewater treatment within the entire Emscher river basin. While some of the cities and municipalities within the Emscher river basin manage their own network for transporting wastewater, all facilities for sewerage treatment are operated by the *Emschergenossenschaft*. For analysing environmental costs, the treatment of wastewater is of primary importance, as it prevents or mitigates Type I EC caused by the water service wastewater management. Therefore, a focused analysis of the *Emschergenossenschaft* allows for studying all relevant effects.[321] Furthermore, in the case of the Emscher river basin, the hydromorphological restructuring of the river course due to the provision of wastewater service, which was historically determined by the mining activities in the region, is of particular relevance for the assessment of Type I EC

321 Nevertheless, activities by municipalities and third parties, which may have an influence on the assessment of environmental costs, are referred to for future reference in the analysis.

and may also be studied through an analysis focusing on the *Emschergenossenschaft* (as outlined in Section 4.2.1.1 above). In addition, the close historic relationship between the activities of the *Emschergenossenschaft* and the mining activities in the region offers valuable insights for the analysis of an adequate contribution in Step 9 (cf. Section 4.3.6.2).

The case of the Emscher river basin: provision of wastewater services

Due to the special historic evolution of the provision of wastewater services in the Emscher river basin (cf. Section 4.2.1.1), the environmental costs linked to service provision, i.e. Type I EC, are comparatively well-known and documented (MUNLV 2008A). The provision of wastewater services in the Emscher river basin is deeply influenced by the region's history of 150 years of industrial development and intensive mining. The political decision at the time of using the Emscher and its tributaries as an open sewer system for the transport of wastewater and rainwater had a major impact on the river basin's water quality and hydromorphology as well as on its flora and fauna (aquatic species and biodiversity).

The Article 5 report for the Emscher river basin accounted for 4,200 discharges of varying dimension and from diverse sources with about 780 municipal or industrial discharges still flowing directly into the Emscher (STUA HERTEN 2004).[322] The special situation of the Emscher river basin is also highlighted by the fact that about 38 percent of the territory is artificially drained through pumping stations (EMSCHERGENOSSENSCHAFT 2009A). Overall, the river's hydromorphology has been severely affected by mining activities and the corresponding decision to establish an open sewer system for the provision of wastewater services: the Emscher was diked, straightened and fitted with concrete formwork. Among others, the estuary of the Emscher into the river Rhine was twice relocated in order to improve flood protection due to newly developed subsidences. As a consequence, the *'Old Emscher'* and the *'Small Emscher'* were separated from today's main river Emscher. Furthermore, the Emscher flows in large parts outside of its original bed and on different geographical heights.

Against this background, the Article 5 report for the Emscher river basin came to the conclusion that about 95 percent of all water bodies will not reach good water status by the year 2015 (STUA HERTEN 2004). In the draft RBMP more than 90 percent of the river basin's water bodies have been designated as heavily modified or artificial (EMSCHERGENOSSENSCHAFT 2009A). While a significant part of environmental damages in the Emscher river basin were not directly

322 Data from 2002.

caused by the provision of wastewater services, but rather motivated by earlier decisions to pursue mining activities and meet flood protection concerns, a rehabilitation of the Emscher river basin will only be possible, once the river has been freed from its use as an open sewer system.

The following environmental damages in the Emscher river basin can be linked to the provision of wastewater services in the region (MUNLV 2008A, EMSCHERGENOSSENSCHAFT 2006, 2009)[323]:

- The river basin's hydromorphology was significantly altered in order to provide optimal conditions for the transportation of wastewater (under the impact of industrial and mining activities). Therefore the Emscher was reconstructed to a hydraulically optimized and straightened river bed with concrete formwork. In its reconstruction as an open sewer system, the bed of the Emscher was reduced to an absolutely necessary minimum.
- Despite the considerable number of already implemented rehabilitation projects and measures undertaken since the beginning of the 1990s (cf. discussion below, Step 4.1), discharges from municipal sewerage treatment plants still have a strong impact on the ecological and chemical status of the Emscher and its tributaries[324].
- While quantitative pressures are not present at the moment due to the artificial reconstruction of the Emscher, the quantity of some discharges could prove to be problematic, once the system has been restored to near-natural ecological conditions: some discharges amount to a multiple of average natural runoff. Yet, these effects have been anticipated and are tackled as part of the reconstruction of the Emscher river basin.

In sum, the Emscher is still in part used as an open sewer system leading to severe consequences for water quality, hydromorphology and river continuity. It also results in important effects on aquatic species and river habitats. Nevertheless, in discussing these effects, the underlying historic reasons for these developments need to be taken into consideration. They are studied in detail under Step 4.4.

According to present planning, the *Emschergenossenschaft* aims to free up the entire river basin from wastewater flows by the end of 2017 (EMSCHERGENOSSENSCHAFT 2009A:58). It is expected to take up to the year 2020 for completing

---

323 However, it should be noted that other water uses contributed as well to these damages, or were the ultimate causing factor for these developments, most notably, the mining sector (cf. discussion below).

324 Already rehabilitated parts of the river basin, such as the *Deinighauser Bach* and selected upstream tributaries, which had not formed part of the open sewer system, constitute an exemption.

the ecological and hydromorphological restructuring of all water bodies. As experience gathered so far with the implementation of measures in the Emscher river basin documents, it can be expected to take another ten years for the river basin's ecology to react to these improved conditions. In addition, the draining and discharge of mining water is expected to continue up to the year 2027. From today's perspective, the *Emschergenossenschaft* considers it an ambitious but realistic objective to reach good water status or good ecological potential within the entire Emscher river basin by the year 2027, once the third WFD implementation cycle has been completed (EMSCHERGENOSSENSCHAFT 2009A).

Step 4.1: Taking stock and investigating present internalization mechanisms

The first step of the analysis focuses on those environmental costs that are directly linked to the provision of water services, i.e. Type I EC. It is assessed in how far existing measures and instruments already address these environmental costs (cf. Section 3.2.4.1).

Measures

As outlined in Chapter 3, Type I EC may be internalized through measures aimed to rectify or prevent environmental damages. In the Emscher river basin, the environmental costs linked to the provision of wastewater services have long been recognised and tackled through extensive and diverse programmes of measures.

With the continuous closure of collieries since the end of the 1960s and the shifting of mining activities further to the North since the 1980s, the cause for subsidences in the Emscher river basin was eliminated. Hence, the former need for an open sewer system was no longer present. Against this background, the first rehabilitation project was implemented in the Emscher river basin in 1982, which aimed to tackle the environmental costs caused by the open sewer system.[325] Soon the need for an integrated and long-term rehabilitation of the

325 Higher standards for wastewater treatment provided the initial motivation to question the entire open sewer system within the Emscher river basin. At the time, only the wastewater treatment plant Duisburg-Old Emscher, which became operational in 1988, complied with the relevant parameters, but not the wastewater treatment plants Emscher-Estuary and Duisburg-Small Emscher. As a consequence, the effluent charge (cf. Table 23) payable by the *Emschergenossenschaft* would have increased tenfold from approximately 3 million Euro (6 million Deutsche Mark) to 30 million Euro (60 million Deutsche Mark) annually, in particular due to non-compliance with the parameter nitrogen. The effluent charge however

entire Emscher river basin became apparent. In 1991, an extensive Programme of Measures was finally decided on by the *Emschergenossenschaft* for the river basin's reconstruction, with an expected implementation schedule of more than one generation and an anticipated budget of 4.4 billion Euro (EMSCHERGENOSSENSCHAFT 2009A). The measures included in this reconstruction programme aim to remedy negative environmental effects of the open sewer system as well as the adverse impact of intensive mining activities over a period of years in the Emscher region. They have been planned and implemented under the umbrella of the so-called *'Emscher Masterplan'*, which takes an integrated perspective on the rehabilitation of the Emscher river basin and simultaneously addresses aspects of the region's spatial and economic development (EMSCHERGENOSSENSCHAFT 2006).

Until recently, the measures implemented within this framework focused more on the construction and retrofitting of wastewater treatment plants along with the construction of a state-of-the-art subsurface sewerage network and less on improving the river basin's hydromorphology (EMSCHERGENOSSENSCHAFT 2009A, NAFO 2009).[326] The most pressing problem was to improve wastewater treatment in the Emscher river basin in order to alleviate the pressure on the Emscher river basin's water status triggered by these environmental costs. Hence, the first measures included the construction of two new river wastewater treatment plants with nutrient elimination techniques and the technical upgrading of the wastewater treatment plant at the Emscher estuary.[327]

By the end of 2008, about 203 km of a state-of-the art subsurface sewerage network were constructed (including associated rainwater management infrastructures) making up about 48 percent of the required canal system for disentangling the Emscher river basin from sewerage flows (cf. Table 22). An ecological rehabilitation has been undertaken for 53 km of watercourse, which equals about 16 percent of the required work (EMSCHERGENOSSENSCHAFT 2009A). The effectiveness of these measures in mitigating (historic) environmental costs of wastewater services provision can be derived from recent monitoring results.

---

provided the possibility to substitute payments with investments in the extension and construction of wastewater treatment plant, thereby setting important incentives for investments and contributing to the decision for reconstructing the Emscher river basin. (EMSCHERGENOSSENSCHAFT 2009A)

326 This staged implementation process has been chosen, as an ecological rehabilitation of the river basin will only be feasible, once wastewater flows have been disentangled from the river basin.

327 The wastewater treatment plant in Dortmund-Deusen became operational in 1994; an additional plant started its operation in Bottrop in 1997 (EMSCHERGENOSSENSCHAFT 2009A).

They demonstrate in particular the improvements of water quality linked to the construction and retrofitting of wastewater treatment plants (EMSCHERGENOSSENSCHAFT 2009A). The measures implemented by the *Emschergenossenschaft* are supported by additional measures conducted by the municipalities within the Emscher river basin, which are detailed in the sewerage management plans of the municipalities (cf. EMSCHERGENOSSENSCHAFT 2009A and as an exemplary case STADTWERKE ESSEN 2003).

Table 22: Status of implementation: reconstruction of the Emscher river basin

Measures	Target value	Status of Implementation 2008	% already implemented
Sewerage network (subsurface canal)	418 km	203 km	48.1
Infrastructure for rainwater management	485,000 km³	257,000 m³	53.0
Ecological rehabilitation of watercourses	343 km	53 km	15.5
Retention	4,650,000 m³	1,793,000 m³	38.6

Source: Emschergenossenschaft (2009a).

By mid-2009, around 50 percent of the overall budget of 4.4 billion Euro have already been invested in measures aimed at eliminating the open sewer system and accompanying mitigation and rehabilitation measures (VAN BÖMMEL 2009). However, not all of these measures can be entirely attributed to the internalisation of Type I EC caused by the provision of wastewater services. For example, measures intended to improve the river's hydromorphology not only rectify damages caused by the provision of wastewater services, but simultaneously remedy the impact of historic mining activities and other water uses (e.g. flood protection).[328] Therefore, the extraction of cost information as an indication of Type I EC is not straightforward and requires expert judgement, despite the fact that information on the costs of these measures can generally be extracted from existing accounts, as they have already been implemented or are under construction.

328 Historically, the mining sector may to a large extent also be considered as the ultimate cause for the alterations and the installation of the open sewer system, necessitating an extensive Programme of Measures today (cf. Section 4.3.6.2). In addition, part of the measures may aim at side services, such as rainwater management (which is vital in case of the Emscher river basin and strongly linked to the provision of wastewater services) or flood protection.

Another difficulty is linked to the fact that the reconstruction of the Emscher river basin is financed through various sources (cf. Figure 23 below). While more than four-fifth of the costs are financed by the *Emschergenossenschaft* itself, the *Land* North Rhine-Westphalia contributes about one-fifth of the required financial means. This comes partly through European funding programmes and revenues stemming from the federal effluent charge (cf. Table 23 below) operated as an internalization instrument for environmental costs of wastewater services provision. In addition, the mining sector contributes additional financial means in excess of its regular member fees in line with its historic contribution to the present pollution situation within the Emscher river basin (cf. also Step 4.4 and Step 9 on '*adequate contribution*'). Box 9 summarises information on the financing scheme for the reconstruction of the Emscher river basin.

Figure 23: Overview on financing scheme for existing internalization measures

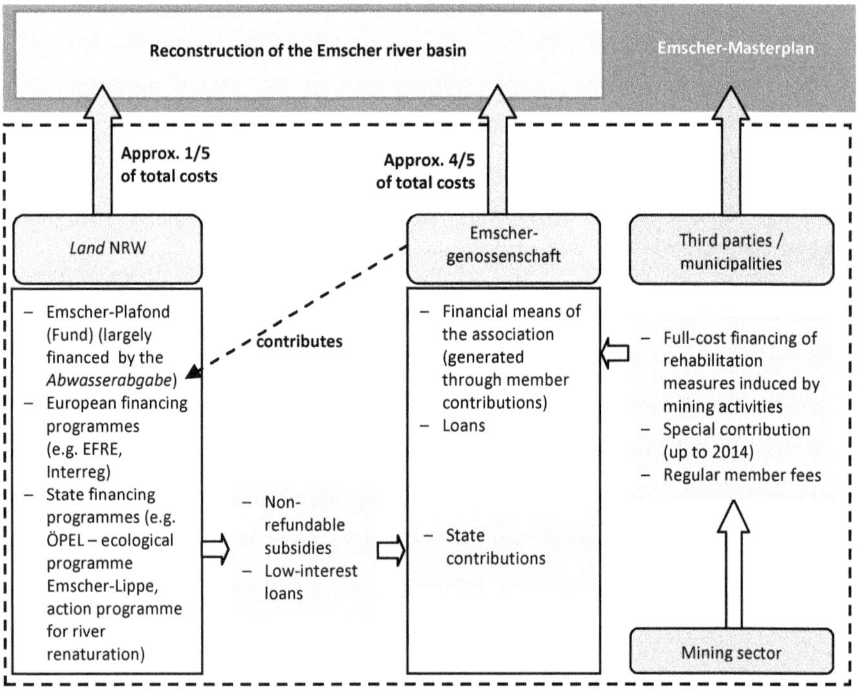

Source: based on Emschergenossenschaft (2009a), van Bömmel (2009).

Box 9: The financing scheme for the reconstruction of the Emscher river basin[329]

Contributions by the Land North Rhine-Westphalia

As outlined in Figure 23, the *Land* North Rhine-Westphalia contributed about one-fifth to the costs of measures already implemented in reconstructing the Emscher river basin. These contributions either take the form of non-refundable grant/contribution or low interest loans, with the latter being the major instrument.[330] The financing programmes can be invoked for supporting measures in the field of constructing and retrofitting of wastewater treatment plants, water management, ecological improvements of watercourses as well as overall planning, scientific research and consulting studies. Currently, the following financing programmes are in place (EMSCHERGENOSSENSCHAFT 2009B):

- The so-called *'Emscherplafond'* is a fund largely provided by the effluent charge, the so-called *Abwasserabgabe* (cf. discussion below, Table 23). As the *Emschergenossenschaft* pays a considerable sum of *Abwasserabgabe*, it likewise contributes to the financial means within this funding programme. From the *Emscherplafond*, the *Emschergenossenschaft* may apply for low interest loans (1 percent for the first 10 years and 2 percent for the next 10 years). It constitutes a central financing programme in terms of the overall financial contribution by the *Land* North Rhine-Westphalia.
- Financial contributions can also be provided from the *'Ecological Programme Emscher-Lippe (ÖPEL)'*. It supports measures designed to improve the ecological quality of watercourses within in the Emscher and the Lippe river basin. In addition, the construction of hiking and cycling paths and other improvements in leisure and recreation activities along the watercourses may be partially financed under this programme.
- The *'Action programme for river renaturation'* focuses on approved measures embedded in an overall concept, which seek to develop near-natural watercourses (such as the purchase of riparian land along with its cultivation and maintenance). The programme's duration will terminate by the end of 2011.

---

329 Please note that part of the information in this Box (on financing programmes and on the contributions by the mining sector) has been provided by the consulted experts (NAFO 2009, SCHÖN 2009, SCHULZ 2009, VAN BÖMMEL 2009) for the validation of the methodological framework. Due to confidentiality concerns, a clearance of these figures for future publication is still pending.

330 The nature of the contribution by the Land North Rhine-Westphalia has changed over the years. By now, the largest part consists of low interest loans instead of direct financial contributions to project costs.

- The *'Directive on financing measures of water construction'* targets constructive measures but also at the ecological rehabilitation of water bodies.
- In addition, one can apply for European financing programmes, such as Interreg IV B and EFRE. However, these deliver a comparatively minor contribution to the overall financing scheme.

Next to these financing programmes, the *Emschergenossenschaft* may apply for low-interest loans from the *'Investment Programme Wastewater North Rhine-Westphalia'* (2007 to 2011), which offers an interest rate of 3 percent for the first 10 years and 2 percent for the following 10 years.

Contributions by the Emschergenossenschaft

The *Emschergenossenschaft* contributes about 82 percent to the financial means required for the reconstruction of the Emscher river basin. Through 2007, about 125 million Euro were invested annually (including the contribution by the *Land*). At the end of 2007, the Association Council and the Association Assembly decided on a 5 percent increase in member fees in order to be able to spend about 200 million Euro annually on reconstruction and rehabilitation measures from 2008 up to the year 2017, when the construction of the subsurface sewerage network is to be completed (EMSCHERGENOSSENSCHAFT 2008, cf. Step 4.3). After their own financial sources the *Emschergenossenschaft* relies upon low-interest loans as much as possible.

Contributions by the mining sector

Due to its decisive historic impact on the evolution of the Emscher river basin, the role of the mining sector in financing rehabilitation and reconstruction measures is briefly investigated. An in-depth assessment of the adequacy of its contribution is however deferred to Step 7 of the assessment in Section 4.3.6.2.

Most importantly, the mining sector contributes to the financing of the reconstruction of the Emscher river basin through its regular member fee. In addition, the sector pays several million Euro annually through 2014 for recovering increased operation and maintenance costs, which are linked e.g. to subsidences.[331] According to the *'Veranlagungssätze'* of the Emschergenossen-

331 This annual contribution is divided between the different owners of the mines in line with the degree to which their coal mining activities have in the past or will in the future lead to subsidences and deteriorations in run-off capability. The owners of inoperative mines also have to make a payment due to the costs of historic environmental damages.

schaft, this contribution is to be used for recovering 30 percent of the costs encountered for operating the open sewer system (EMSCHERGENOSSENSCHAFT 2003). The remaining amount is to be employed towards the operation and maintenance of the sewerage network and disentangled water courses. After the year 2014, the mining sector will still contribute to the operation and maintenance costs of disentangled watercourses.

Measures aimed to prevent or remedy environmental costs directly linked to mining activities are fully recovered from the mining sector. To this end, the costs of rehabilitation measures are included in the '*Vermögensplan*' as an own position and levied on the mining sector. The projected costs of these measures have to be approved by the Association Council. Should project costs evolve differently from the anticipated budget, the differences in cost estimates are accounted for in the upcoming years. For the majority, these measures focus on the negative effects of pumping stations and water flow regulation. They do not address the sewerage network or the general ecological rehabilitation of water bodies. (VAN BÖMMEL 2009)

Judging the degree of internalization of Type I EC in the case of the Emscher river basin requires identifying those measures, which have already been financed by the *Emschergenossenschaft* during the Emscher reconstruction process, that aim to prevent or mitigate Type I EC.[332] Hence, specific rehabilitation measures, which are fully financed by the mining sector are to be excluded from the analysis, as well as those costs related to tackling Type II EC (cf. Step 4.4 below). In addition, measures displaying multiple uses (i.e. that not only address environmental costs linked to the provision of wastewater services) need to be given special consideration in the attribution of costs to the category of Type I EC.

As confirmed by expert interviews (NAFO 2009, VAN BÖMMEL 2009), measures undertaken within the Emscher reconstruction process do not include expenses for maintaining or improving existing sewerage networks. They qualify as operation and maintenance costs under financial costs. Likewise, costs encountered by municipalities (e.g. by the city of Essen) for maintaining their infrastructure systems for wastewater transportation are not accounted for. Furthermore, the costs of rehabilitation measures that only become necessary due to activities of the mining sector within the Emscher river basin are recovered on a full cost-recovery basis from the mining sector in line with the polluter-pays principle

332 In a complementary analysis, Step 4.4 investigates whether the *Emschergenossenschaft* also finances measures which rather qualify as Type II EC, as they are required due to the impact of other water uses.

and are not considered within the budget for the reconstruction of the Emscher river basin (cf. detailed discussion in Section 4.3.6.2 below).

Measures so far have largely concentrated on disentangling sewerage flows from the Emscher river basin, while hydromorphological measures have been less important in terms of budget shares.

Figure 24: Analysis of the degree of internalisation of Type I EC with present measures

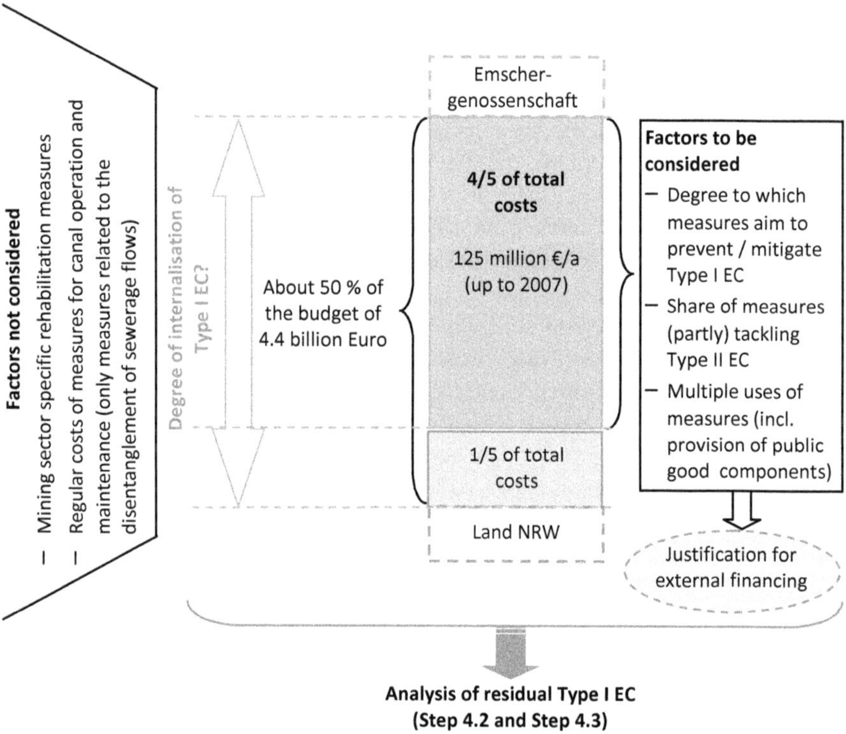

In Step 4.1 of the assessment of Type I EC, the analysis focuses on determining whether measures (and instruments, cf. discussion below) are already operational, that aim to internalise Type I EC. The following steps then investigate whether these are sufficient for addressing Type I EC linked to the provision of wastewater services in the Emscher river basin or whether residual costs remain that are presently not recovered. In addition, it has to be investigated how measures are currently financed (i.e. who pays for what), in order to assess the degree to which Type I EC are in recovered internally from water service users/polluters. Figure 24 above summarises the relevant considerations for Step 4.1.[333]

The measures that have so far been implemented within the reconstruction process of the Emscher river basin focus on mitigating the environmental costs related to the provision of wastewater services through an open sewer system. With a contribution of more than 80 percent to the costs of measures by the *Emschergenossenschaft*, a higher degree of internal cost-recovery can be attested for present Type I EC internalisation measures, in particular as these measures also address environmental damages caused by other water uses or aim in part at the provision of public services. They cover in part general water management issues such as flood protection, the provision of public good components in rainwater management and the historic costs of mining activities. Against this background and considering the historic reasons for the evolution of the open sewer system, a financial contribution by the *Land* North Rhine-Westphalia to the cost of the reconstruction of the Emscher river basin can be justified from the perspective of Article 9 requirements.[334] The decision to establish an open sewer system was a political one, taken against the background of promising economic development and a lack of technical alternatives at the time.

Next to an analysis of already implemented measures addressing Type I EC, an assessment of complementary internalisation instruments completes the analysis of Step 4.1.

Instruments

In the Emscher river basin (as in the whole of Germany), an instrument is operational that aims to internalise the environmental damage costs of polluted discharges into surface water bodies. The instrument '*Abwasserabgabe*' was instituted by the Effluent Charges Act (*Abwasserabgabengesetz* – AbwAG) at the federal level in 1976 and takes the form of a direct payment based on estimates or actual measurements of the quantity and quality of pollutants discharged

333 Whether the contribution by other water users is sufficient in terms of their compliance with the polluter-pays principle is studied at later phases of the assessment in *Step* 7. It first requires an assessment of the degree to which other water uses heighten the costs of service provision, as is undertaken in *Step* 4.4.

334 Furthermore, it has to be noted that part of the financial contribution by the *Land* is generated through the *Abwasserabgabe* to which the *Emschergenossenschaft* also contributes (cf. Box 9 and Table 23 below).

directly into a water body.[335] The effluent charge is calculated based on the amount of certain pollution parameters discharged within a specified period of time.[336] Although the charge is not based on actual estimates of damage costs, which would for example imply varying charges at regional level depending on the sensitivity of receiving water bodies, it can be considered as a step towards the implementation of the polluter-pays principle for wastewater services provision and a cost allocation mechanism for Type I EC, despite inherent drawbacks outlined in Table 23 below.

The following Table 23 summarizes the relevant information on the *Abwasserabgabe* as an internalization instrument of environmental costs in the Emscher river basin in line with the requirements outlined in Section 3.2.4.1 (Box 6).

In 2008, the *Emschergenossenschaft* was charged 20.9 million Euro *Abwasserabgabe* for its discharges by the State Environmental Agency (*Landesumweltamt*) of North Rhine-Westphalia. In total, the *Emschergenossenschaft* was liable for 461.1 million Euro from 1981 to 2008, of which 202.9 million Euro could be offset against investments in pollution abatement or sewage treatment plants.[337] As outlined in Table 23 and Box 9, the *Abwasserabgabe* constitutes an important source of financing for rehabilitation projects within the Emscher river basin reconstruction process. A significant share of the financial contribution of the *Land* North Rhine-Westphalia to the reconstruction of the Emscher river basin stems from revenues of the *Abwasserabgabe*.

335 The *Abwasserabgabe* is levied on direct discharges of effluents into natural waters. The definition of effluents includes (§ 2 AbwAG): industrial effluents; agricultural discharges; discharges from sewage treatment plants (STP); discharges and leakage from landfills; direct rainwater discharges; minor effluents such as domestic sewage from decentralized treatment facilities.

336 Cf. §18 AbwAG in amendment to §3.

337 The amount was charged for different years (delays were caused by judicial interventions). In 2008, the entire sum was paid, as there were no investments undertaken, which could be offset according to § 10(2) *Abwasserabgabengesetz*. However, judicial intervention was claimed for 35 *Abwasserabgabe* notifications levied by the State Environmental Agency of North Rhine-Westphalia with a total sum of 19.2 million Euro in 2008 (EMSCHERGENOSSENSCHAFT 2008:49).

Table 23: Internalisation instrument 'Abwasserabgabe'

Classification questions (cf. Chapter 3)	**Description of the instrument**
Since when is the instrument operational?	The *Abwasserabgabe* was introduced in 1976 as an area-wide instrument at the federal level through the Effluent Charges Act (*Abwasserabgabengesetz – AbwAG*).
Reason for the instrument's creation	A critical chemical and biological water quality in Germany's river basins led to the introduction of the *Abwasserabgabe*. The instrument aims to internalize environmental damage costs of polluted discharges into surface water bodies in line with the polluter-pays principle.
Which activities are addressed by it?	Direct discharges of wastewater into surface water bodies.
What is the fee/charge levied on?	The level of the *Abwasserabgabe* depends upon the quality of the discharge, i.e. the number of pollution units. The charge of 35.79 Euro (since 1998) per pollution unit is uniform across the entire country. Individual charges are based on the values stated in the discharge permit issued by the responsible water authority and based on the following general formula: $EC = f\,(a_i;\ pu_i)\ 1 \leq i \leq n$ where *EC* is the effluent charge for one year, a_i is the rate for a certain pollutant, pu_i stands for the number of pollution units discharged in one year and *n* is the number of pollutants entering into the calculation. Dischargers conduct self-monitoring and the permit-issuing water authority has the responsibility of inspection. The frequency of compliance monitoring by the water authority depends on the authority, except for sewerage treatment plants, for which the Urban Wastewater Treatment Directive (UWWTD) specifies monitoring intervals.
Is the level of the fee/charge determined by environmental targets or politically?	The level of the charge is determined politically. From 1976 to 1998 it has been gradually increased up to 35.79 Euro per pollution unit. The level of the charge is not differentiated depending on the quality and perceptibility of discharges of receiving waters. The charging level has proven to be set high enough for acting as a deterrent to water pollution and has motivated considerable investments in pollution abatement measures.
Does the level of the fee/charge vary? With which parameters (e.g. seasonally, according to pollution load, etc.)?	There is no seasonal variation for the charge. The amount payable depends on the number of pollution units, i.e. the quality and the amount of polluted discharge.

Classification questions (cf. Chapter 3)	Description of the instrument
Are exemption regulations in effect (e.g. exempting certain sectors or activities from the fee/charge)?	The charge considers only direct discharges. The payable charge may be reduced if monitored values diverge by at least 20 percent from expected pollution values stated in the permit. Pollutants that are subject to the effluent charge but do not exceed specified threshold values or dilution factors are not included in the calculation of the charge. Charges can be offset against investments in pollution abatement or expenditures for improving sewage treatment plants, to promote the removal of dangerous substances from discharges and to support the adoption of BAT.
Is the instrument effective in reaching its objectives?	Since its introduction in 1976 important improvements in water quality could be achieved through the *Abwasserabgabe*: – it motivated considerable investment in pollution abatement measures; – it provided financing means for improvements in infrastructural systems and treatment capacities; – it contributed significantly to the full-scale implementation of the UWWTD in Germany.
Does the instrument set incentives for efficient resource use (*objective 2*)?	According to §10(3) of the Effluent Charges Act investments in improved sewerage treatment capacity can be offset against the charge, if the investment leads to a parameter's reduction by at least 20 percent and a decreased overall wastewater stream. While this regulation set important incentives for reducing the harmfulness of wastewater discharges in the past, the incentive function has been restrained over the year along with improvements in wastewater treatment. High standard of wastewater treatment in Germany and the need for reducing instead of extending networks and treatment capacities in most parts of Germany (e.g. in response to demographic change) have eroded the incentive effect. The incentive function of the instrument could be further improved by levying the charge on actual monitoring results (quality and quantity of effluent), instead of basing it on discharge permits. Reductions in the harmfulness or quantity of discharges would then lead to lower payments without necessitating a 20 percent reduction in one parameter, which is increasingly becoming technically infeasible for wastewater service providers. This would also strengthen the polluter-pays principle.[338]

338 The incentive function of the Abwasserabgabe is currently studied in a research project on behalf of the German Federal Environmental Agency; results are expected by the end of 2010 and may induce a reform of the charging system.

Classification questions (cf. Chapter 3)	Description of the instrument
Are the revenues of the instrument earmarked? How are they employed (*objective 1 and objective 3*)?	According to Article 13 of the Effluent Charges Act, revenues from the *Abwasserabgabe* are earmarked for a range of water management activities, such as contributions to investments in water pollution control and other quality improvement measures. The revenue is thus used to maintain and improve water quality and also fulfils a financing function. Revenues are collected on an annual basis. As the *Länder* are the competent authorities for water management and legislation, they are the recipient authorities and in addition responsible for the redistribution of revenues.
Instrument critique	There is no rate differentiation depending on the quality of receiving waters, i.e. the same charge is levied on discharging a certain amount of effluent into a sensitive water body with existing impairments in water quality as on discharging the same amount and quality of effluent into a water body with sufficient carrying capacity to accommodate the polluted load. The level of the charge is not determined through actual measurement of pollution loads but rather through permits, which constitutes a second-best solution with regard to incentive setting and the implementation of the polluter-pays principle.

Source: Effluent Charges Act (2005).

Step 4.2: Rough analysis for identifying potential additional Type I EC (residual damage)

Despite the extensive rehabilitation projects implemented in the Emscher river basin since the beginning of the 1990s, severe environmental problems continue to persist. They are partly linked to the (past) provision of wastewater services and prevent the timely implementation of the WFD objectives in large parts of the Emscher river basin.[339] As outlined in Chapter 3, the distance to the target *'good water status'*, which can be attributed to the provision of wastewater services provision is an indication of the magnitude of remaining Type I EC. In the Emscher river basin, particular problems are linked to the following aspects (EMSCHERGENOSSENSCHAFT 2009A):

- In the past, the river's hydromorphological development was superseded by the transportation of wastewater. As a result, the Emscher river basin was gradually straightened, deepened, diked and fitted with concrete formwork, which resulted in the adverse hydromorphological effects described above.

339 As a reminder, environmental costs linked to other water uses (e.g. the mining sector) are not considered at this point of the analysis, but studied in Step 4.4.

While the *'Emscher-Masterplan'* remedied part of these effects already, additional work remains.

- The polluted discharges of the open sewer provision of wastewater management services and the accompanying changes in the Emscher and its tributaries' hydromorphology resulted in a number of problems. These included a loss in biodiversity, negative effects on aquatic species and the elimination of large parts of the original wetlands;
- Chemical contaminations have accumulated through the long period of heavily polluted discharges, stemming partially the provision of from wastewater services.

More specifically, the monitoring results for the Emscher river basin revealed that due to impairments in hydromorphological conditions water bodies within the Emscher river basin fail to reach good status (EMSCHERGENOSSENSCHAFT 2009A). In this context it must be noted that even for already rehabilitated watercourses that had been used over decades as open sewer systems, it will take at least ten years to develop a stable and self-dependent biocoenosis. There are few fish and aquatic macrophytes in the Emscher river basin today, contributing to a negative assessment for most water bodies in the monitoring exercise. The chemical assessment of the Emscher river basin shows significant impairment in seven out of ten groundwater bodies. This was expected given the historic open sewer system. Also the near-natural upstream tributaries of the Emscher system reveal impairments for certain parameters, such as nitrogen and sulphate or heavy metals.

Fact Sheet 13 below provides a rough analysis of residual Type I EC, for which a more detailed analysis has to be undertaken in Step 4.3.

Step 4.3:Taking stock of residual damage – unaccounted for environmental costs

As highlighted in Section 2.4.4, a cost-based approach can be followed for assessing (residual) Type I EC. During the WFD implementation process extensive information is collected on both, environmental damages, i.e. expressed as deficits in water status, as well as on the related causing activities, i.e. though an analysis of pressures and impacts in addition to extensive monitoring programmes.[340] The draft Programme of Measures for the Emscher river basin outlines, which (cost-effective) measures have to be implemented for rectifying

340 In the Emscher river basin, the local government districts Münster is responsible for conducting the monitoring for WFD implementation with support from the government districts Düsseldorf and Arnsberg.

the identified environmental damages. Furthermore, indicative information on the costs of these measures is also available and will further improve as implementation progresses. For the analysis, additional information is required on how measures are to be financed, i.e. on the cost allocation between different user groups, and the degree to which financing mechanisms are in line with the polluter-pays principle.[341]

Fact Sheet 13: Rough analysis for detecting Type I EC (Emscher)

	Type I environmental cost categories														
	Overabstraction			Morphological changes (river regulation and flow diversion)			Polluted discharges			Other					
Water service	Adverse effects on aquatic species	Loss of biological diversity	Adverse effects on water dependent land ecosystems (wetlands, forests)	Adverse effects on aquatic species	Loss of biological diversity	Adverse effects on water dependent land ecosystems (e.g. wetlands, forests)	Water quality degradation (impairment of natural rate of recovery)	Loss of biological diversity	Adverse effects on aquatic species	Land erosion	Degradation of soil quality, loss of soil fertility	Eutrophication	Chemical contamination	Salination[342]	Water logging
Wastewater management				x	x	x	x	x	x				x		

"x" indicates relevance of environmental cost categories

For a comprehensive analysis of Type I EC, it also has to be investigated, whether a deadline extension or lower objectives are requested (so-called

341 During the first implementation cycle, detailed information on financing schemes of measures and on the implementation of the polluter-pays principle will often not be publicly available and only be accessible through expert consultation. The information is required for assessing the implementation of the polluter-pays principle and the adequacy of contribution from water uses in Step 9 below (cf. Section 4.3.6).

342 While salination constitutes an environmental problem in the Emscher river basin, it is not linked to the provision of wastewater services, but rather to the discharge of draining waters from the mining sector and hence discussed under Step 4.4.

exemptions, cf. Section 1.3.2) that are linked to the environmental damages caused by the provision of the water service under investigation. In that case, the outlined Programme of Measures for the first implementation cycle would not contain all measures required for rectifying the identified environmental damages and thus could not serve as a complete indication of environmental costs.

In the case of the Emscher river basin, detailed information on measures and in part on related costs can be extracted from the following publicly available documents:

- Draft RBMP for the North Rhine-Westphalian parts of the rivers Rhine, Weser, Ems and Maas (2008);
- Draft Programme of Measures for the North Rhine-Westphalian parts of Rhine, Weser, Ems and Maas (2008);
- Brief statements on the measures that are to be implemented within the planning units of the Emscher river basin (so-called *'Steckbriefe'* (2008));
- RBMP for the Emscher river basin (*'Flussgebietsplan Emscher'*), prepared by the *Emschergenossenschaft* (2009A);
- Annual report of the *Emschergenossenschaft* (2008);
- *'Emscher:Zukunft'* – Masterplan Emscher (2006).

Furthermore, additional information and expert opinion were gathered through interviews with representatives from the Ministry of Environment of North Rhine-Westphalia (MUNLV) and the *Emschergenossenschaft*.

For the Emscher river basin, the draft Programme of Measures and the accompanying *'Steckbriefe'* already outline all measures that are (from the present planning status) required for reaching good water status or good ecological potential, including those that will only be (fully) implemented in upcoming implementation cycles (*time exemptions*). Hence, the information on measures required for reaching good water status or good potential by 2027 is complete. Although one should note that, as implementation progresses, cost data will become more accurate and measure selection for upcoming implementation cycles may be amended or adapted.

For the following Table 24 those groups of measures have been selected from the overall draft Programme of Measures for 2008, which aim (in part) at rectifying remaining Type I EC of wastewater services provision.

Table 24: Selected measures from the Programme of Measures for the Emscher river basin

<table>
<tr><th>(Group of) measures</th><th>Main aim of the measures</th><th>Timeframe for implementation</th><th>Important aspects for the analysis of Type I EC</th></tr>
<tr><td>Construction and adaptation of infrastructure assets for transporting, treating and retaining combined sewer</td><td>Separating water streams and eliminating wastewater flows from the watercourse</td><td>Finalisation expected by 2017.</td><td>Includes investment costs in infrastructural assets that form part of 'capital costs'. Risk of double counting.
Given the specific historic development of the Emscher as an open sewer system, also investments in wastewater transportation aim to reduce Type I EC.</td></tr>
<tr><td>Extension and adaptation of wastewater treatment plants (elimination of nitrogen compounds)</td><td rowspan="2">Improving wastewater treatment</td><td>Expected to be completed by 2015.</td><td rowspan="2">Includes investment costs in infrastructural assets that form part of 'capital costs'. Risk of double counting.</td></tr>
<tr><td>Eliminating infiltration water for reducing nitrogen and phosphorous input</td><td>Expected to be completed by 2017.</td></tr>
<tr><td>Ensuring linear continuity of infrastructural works</td><td rowspan="6">Improving hydro-morphology and establishing river continuity</td><td rowspan="6">Implemented in the 1st and 2nd WFD implementation cycles; expected to be fully operational over the 3rd implementation cycle.</td><td rowspan="6">Hydromorphological damages are not only caused by wastewater service provision.</td></tr>
<tr><td>Improving river habitats through changes in watercourse structures</td></tr>
<tr><td>Improving habitats in watercourses incl. the development of wetlands</td></tr>
<tr><td>Improving habitats at the waterside (e.g. grove development)</td></tr>
<tr><td>Vitalizing watercourses within existing river profiles</td></tr>
<tr><td>Reducing flow peaks caused by water uses</td></tr>
</table>

Source: MUNLV (2008a), MUNLV (2008b), MUNLV (2008c).

The relevance of the selected groups of measures outlined in Table 24 above for the analysis of Type I EC could be validated through expert consultations, in line with the procedures described in Chapter 3.[343]

Description of measures

As discussed in Step 4.1 of the analysis, an overall improvement of water status is expected from an integrated and area-wide rehabilitation of the Emscher river basin. The open sewer canal is to be replaced by a subsurface sewerage network that separates groundwater and surface water streams from wastewater (and rainwater). The following groups of measures have been selected for rectifying the identified environmental damages in the Emscher and its tributaries (cf. Table 24 above):

- Construction of separate sewer systems and rainwater management infrastructure;
- Extension and adaptation of existing sewerage treatment plants;
- Rehabilitation of the Emscher river basin's hydromorphology to a functioning and healthy ecosystem.

The sewerage treatment plants that are currently still operated as river treatment plants[344] (Dortmund-Deusen (up to the end of 2009), Emscher-Estuary and Bottrop) will be refitted to sewerage treatment plants with an underground reception of wastewater flows. According to the draft Programme of Measures for the Emscher river basin, the final extinction of the open sewer system will be achieved by the end of 2017. Those parts of the Emscher river basin, which have been freed from wastewater flows, will be rehabilitated to their (near) natural conditions on a step-by-step basis as far as local structural circumstances of the landscape permit. The concrete formwork will be removed and river beds will be given more space and room to meander. While rehabilitation work on the Emscher river basin's hydromorphology is to be completed by the year 2020, some of the measures will only take full effect with a time lag, i.e. at the end of the third implementation cycle.

For the rehabilitation of the Emscher river basin's ecology, near-natural bed and shore structures have to be constructed (cf. Table 24), which will allow for an independent and dynamic development of the Emscher. In addition, the

343 According to Chapter 3.2.4.3, the degree of certainty with which relevant measures could be singled out from the overall Programme of Measures and attributed to the provision of wastewater services is to be assessed on the basis of expert judgement.

344 Cf. Footnote 24 above.

reconstruction of wetlands and the establishment of continuity in the Emscher tributaries are required for ensuring the redevelopment of habitats in the Emscher itself.

The ecological concept elaborated for the Emscher river basin allows for the development of a network of biotope structures to ensure the ecology of the river basin. Due to the restrictive structural conditions in part of the Emscher river basin (e.g. a high population and habitat density), the concept not only relies on a significant extension of the Emscher's river bed, but also aims to link existing structures and habitats to establish a biotope network of ecologically complementary and supporting habitats. The river bed will be rehabilitated in such a way, that the Emscher will flood the newly constructed wetlands for several weeks a year, thereby creating the necessary conditions for a typical and independent development of wetlands. Improved interlinkages of the Emscher with its accompanying wetlands are to be further facilitated through river bed lifting, which will simultaneously contribute to the perceptibility of the Emscher within the region.

Overall, the selected measures for rectifying existing environmental costs of wastewater service provision aim to reconstruct the Emscher river basin as a functioning aquatic eco-system with typical hydromorphological structures and habitats.

To summarise, for the first WFD implementation cycle, the *Emschergenossenschaft* anticipates completing the implementation of about 350 measures as part of the Emscher reconstruction project (EMSCHERGENOSSENSCHAFT 2009A). By 2015, all four wastewater treatment plants in the Emscher river basin will utilise state-of-the-art technology, 350 km of subsurface sewerage network will have been constructed and about 200 km of watercourse will be ecologically rehabilitated (EMSCHERGENOSSENSCHAFT 2009A).

Costs of measures

For the implementation of these measures, planners anticipate a budget of about 200 million Euro annually (EMSCHERGENOSSENSCHAFT 2008). The *Emschergenossenschaft* will cover about 80 percent of these anticipated costs. To raise the required funds, a 5 percent increase in revenues obtained from member fees was decided on by the *Emschergenossenschaft* in 2007.[345] As outlined

345 Before, 125 million Euro had been invested annually in the reconstruction process of the Emscher river basin to which the *Emschergenossenschaft* contributed about 82 percent (cf. Figure 24 above). The 5 percent increase in revenues from member fees has been decided on

under Step 4.1, the *Land* North Rhine-Westphalia supports the reconstruction of the Emscher river basin financially through low-interest loans, direct financial contributions for selected measures as well as through (comparatively minor) contributions from European financing schemes (cf. Box 9). For example, for the years 2009 to 2013 about 25.6 million Euro will be provided through the financing programme EFRE, mainly for measures aimed at ecological and hydromorphological reconstructing (MUNLV 2009).

While the overall budget required for the reconstruction of the Emscher river basin is known (i.e. 4.4 billion Euro of which approximately 50 percent have already been invested, cf. Step 4.1), including all measures aimed to eliminate negative environmental effects of the open sewer system, there is no publicly available information at this stage of the implementation process on the budget's allocation to different groups of measures. Therefore, one is currently unable to conduct a complete assessment of the costs which should actually be borne by the *Emschergenossenschaft* as a means of internalising Type I EC.

One can argue, however, that the financial means required for the internalisation of Type I EC will be lower than the overall budget assigned for the reconstruction of the Emscher river basin as some measures display multiple uses or have to be implemented due to multiple pollution causes. Hence, the overall budget provided for the reconstruction of the Emscher river basin has to be considered as an overestimated upper limit for Type I EC, to which the *Emschergenossenschaft* contributes more than 80 percent.

Given the information restrictions at this stage of the implementation process, information on the overall budget assigned for reconstructing the Emscher river basin together with a qualitative description of those measures aimed specifically at reducing Type I EC (Table 24) along with their expected impact on water status may be considered sufficient for assessing whether Type I EC are addressed through present financing mechanisms for water service provision or largely left unaccounted. Based on the 80 percent contribution of the *Emschergenossenschaft* to the overall budget, it can be concluded that Type I EC are largely internalised in line with the requirements of WFD Article 9 (i.e. an internal cost-recovery approach for environmental costs caused by the provision of water services). Yet, given the restraints on cost allocation data, the assessment remains indicative in nature as important aspects remain unresolved. As soon as more detailed cost information becomes available during the implementation process, the analysis must with respect to the following points:

---

by the Association Council and the Association Assembly of the *Emschergenossenschaft* on 22 November 2007 (EMSCHERGENOSSENSCHAFT 2008).

- The relevant share of the costs of measures with multiple uses and polluters (e.g. hydromorphological measures) that is encountered for alleviating the negative environmental effects of wastewater services provision has to be assessed with the help of expert judgement;
- On this basis, a more detailed analysis of the financing scheme can be undertaken, which studies more closely whether the financial contribution of the *Emschergenossenschaft* to the reconstruction of the Emscher river basin is in line with the specifically identified share of measures aimed at internalising Type I EC.

Finally, in an integrated assessment over all steps of the methodology, special attention should be given to the risk of double-counting costs of measures (e.g. investments in sewerage treatment plants) under financial costs and under Type I EC in line with the guidelines provided in the *Fact Sheets* (cf. Sections 3.2.2 and 4.3.1).

Overall, it can be concluded that the problems encountered in applying the methodological framework were linked to confidentiality concerns and the early stage of implementation. These difficulties will subside as implementation progresses.

Step 4.4: Assess Type II EC – environmental costs caused by other water uses

The river basin management plan and in particular the analysis of pressures and impacts and the monitoring programme provide information on the origin of environmental damages and pollution pathways. For the Emscher river basin, the following environmental damages can be identified that are not directly linked to the provision of wastewater services in the Emscher river basin, but may trigger higher costs of service provision (derived from EMSCHERGENOSSENSCHAFT 2009A):

- Water from the draining of mines, which is contaminated by past pollution and has a high salt concentration, constitutes a main pressure on water status.[346] It has an impact on the functioning and effectiveness of river treatment plants and negatively affects the river basin's pollution carrying capacity.

346 In 2007, the seven mine water drainages in the Emscher river basin discharged around 22.6 million m³ of draining water into the Emscher with a chloride contamination of approximately 265,000 t (EMSCHERGENOSSENSCHAFT 2009A). As long as active coal mining is undertaken in the Emscher river basin, draining water from the mines will have to be discharged into the Emscher.

- Due to subsidences caused by mining activities, the Emscher and many of its tributaries had to be diked. The missing run-off capability was replaced by pumping stations, leading to higher costs of service provision.
- The laying of pipes is more cost-intensive due to contaminated sites (brownfields)[347]. Furthermore, the geo-structure and urban development setting within the Emscher river basin that resulted from the region's evolution from a predominantly agricultural are to a mining and finally a post-mining region implies special building techniques in laying the subsurface sewerage network. While sewerage networks are normally constructed along with the development of settlements, the history of the open sewer system implies higher construction costs today, because it is an area characterized by high population and use density (cf. Section 4.3.5). Generally, the pipe network has to be installed at considerable depth resulting in higher pumping costs and long transportation distances with associated high requirements on corrosion prevention.

The following *Fact Sheet* 14 summarises the effects of Type II EC related to the provision of wastewater services in the Emscher river basin.

Fact Sheet 14: Identifying Type II EC of wastewater services provision (Emscher)

Water service	Effect	Related water uses	Affected cost category
Wastewater services	Reduced self-cleaning capacity of receiving water bodies	Mining activities (draining water) Brownfields Industry	Operation and maintenance costs (e.g. material costs, energy costs, monitoring costs) Capital costs
	Subsidences necessitating a dyking of watercourses	Mining sector	Operation and maintenance costs (higher energy costs, more expensive building techniques) Capital costs
	Cost-intensive laying of pipes due to geo-structural and urban development conditions, as well as contaminated sites	Households (high population density with associated structures) Industry and mining activities → Framework conditions in the region	Operation and maintenance costs Capital costs

347 There are about 4,800 registered brownfields or expected brownfields in the Emscher river basin (EMSCHERGENOSSENSCHAFT 2009A).

As described in Chapter 3, the cost information for Type II EC is generally included in the financial accounts of water service providers (i.e. by definition Type II EC refer to 'higher costs for water service provider'). In the case of the Emscher, the mining sector constitutes the most relevant cause for Type II EC. The additional costs for water service provision caused by mining activities are in part addressed through a special charge in the Emscher river basin.

As already outlined under Step 4.1 (cf. Box 9), the mining sector helps finance the reconstruction of the Emscher river basin through its regular member fee and pays a special fee of several million Euro annually through 2014 (EMSCHERGENOSSENSCHAFT 2003). The fee seeks to recover increased operation and maintenance costs, which are directly linked e.g. to subsidences (EMSCHERGENOSSENSCHAFT 2003).[348] After 2014, the mining sector will still contribute to higher operation and maintenance costs of disentangled watercourses caused by its past activities. Furthermore, the mining sector fully pays for reconstruction measures that are directly necessitated by its activities. Hence, a polluter-pays approach for Type II EC is already in place in the Emscher river basin. An assessment of the adequacy of the present contribution of the mining sector is undertaken in Step 9 (cf. Section 4.3.6.2).

In order to ensure the security of old pits, the owners of mines are required to continue draining mining water, which then has to be discharged into the Emscher. In terms of WFD implementation cycles, it can be expected that the discharge of draining water will continue through 2027.

The assessment of Type II EC is taken up again in later phases of the investigation, namely when the adequacy of the contribution to cost-recovery of water uses is studied and the implementation of the polluter-pays principle is assessed.

4.3.4 Testing Step 5: assessing resource costs and their recovery

Step 5 of the methodology, the assessment of resource costs and their impact on the degree of overall (internal) cost-recovery is also illustrated on the case of the provision of wastewater services in the Emscher river basin. As discussed in Chapter 3, it may often be argued that no alternative higher value uses exist for

348 The owners of the mines contribute to this fee in relation to the degree to which their ongoing or expected mining activities had led or are expected to lead to subsidences and associated deteriorations in run-off capability. Also the owners of inoperative mines continue to contribute to the costs of long term damages of mining activities caused by the deterioration of run-off capabilities and subsidences.

the provision of water supply and wastewater services, due to their high societal preference. While this will in many cases be true, the following illustration of the Emscher river basin demonstrates the need for an unobstructed view on the specificities of each case. The severe environmental impact of the provision of wastewater services in the case of the Emscher river basin are linked to important obstructions of non-value uses, which could be prevented or minimized with a different organisation of service provision, i.e. as envisaged by the removal of the open sewer system and the accompanying rehabilitation projects foreseen for the Emscher river basin (cf. Section 4.3.3). Nevertheless, it is again vital to emphasise the link to the assessment of the polluter-pays principle (cf. Section 4.3.6.2) and the adequacy of contribution by other water users. While important obstructions and environmental damages (cf. analysis above) as well as resource costs may be linked to the present provision of wastewater services, its implementation was historically necessitated by the mining sector.

However, in the case of the provision of wastewater services, the analysis of alternative economic activities can be ignored due to the unsurpassed high societal preference for the service's provision. Hence, the following paragraphs investigate the impact that the provision of wastewater services exposes on ecosystem service in the Emscher river basin in greater detail. Finally, the illustrative testing on the Emscher river basin is complemented with information on the region Emilia-Romagna.

The case of the Emscher river basin: provision of wastewater services

Step 5.1: Screening for potential alternative water uses

The first screening of potential resource costs draws on the river basin management report, as well as on accompanying documents to the Emscher rehabilitation process (cf. Section 4.3.3). These outline the impact of the present mode of provision of wastewater services on the ecosystem. In line with the assessment framework proposed in Chapter 3, the following *Fact Sheet* 15 summarises a rough analysis of those ecosystem services, which may constitute relevant alternatives in the Emscher river basin giving rise to resource costs.

Fact Sheet 10 demonstrates the close connections between the assessment of Type I EC and resource costs. The assessment of Type I EC also outlined effects of service provision on ecosystems as far as they coincided with the achievement of the target situation *'good water status'*. However, it did not take into consideration whether the service provided by the ecosystem would

Fact Sheet 15: Rough analysis of resource costs stemming from ecosystem services (Emscher)

Ecosystem service	Affected by wastewater service provision	Description of impact
Fresh water provision	xxx	The open sewer system has lead to a severe degradation of water quality within the Emscher river basin. Water supply has to rely on alternative sources.
Water regulation	xxx	The modifications of the Emscher and its tributaries affected the natural water storage potential of the system: wetlands were removed; run-off was altered; river flow changed; and the river bed was constricted and straightened.
Erosion regulation	(x)	The modifications of the Emscher river basin led to changes in land cover.[349]
Water purification and treatment	xxx	The river basin's pollution carrying capacity has been strongly surpassed by the open sewer system.
Disease regulation	(positively affected)	(One main reason for the establishment of the open sewer system in the Emscher river basin was an insufficient natural disease regulation, cf. Section 4.2.1.2)
Recreation and ecotourism	xx	The open sewer system impacts in multiple ways on the region's recreation and (eco)tourism potential: odour problems linked to the open sewer system; hydromorphological alterations with negative impacts on the attractiveness of the overall landscape; severe impact on habitats; very low fish population (restrictions on angling). Overall, these impacts render the region less attractive for recreation and (eco)tourism.

349 These changes were however minor in comparison to those changes induced the by mining sector and can therefore be neglected in the upcoming analysis.

be more valuable than the water service's provision. The number of 'x' in *Fact Sheet* 10 clarifies the severity of the water service's impact on the ecosystem service. Generally, it is difficult to assess if the alternative ecosystem service would be more valuable than the present provision of the water service, specifically because of the high societal preference for wastewater services provision. However, in the case at hand, a reduction in resource costs (i.e. in the impact of service provision on ecosystem services) can be achieved by an extensive reorganisation of the mode of provision of wastewater services, i.e. by establishing of a separate sewer system and rehabilitating the river basin's hydromorphology.

The analysis of the degree to which other water uses should contribute to remedying these resource costs (i.e. the appropriate share of the mining sector) is deferred to Step 4.3.6.1 of the assessment, i.e. the adequacy of contribution by other water uses.

Step 5.2: Assessing the relevance/magnitude of resource costs and existing compensation schemes

As outlined in Chapter 3, for the purpose of the assessment at hand only an indication of the relevance/magnitude of resource costs is required and not an exact specification in monetary terms. The assessment should allow for a judgement on the degree to which uncovered resource costs impact on the overall cost-recovery level. To this end, a three-tiered classification scheme based on expert judgement was proposed, which assesses the magnitude/relevance of resource costs on a scale from 1 to 3:

(1) (Signalling a low magnitude/relevance of resource costs;
(2) Indicating a medium magnitude/relevance of resource costs, translating into a downward trend in cost-recovery levels in the integrative assessment (cf. Step 3);
(3) Indicating a significant magnitude/relevance of resource costs, necessitating a downgrading of cost-recovery levels in the integrated assessment (cf. Step 3) and an analysis of potential alternatives approaches for the provision of these water services in the future (cf. Section 3.3.3).

In Step 5.1, four ecosystem services were identified on which the provision of wastewater services displayed relevant effects:

- **Freshwater provision**: as the analysis of Type I EC above revealed, the provision of wastewater services through an open sewer system led over time to a severe degradation of water quality. Early on, it was decided to neglect the capacity of freshwater provision of the Emscher river basin (cf. Section 4.2.1.2) in order to (over)exploit the river's pollution carrying capacity. Hence, at the time the decision was made, the benefits to society as a whole of the open sewer system were considered to outweigh its negative consequences on the water environment. Hence, the region's water supply had to rely on alternative sources, i.e. from the river Ruhr. The extensive Programme of Measures outlined in Step 4.3 (Section 4.3.3) required for improving water status in the Emscher river basin and for remedying the effects of the open sewer system on water quality provide an indication of the effects on the ecosystem service of freshwater provision.

- **Water regulation**: the reconstruction of the Emscher and its tributaries affected the natural water storage potential of the system: wetlands were removed, the river's run-off and flow were altered, and the river bed was constricted and straightened. To remedy these effects, measures will have to be implemented over the upcoming three WFD implementation processes, as outlined in Step 4.3 above. These will require significant financial means.
- **Water purification and treatment**: The Emscher river basin's natural carrying capacity for polluted discharges and its natural treatment potential has been strongly surpassed by the open sewer system. Again, the extent of the measures (and related costs) outlined under Step 4.3, which aim to separate wastewater streams from the Emscher river and its tributaries by constructing a subsurface piped sewer system, can be taken as an indication of the magnitude of the effect on the ecosystem. In addition, those measures that are required for restoring water quality within the Emscher River and its tributaries have to be considered.
- **Recreation and tourism**: The alterations to the Emscher river basin, which were historically induced by the mining sector, but heightened by the resulting open sewer system for wastewater services provision, display an important impact on the river basin's recreation and tourism potential. The open sewer system created odour problems, led to significant landscape alterations, severely impacted river habitats, and led to a very low fish population (with the associated restrictions on angling). Overall, these changes rendered the region less attractive for recreation and ecotourism, resulting in a negative impact on the region's quality of life. Furthermore, they can be regarded as a competitive disadvantage for the region, hampering its potential for economic development. Investigations within the project *'Emscher.Zukunft'* outline the positive effects to be expected from the river basin's reconstruction, e.g. in terms of rising house prices and an improved standard of living for the local population and surrounding areas due to an improved recreation potential (EMSCHERGENOSSENSCHAFT 2009A). Next to the

river basin's ecological reconstruction a wide variety of additional land use planning measures has been envisaged. For example, hiking and cycling paths will be established along the Emscher to contribute to the recreation potential of the region. These effects have so far not figured prominently in the analysis, e.g. under *Step 4*.

Overall, the magnitude and relevance of resource costs is evaluated as medium (2), which translates into a downward trend of the cost-recovery level for the integrative assessment. While some of the effects on ecosystem services qualify as significant (and would thus result in a classification of (3)), the interlinkages with the assessment of environmental costs is very strong, so that these effects have in part – as far as they relate to water status – already been controlled for through the analysis under Step 4. Other effects, such as those on recreation and tourism have not yet been addressed.

For the overall assessment, it should be positively noted that a (future) alternative for the provision of water services has already been identified for the Emscher river basin and is currently being implemented.

Step 5.3: Indicate the impact of uncovered resource costs on overall environmental and resource cost-recovery rate

As the analysis in Step 5.2 revealed, resource costs are encountered in the provision of wastewater services, which should be considered in the assessment of the overall environmental and resource cost-recovery rate. While alternative economic activities are not relevant in the case of wastewater services provision in the Emscher river basin, due to the high societal preference given to the service's provision, ecosystem services are significantly affected by the service's present mode of operation. Hence, one needs to investigate how far a different collection and treatment technology may reduce the service's negative effects on ecosystem services, thereby generating a higher total economic value: this implies to compare the situation of wastewater services provision through an open sewer system, with the situation of a state-of-the-art sewerage collection and treatment, which simultaneously enables the provision of ecosystem services. The analysis in Step 4 and Step 5.2 has shown, that the effects of the present system on ecosystem services is medium, but that these effects may be mitigated or completely remedied through appropriate measures. As the empirical illustration case underlined, the analysis of environmental costs constitutes an important input to the assessment of resource costs related to ecosystem services.

Ultimately, the overall magnitude and relevance of resource costs is evaluated as medium (2). Therefore, the location of the Emscher river basin within Figure 13 (Chapter 3.1) does not have to be adjusted downward to a lower degree of internal environmental and resource cost-recovery through a move to the left on the 'x'-axis, but rather a downward trend has to be considered. It signals that resource costs are not yet fully accounted for, which has been thoroughly document through this analysis, but may be neglected in the further assessment of cost-recovery. Nevertheless, an analysis of the adequacy of contributions by different users should also account for these underestimated costs.

Extending the information base: Emilia-Romagna – irrigation water service

To extend the evidence gathered through the Emscher illustration case, *Step 5.1* of the analysis is also applied to the provision of irrigation water services in the region Emilia-Romagna (CBRO). The following *Fact Sheet* 16 provides an overview of the service's effect on the provision of ecosystem services as potential alternative uses.

As a result of this rough analysis, the ecosystem services *'water regulation'* and *'erosion regulation'* would warrant a further in-depth investigation.

4.3.5 Testing Step 7: evaluating the relevance of exemptions from Article 9

As outlined in Section 3.3.2, obtaining an exemption from the cost-recovery objective on the basis of Article 9 (4) requires a transparent procedure. The following section highlights the relevant aspects of the procedure proposed in Chapter 3 for the justification of an exemption from the cost-recovery objective based on environmental conditions and geographic and climatic conditions on the Emscher river basin, as these two justification pathways are less well defined in terms of indicators or guiding analytical questions. For the same reason, a testing of the proposed procedure for justifying an exemption from the cost-recovery objective due to social effects and economic effects is neglected as well-documented and established indicators were proposed for the assessment in Chapter 3.[350]

350 For an application of indicators related to social effects of water pricing and cost-recovery for water services, cf. OECD (2003), COURTECUISSE (2005) and KLAUER ET AL. (2007). On the economic effects of increased water prices in the Italian agricultural sector cf. BERBEL AND GUTIÉRREZ (2004) and BAZZANI ET AL. (2005).

Fact Sheet 16: Rough analysis of resource costs through irrigation water service provision (Emilia-Romagna)

Ecosystem service	Affected by irrigation service provision	Description of the impact
Fresh water provision	x(x)	Freshwater provision is locally influenced through nitrate and pesticide pollution from agriculture.[351]
Water regulation	xxx	Extensive canal networks are operated for providing irrigation water throughout the region, either through communal open canals or through pressure pipes. Significant water diversions from the Po river in the case of the C.E.R. and other rivers is undertaken; river flow is regulated to enable stable service provision; river beds have been adapted; historically, irrigation has fostered the extension of agricultural land use with effects on original land covers.
Erosion regulation	xxx	Changes in vegetation cover and work on water sides lead to problems with erosion and landslides.
Water purification and treatment	x	Partly affected through contaminated run-off from agricultural sites.
Disease regulation		(not relevant)
Recreation and ecotourism		(The cultivated agricultural landscape is considered an asset for recreation and ecotourism.)

Source: CBRO (2009), Bartolini et al.(2007), Massarutto (1999).

351 GALLERANI ET AL. (2004) note that *"environmental improvements would be desirable in the area, which lies within a region with relatively high pollution (despite efforts to reduce pesticides in recent years and to introduce more natural forms of pest management and organic farming methods). Present pollution pressures are 90 kg/ha per year of nitrogen and 10 kg/ha per year of active matter of pesticides (21 kg/ha per year of total products)."*

Emscher River Basin – wastewater services provision

Environmental conditions

As outlined in Section 3.3.2.2, unfavourable environmental conditions may lead to higher costs of water service provision. For example, if an unfavourable topography or a restricted carrying capacity of receiving waters requires the establishment of cost-intensive wastewater collection and treatment techniques, it may render a strong internal cost-recovery approach socially or economically unacceptable. In this case, a phased implementation of a strong internal cost-recovery approach could be justified (i.e. a time prolongation), also under the policy principle of regional equality. In addition, an exemption may be justified if unfavourable environmental conditions imply that cost-intensive measures have to be implemented for reaching good water status. This would have to be financed on top of the costs of water services provision within a river basin, leading to social hardship for certain groups of the population.[352]

In the Emscher river basin, an extensive and cost-intensive Programme of Measures has to be implemented for reaching good water status or good ecological potential. The Emscher river basin's water status has been severely affected by the river's long term use as an open sewer system and the associated hydromorphological alterations (cf. Step 4). As outlined in Section 4.3.3, a large variety of measures has to be undertaken in the upcoming three implementation cycles for reaching good water status / good ecological potential by 2027 in the Emscher river basin, whose implementation will require substantial financial means. Against this background, part of the Emscher river basin has been notified for time exemptions in line with WFD Article 4 (4) for reaching the Directive's environmental objectives (cf. MUNLV 2008A).

As a wastewater services provider, the *Emschergenossenschaft* and its members contribute substantially to the financing of measures. Hence, additional costs are invoked for the provision of water services due to the environmental conditions in the river basin. While these costs constitute in part an internalization of historic environmental costs caused by the provision of wastewater services in the past (cf. Section 4.3.3), existing environmental conditions also

352 In this case one could also argue, whether a time exemption should be sought for the implementation of the cost-intensive Programme of Measures (according to WFD Article 4 (4)). However, in a hierarchy of WFD objectives, reaching good water status figures more prominently and is given preference as the overall WFD objective compared to a timely implementation of an internal cost-recovery approach in line with Article 9.

lead to higher costs of water service provision (cf. discussion below). Increases in water service costs may induce affordability concerns for certain user groups within the service area, especially as the average available income in the Emscher region is below the national average.[353]

Hence, financing the required measures for reaching good water status could conflict with the simultaneous goal of establishing strong internal cost-recovery of water services in the Emscher river basin by 2010, especially when one includes all historic environmental costs. Therefore, other financing mechanisms (e.g. financial contributions from the *Land* North Rhine-Westphalia) may be invoked for financing part of the required measures and thus for internalising (historic) environmental costs of wastewater service provision. In the case of the Emscher river basin, a divergence from the strong internal cost-recovery approach and a prolonged implementation schedule could therefore be justified under Article 9(4). For this justification, it would however also be relevant to assess, whether other water uses contribute their adequate share to the costs of service provision, including environmental and resource costs (cf. Step 9 below).

Climatic and geographic conditions

In the Emscher river basin, local geographic conditions considerably influence the costs of water service provision. The impact of mining activities on the river bed's location and the flow of the river determined the infrastructural set-up for service provision, i.e. the technological options available for service provision were predetermined by the manmade alterations to the geographical surroundings. Substituting the open sewer system today by an environmentally viable and state-of-the-art closed sewer system requires substantial investments. Unfavourable geo-structural conditions resulting from a high density of urban development and multi-functional uses within the river basin further heighten the costs of investment.

First, a substantive number of registered potentially contaminated sites within the Emscher river basin (i.e. more than 4,000 sites, EMSCHERGENOSSENSCHAFT 2009A) leads to higher costs of network construction, e.g. as one must dispose

353 In the Emscher river basin, the costs of wastewater services for households are still below the North Rhine-Westphalian average (cf. Figure A4.1 in the Annex), rendering affordability concerns less prominent. Nevertheless, an investigation by MUNLV (presented by KLAUER (2008) on a European Commission Workshop in Copenhagen, internal workshop document) revealed that the lowest income deciles may still have to spend more than 3 percent of their disposable income on the provision of wastewater services, thus reaching the threshold of affordability established by the OECD in 2003.

of excavated contaminated soil. Second, the costs for constructing the pipe system cannot be compared to the costs that would be encountered in a '*green field*' approach (NAFO 2009). Due to the open sewer system established at the time in response to mining activities, a subsurface sewerage network was not developed along with urban development. Nowadays, special building techniques are required in laying the piped sewerage system. Due to the extensive urban development within the entire river basin, there is little unobstructed space available and construction costs are considerably higher, e.g. in order to protect existing structures and buildings during the construction process. These factors also impact the network's operation and maintenance costs, as the pipe system has to be constructed at greater depth, requiring expensive building techniques and, among others, higher pumping and monitoring costs in operation.

Hence, due to the geo-structural conditions resulting from a high density of urban development in the Emscher river basin, the costs of constructing and maintaining the relevant infrastructural assets are considerably higher than in a standard municipal system (EMSCHERGENOSSENSCHAFT 2009A). These higher costs add to the considerable financial burden encountered for internalising Type I EC of water service provision (cf. Step 4). A divergence from a timely implementation of a strong internal cost-recovery approach by 2010, e.g. due to financial contributions from external sources, could be justified against this background.

4.3.6 Testing Step 9: analysing the complementary objectives of Article 9

The following section verifies the proposed framework for testing compliance with the complementary objectives of Article 9, namely

- To set incentives for efficient resource use (*objective 2*);
- To ensure an adequate contribution by other water uses to the cost-recovery of water services in line with the polluter-pays principle (*objective 3*); and
- To promote sustainability of infrastructural systems for water service provision and transparency of financial flows in the water sector.

For each objective one illustration case is studied to exemplify the proposed methodological framework in a practical application.

4.3.6.1 *Incentives for efficient resource use*

Berlin – drinking water service provision

With the analysis of the tarification system of Berlin in Section 4.3.2, part of the required information for assessing the system's incentive structure has already been compiled. The *Fact Sheets* on endogenous revenue generation mechanisms indicated to which degree the present charging scheme accommodates individual consumer behaviour through variable tariff components and discourages excessive water use. From the analysis above, the following information can be extracted:

- A variable price component, dependent upon water demand makes up for a significant share of revenue generation[354], thus setting incentives for consuming less water at the level of each metering unit (i.e. the individual household or apartment, or to less extent at the level of a multi-storey apartment building[355]).
- There are no exemptions from the general tarification scheme for specific water service user groups. The concerns of vulnerable groups in water pricing are addressed through the general social security scheme.
- The fixed tariff component varies by meter size, thus promoting a user-pays related contribution to cost-recovery. Furthermore, it sets the incentive to avoid over-dimensioning of water meters and thus allows better projections of water demand and an adapted infrastructure.
- Consumption is fully metered in Berlin. The size of the water meter is adapted to the average consumption projected per inhabitant in the apartment block; and the overall water bill is determined by individual metered water consumption.
- A projected elasticity of water demand is taken into account in the construction of the tarification scheme.

An additional tariff component, the water abstraction charge, has not been studied in the analysis so far, as it qualifies (in part) as an internalisation mechanism for environmental costs of water service provision (i.e. regional scarcity

354 A fixed tariff component was only introduced in 2007 to guarantee a sustainable service provision and financing (cf. Section 4.2.1.3).

355 Data on the percentage of apartments in Berlin equipped with an individual meter are not available. In the case of a large apartment building, the individual household is unable to directly influence the costs of water supply, i.e. if a household saves water the effect on its bill may also be non-existent or negligible; the incentive effect is also dependent on how the house owner divides the costs among apartments.

costs and negative environmental effects of water abstraction), which has been extracted from the analysis of *Step 2* and referred to *Step 4*. While the level of the charge is determined politically instead of by environmental concerns, with e.g. varying levels for different abstraction zones depending on the sensitivity of the groundwater body, it was intended for internalising environmental costs by setting incentives for an efficient resource use.

The water abstraction charge, which has to be paid by the *Berliner Wasserbetriebe* per cubic meter abstracted[356], further intensifies the incentives for water saving provided by the variable pricing component. It amounts to 0.31 Euro/m³ and is by far the highest water abstraction charge in the whole of Germany.[357] The Berlin Water Law mandates that all of the water needed to cover Berlin's public water supply has to be taken from Berlin territory (§37 a (4) of the Berlin Water Law, *Berliner Wassergesetz*), unless an exemption is granted by the responsible administrative body of the Berlin Senate. Hence it is not possible for the *Berliner Wasserbetriebe* to substitute water abstraction in Berlin with water abstraction in Brandenburg at a larger scale, where a far lower water abstraction charge is operational of 0.102 Euro/m³.[358] The *Berliner Wasserbetriebe* operate one water work in Brandenburg (Stolpe) and are allowed to abstract 20.7 Mio m³ annually for water provision in Berlin. With a water abstraction of 138.4 million m³ in 2007 (private households), the *Berliner Wasserbetriebe* paid 38.3 million Euro of water abstraction charge to the *Land* Berlin and 1.5 Mio Euro to the *Land* Brandenburg for the provision of water to private households.[359] For the entire water supply segment of the *Berliner Wasserbetriebe*, the water abstraction charge amounted in 2007 to 11 percent of total costs (BWB 2007).

The revenue from Berlin's water abstraction charge is not specifically earmarked. The Berlin Water Law only mandates in § 13a (1) that the financial means generated by the charge should *primarily* be employed for the protection of the quality and quantity of groundwater and the prevention or mitigation of dangers to groundwater bodies. According to an interpellation by mem-

356 According to § 13a (1) Berlin Water Law, abstractions for groundwater management purposes are exempted from the charge. In addition, no water abstraction charge is levied on abstractions below 6000 m³ per year.

357 The water abstraction charge is paid per calendar year in estimated quarterly installments (§13 a (2) Berlin Water Law).

358 Under strict conditions, an exemption from this general prohibition may however be granted.

359 Assumption: 72 percent of 20.7 Mio m³ (14.9 Mio m³) are abstracted in Brandenburg (0.102 Euro/m³ water abstraction charge), while the rest of the water supply is guaranteed through water sources in the Berlin area (0.31 Euro/m³ water abstraction charge).

bers of the Berlin House of Representatives to the Berlin Senate as of 11 August 2008, only a comparatively small share of total revenues from the charge had been invested in groundwater management in the year 2007 with 1.3 million Euro.[360]

In addition, § 37 a (3) of the *Berlin Water Law* mandates the *Berliner Wasserbetriebe* to encourage its water service users (e.g. through targeted information campaigns) to use water rationally and to reduce its leakage rate to the economically achievable minimum.

As *exogenous revenue components* and *other revenue components* play a minimal role in the overall financing scheme of water service provision to private households in the case of Berlin, their impact on incentive setting can be considered negligible.

Problems related to the present incentive structure for sustainable water service provision

Due to the high variable component in the pricing scheme, individual households face strong incentives to safe water thereby lowering their individual water bill. However, on a broader perspective reductions in water demand lead to higher costs of water provision per cubic meter due to the sector's high share of fixed costs. Furthermore, with the over-dimensioned water supply system constructed in Berlin (due to the false projections after the political change that Berlin's population would double to six million inhabitants), further reductions in water demand led to technical problems in infrastructure management and maintenance, requiring costly compensatory measures. Overall, saving water may therefore not lead to a reduced water bill.

Next to technical-operational problems, sinking water demand levels also have a local impact on groundwater management, as rising water tables pose problems in certain areas of Berlin, which in turn lead to additional costs for society as a whole. To cushion the effects of decreased water demand in these areas, water is abstracted from groundwater bodies and discharged into surface water bodies. These costs are not directly linked to water supply, but could be avoided at higher water use rates as a side-benefit for the inhabitants of these regions, in case of avoided damp cellars, and society as a whole, in the case of preserved woods and avoided swamps. Nevertheless, other areas of Berlin are rather prone to (seasonal) water scarcity.

360 From 1998 to 2007, the *Land* Berlin obtained total revenue from the water abstraction charge of 572,651,506.54 Euro (Interpellation as of 11 August 2008).

Next to the information provided under the analysis of revenue generating mechanisms in Step 2, other scientific investigations have studied the incentive effects of pricing schemes for the agricultural sector in the region Emilia-Romagna.

BAZZANI ET AL. (2005) show in a modelled analysis for the region Emilia-Romagna based on a representative farm, which was constructed through a cluster analysis on a sample of 1969 farms, that water demand is inelastic to price changes until water prices are higher than the current price. The amount of water used by fruit crops is affected by changing water prices only above 0.5 Euro/m^3. As a consequence, pricing policies would require very high charges in order to produce relevant changes in water use. The authors demonstrate that demand is more elastic in the long-run, but only for relatively high water prices. They stipulate that in the long-run, actual farm reaction to an increase in water price is likely to depend on the change of the local social pattern, in particular on the demographic changes (age and retirement) affecting farmers. This is because changes in technological options are slow due to the need for long-run investment associated with tree crops and the weight of a strong commercial chain on farm decision making.

VIAGGI (2003) reports for the region Emilia-Romagna relevant incentives effects of pricing structures, particularly in those cases where volumetric pricing is adopted. According to interviews carried out among farmers, the price effect discourages the use of water for irrigation of water extensive annual crops (such as sugar beet) that were traditionally irrigated when water was taken from open canals with a fixed price. This seems to demonstrate that, at the present price, irrigation is still highly profitable for fruit and horticulture, but not for other annual crops. VIAGGI assumes that in cases where irrigation is still practiced on annual crops, famers already own the required irrigation machinery, as new purchasing or hiring would no longer be profitable. The study also revealed that farmers show different attitudes towards water use and attention to price. While older farmers do not adjust their management practices depending on the water price, younger farmers react more sensitive to the water price, as well as to other variables affecting their profit. This may be explained by the fact that younger people have a higher opportunity cost of labour and can shift more easily to other work than older farmers.

4.3.6.2 *An adequate contribution of water uses in line with the polluter-pays principle*

Emscher – wastewater service provision

The illustrations provided on the Emscher river basin in this chapter documented the immense influence of the mining sector on the (historic) development of the Emscher region in economic terms as well as on the status of water quality and the river basin's hydromorphological condition. As outlined in Step 4 of the analysis (Section 4.3.3), the open sewer system is linked to important environmental costs. However, it was installed due to a lack of technical alternatives given the ongoing mining activities within the region. Hence, while the environmental costs observed in the river basin today largely stem from the provision of wastewater services via an open sewer system, the system's installation was necessitated by mining activities.

Consequently, it has to be investigated, whether all environmental costs should in fact accrue to the water service of wastewater collection and treatment as Type I EC, or whether the causal link to the mining sector has to be considered in the refinancing of these environmental costs (i.e. as a '*historic Type II EC*') by means of the sector's adequate contribution.

Historically, the mining sector financed the open sewer system. To the system's reconstruction today, the mining sector contributes with its regular member fee and by bearing the full costs of rehabilitation measures that are directly linked to mining activities. This contribution is calculated on an annual basis and determined by the actual costs for rehabilitation measures. It implements the polluter-pays principle for past and present activities of the mining sector in the river basin. The revenue from this fee is strictly earmarked for rehabilitation measures. While difficulties in calculating the required financial means to be generated may relate to delineating the proportion of measures that is required due to mining activities alone, e.g. in the case of measures fulfilling multiple purposes or addressing several causes[361], these are not considered significant in practice. In addition, the mining sector pays a special fee through 2014 of several million Euro annually. The fee seeks to recover increased operation and maintenance costs, which are linked e.g. to subsidences (EMSCHERGENOSSENSCHAFT 2003). After the year 2014, the mining sector will still contribute to higher operation and maintenance costs of disentangled watercourses caused by its past activities.

361 For example, if habitat improvements are achieved by a measure, which were not only required due to mining activities but also for rectifying historic environmental costs.

In order to investigate the degree to which the polluter-pays principle is implemented with regard to historic environmental costs caused by mining activities, information on how the decision to construct the open sewer system was taken and whether the negative consequences of the aquatic environment were already taken into consideration at the time. It was a political decision to develop mining activities in the region in which the *Land* North Rhine-Westphalia was a key actor. Based on the decision to develop the mining sector, the region prospered and had an immense impact on the economic development of the whole of Germany. Today, these past activities pose environmental problems, which have to be rectified and financed in line with the polluter-pays principle as part of the Emscher reconstruction process.

The *Land* North Rhine-Westphalia contributes about one-fifth of the total costs to the reconstruction of the Emscher river basin. If all environmental costs related to the open sewer system were considered as Type I EC of wastewater services provision, this contribution would have to be classified as a subsidy.[362] However, considering the causal effect of the mining sector on the evolution of the present environmental costs, a different viewpoint is warranted. As a key actor in the political decision-making process that led to the development of the mining sector in the Emscher region, the *Land* North Rhine-Westphalia thereby contributes to the recovery of Type II EC that are caused by the mining sector. Whether contribution of about 20 percent of the total reconstruction costs is adequate for remedying the *Land*'s causal link in the generation of environmental costs remains therefore debatable.

Within this set of questions it should, however, also be noted that the mining sector already receives considerable state subsidies. Hence, a divergence from a more polluter-pays oriented financing of historic costs may in part be justified by social and economic reasons (cf. Section 4.3.5). Nevertheless, it should be noted that ability-to-pay considerations are also warranted for low income households, on which higher costs of wastewater service provision are now alternatively levied.

In conclusion, a political consensus between actors has been reached for this cost attribution. However, under the chosen approach, the polluter-pays principle is only partly implemented with regard to historic costs. A higher contribution by the mining sector and the *Land* North-Rhine Westphalia to the mitigation of historic Type I EC linked to the operation of the open sewer system would be warranted according to the assessment in line with the requirements of WFD Article 9.

362 Note: Step 4.1 and Step 4.3 discuss in how far the costs of measures for the reconstruction of the Emscher river basin qualify as Type I EC.

4.3.6.3 *Sustainability of infrastructure*

Berlin – drinking water service provision

In the following paragraphs, the criteria outlined in Chapter 3 for assessing the sustainability of service provision are applied to the case of drinking water provision to private households in Berlin:

- The depreciation approach adopted by the *Berliner Wasserbetriebe* is based on the replacement value of existing assets and is thus well in line with the proposed accounting framework. It enables the accumulation of sufficient financial means for reinvestment after the infrastructure's economic lifetime has expired.
- The annual rate of network renewal is quite high, considering the diverse age structure of the pipe network, with about 0.8 percent being renewed on an annual basis. The low percentage of pipe breaks of 0.1 percent testifies to the proactive maintenance and renewal of the pipe system.
- Security of supply is high, both in terms of disconnections as well as in terms of the network leakage rate, which is exceptionally low at 5.18 percent of abstracted water resources (OELMANN ET AL. 2009).
- There is full compliance with the European Drinking Water Directive. According to OELMANN ET AL. (2009), only in one in one thousand investigations of drinking water quality is the threshold value exceeded. In comparison, for the whole of Germany, four investigations in one thousand provide evidence of an exceeded threshold value.
- As the analysis under *Step 2* of the assessment framework (Section 4.3.2) demonstrated, exogenous revenue sources do not play a role in water service provision to private households in Berlin and therefore will not impede the long-term sustainability of infrastructure financing.
- The legislative framework for water pricing mandates the *Berliner Wasserbetriebe* to levy cost-recovering water prices on their customers and allows for conducting full cost pass-through. Under or over-coverage of costs in one period has to be balanced over the next two accounting periods.
- From 1999 to 2005, the *Berliner Wasserbetriebe* invested about 663 million Euro into the company segment water supply and abstracted 1.466 million cubic meter of water, resulting in investments of 0.45 Euro/m³. In comparison, the German average amounts to 0.38 Euro/m³. (OELMANN ET AL. 2009)
- According to OELMANN ET AL. (2009), dividend payments to private investors amounted to an annual average of 6.45 percent between 1999 and 2008 and are thus in line with established water sector averages (BDEW 2008) of

6–7 percent. Looking only at the post 2003 timeframe, annual dividend payments varied from 7.7 percent in 2003 to 13 percent in 2007, which would be high in comparison to the low risk associated with the water sector. OELMANN ET AL. (2009) conclude however in their investigation that these high levels are not likely to be maintained.

The following Figure 25 provides an overview on the financial flows within the system of water service provision to private households in Berlin.

Figure 25: Financial flows in water supply service provision in Berlin

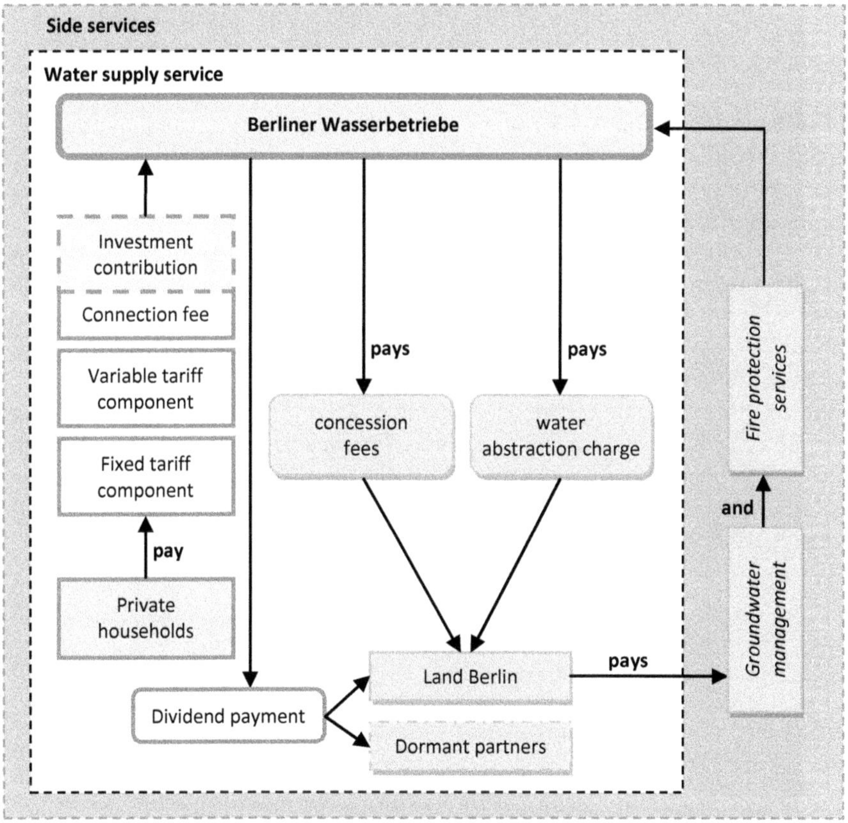

Overall, the accounting framework could be well-applied to the illustration steps of the case study Berlin. No severe divergences were encountered in the application of the cost accounting and revenue accounting framework. The *Fact*

Sheets allowed for documenting divergences, trends and expert judgements. A good quality of data was available for this case study, with most information being publicly available. Uncertainties relate to the exact internal distribution of endogenous revenue components, i.e. which share of revenues has been generated by the fixed tariff component etc. Here, only a rough indication could be obtained through expert interviews. Equal uncertainties relate to the magnitude of administrative and monitoring costs, which were also gathered through expert investigations. To conclude, the accounting framework proved successful in promoting the transparency objective by documenting costs and revenue components of water supply service provision in the case of Berlin.

Extending the information base: Emilia Romagna – irrigation water services

As evaluated in Sections 4.3.1 and 4.3.2, financing of (large scale) investments in the shared infrastructure system of the Bonifica is dependent upon public contributions, while its operation and maintenance is guaranteed through user contributions. Hence, the system may only function sustainably, if there is a recurrent inflow of construction subsidies. In terms of Article 9, the present system thus presents sustainability weaknesses in terms of financing. These weaknesses must be negated, either by describing how the system is intended to evolve over time into a more self-financing state or by outlining, how the necessary financial means are to be secured at the time of reinvestment. In addition, it needs to be outlined in relation to Step 7, how the divergence from a strictly internal cost-recovery approach is justified for the region, e.g. through social or economic considerations. The public good component of large scale shared infrastructure may also serve as a justification for a divergence in this case (cf. Section 2.1.4).

5 Conclusions and Outlook

This chapter provides a summary of the problem statements addressed by this thesis and summarizes main findings of the analysis. It outlines remaining open issues, which were identified in the respective chapters, warranting further concretisation in upcoming WFD implementation cycles. Finally, it discusses further research needs identified through the analysis.

5.1 The problem statements of the thesis revisited

Article 9 of the WFD formulates new and unprecedented objectives for water management in Europe, whose implementation will require the adaptation of existing practices. As the Article's dissection in Chapter 1.4 revealed, four partly contradictory objectives are explicitly referred to or implied by the Directive's text, namely:

Objective 1: recovering financial and environmental and resource costs for water services;

Objective 2: using water prices for setting incentives for efficient water resources use;

Objective 3: making the polluter pay and ensuring an adequate contribution of water uses;

Objective 4: establishing transparency of financial flows within the European water sector, and guaranteeing the sustainability of water services provision and infrastructural systems.

Implementing these objectives in a way that allows for European cross-country comparisons is by no means straightforward. While a certain harmonisation of definitions and procedures could be achieved through an exchange within the economic working groups of the CIS Process at EU level, the definition of and the accounting rules for the cost categories to be considered in the assessment of cost-recovery as well as the accounting for different approaches to water service financing varies considerably across Member States. Against this background, the EUROPEAN COMMISSION (2000) highlighted that *"the adoption of common definitions for key cost variables would facilitate the comparison between costs and prices*". Furthermore an investigation of the understanding of the concept of cost-recovery in different Member States as well as in the scientific literature reveals a notable discrepancy. Hence, a harmonised and comparable implementation of Article 9 requires further substantiations with regard to underlying definitions, the cost categories to be included in the assessment as well as the accounting approaches used for different financing schemes. Furthermore, a set of complementary indicators is needed that allows for compiling the required additional information on framework conditions for water service provision that is needed for a meaningful cross-country comparison of the objective's implementation.

Against this background, this thesis posed the question how a methodological assessment framework could be designed and made operational that allows for comparatively analysing the implementation of the partly contradictory objectives of Article 9 in different EU Member States along with the possibilities for justifying exemptions from their timely implementation. To this end, the thesis investigated in *Chapter 2* the influence of the perception of the different good dimensions of water on the provision and financing structure of water services, examined the cost-recovery understanding of Article 9 and outlined which cost components have to be addressed through an assessment of cost-recovery in line with WFD Article 9. Furthermore, the role of water pricing in implementing Article 9 was discussed and specificities of the household as well as the agricultural sector were highlighted. Overall, the chapter addressed the question of which cost factors should be covered through (irrigation or household) water prices and sewerage charges (or other sources) in order to ensure long-term sustainability of service provision. In addition, the chapter highlighted the significance of clearly defining and delineating water services for the assessment and provided definitions of environmental and resource costs along with potential economic approaches for their measurement.

Chapter 3 then addressed the question of which framework conditions are decisive for a comparative assessment of cost-recovery in different Member States and through which indicators they can be effectively measured. It proposed the methodological assessment framework for assessing cost-recovery as well as for investigating the implementation of the additional objectives of Article 9. Next to a cost accounting framework, which delivers accounting guidelines in line with Article 9 on each cost component to be considered in the assessment, a complementary revenue accounting framework was proposed. It allows analysing how the provision of water services is financed in different countries and in how far different financing schemes are in line with the objectives and requirements of Article 9.

Chapter 4 finally tested the methodology in illustrative cases in order to detect difficulties in applying the assessment framework and to generate first '*lessons learned*'. The methodological framework proposed in this thesis is set-up with the goal of delivering a pragmatic tool that is able to expand over the years as more reliable data become available while also accommodating national particularities. Hence, with its illustrative testing Chapter 4 posed the question where further research should concentrate on to improve practical implementation at later stages.

The following Chapter 5.2 outlines main findings of this thesis, summarising the requirements of WFD Article 9 and the prerequisites for a methodological assessment framework as well as associated '*lessons learned*' from the empirical illustration cases. Chapter 5.3 then concludes the chapter with an outlook on further research needs.

5.2 Main findings of this thesis

This thesis adds new insights on the analysis of costs and associated financing schemes of water service provision by developing an assessment framework in line with WFD Article 9 that is applicable to different European Member States. With this methodological framework, the thesis delivers a pragmatic tool whose application may facilitate an increased transparency of financial flows in the European water sector and sets the basis for meaningful comparisons of the implementation of Article 9 across Europe. More specifically, the thesis provides

- A clear-cut analysis of the underlying objectives of Article 9 and the intended WFD approach to cost-recovery;
- Concise definitions of cost categories and accounting guidelines in line with the specific requirements of WFD Article 9;
- A guiding analytical framework for analysing different financing schemes for water services provision and their respective compliance with the objectives of Article 9;
- A pragmatic approach for the incorporation of environmental and resource costs in the overall cost-recovery analysis;
- A set of guiding analytical questions and indicators for evaluating the justification of Article 9 exemptions and the implementation of the additional Article 9 objectives.

Overall, the thesis proposes an integrated assessment tool for the entire Article 9 implementation process, including its exemption regulations.

5.2.1 The requirements of WFD Article 9 and the prerequisites for a methodological assessment framework

When developing the integrated assessment framework on the implementation of Article 9, the underlying requirements of the Article first had to be investigated, and the prerequisite to which the assessment framework would have to adhere to, had to be singled out. To this end, *Chapter 1* investigated the underlying spirit of the Directive by outlining the evolution of EU water policy up to the WFD and by highlighting the objectives and novelties introduced by the Directive with a focus on its economic elements. Furthermore, the chapter identified the explicit as well as the implicit objectives of WFD Article 9 along with inherent controversies, thereby delineating the analytical frame for the methodological development.

Subsequently *Chapter 2* addressed these objectives in detail by providing the theoretical background for their incorporation in an integrated assessment framework. It was concluded that the perception of water's different dimensions at Member State level has important implications for the objectives assigned to water pricing and may significantly influence the financial structure chosen for financing the provision of water services. Regarding the cost-recovery approach under Article 9, the analysis in Chapter 2 comes to the conclusion that the Directive specifies the costs to society as a whole as the relevant reference frame for decision making, implying a wide cost definition for the assessment of cost-recovery. In other words, the Directive aims at a high degree of internal financial and environmental and resource costs recovery, with a use-/pollution-dependent allocation of costs onto different user groups and no costs being transferred to the next generation. Nevertheless, the analysis revealed that through its exemption regulations the Directive accepts an iterative approximation to this goal. Overall, transparency of how costs are currently recovered and a full cost accounting approach (i.e. which costs are encountered and by whom) are at the centre of attention in implementing Article 9. To facilitate this objective and to set the basis for the subsequent methodological development, the chapter provided definitions on each cost category to be addressed in the assessment of cost-recovery in line with WFD Article 9 and highlighted differences for the agricultural and the household sector. Furthermore, it provided an overview on different approaches to water pricing and highlighted specificities of the different models in terms of their allocative and informative, financing and incentive properties. As a further input to the methodological development, the chapter dissected the definition of water services and water uses along with their respective role in the implementation process of Article 9 and likewise addressed environmental and resource costs.

With these basic settings outlined, *Chapter 3* developed a comparative framework for assessing cost-recovery in line with the identified requirements. To this end, the financial cost categories identified in Chapter 2 were complemented with appropriate guidelines for their accounting in order to allow for a comparable cost basis and to make divergences from the Article 9 requirements transparent. In addition, a revenue accounting matrix was constructed, which allows for documenting all endogenous and exogenous revenue sources along with their incentive properties. On this basis, financial cost-recovery of water services provision can be assessed and divergences from the ideal Article 9 path be mapped out. To meet the requirements of Article 9, the analytical framework was amended with a procedure for assessing environmental costs, both caused directly by water service provision and by other water uses, and their recovery. Finally, Chapter 3 proposed a pragmatic procedure for assessing the role of resource costs, particularly the impact of unaccounted for resource costs, on the overall degree of cost-recovery in water service provision.

With these prerequisites for Article 9 implementation fulfilled, the methodological framework developed in the first three Chapters provides an integrative basis for evaluating the compliance of Member States with Article 9 of the Directive. Likewise, it may serve as a basis for developing a tool that allows customers of water services to assess how their services are being financed, which role accrues to the polluter-pays principle, whether the present burden-sharing is in line with societal preferences and water services are affordable to all.

In *Chapter 4*, the accounting framework was finally put to a practical test in terms of applicability and pragmatisms of the proposed procedures. The results of the illustrative testing of the methodology in selected cases are summarised in the following section.

5.2.2 Lessons learned from the empirical illustration cases

Using the household sector in Germany and the agricultural sector in Italy, *Chapter 4* applied different steps of the assessment framework to empirical examples. The chapter addressed the question of how well the methodological framework performed in its practical implementation in terms of applicability, practicality, adaptability to national particularities and the productivity of valuable and sufficiently detailed results for assessing the implementation of Article 9.

Overall, the methodological framework with its detailed *Fact Sheets* on costs and revenue sources and the accompanying guiding analytical questions could be well adapted to the regional specificities/particularities of each illustration case and prove to be a pragmatic tool. The accounting guidelines were precise enough to enable the identification of diverging approaches but likewise allowed for documenting national or regional particularities in a concise and effective manner.

In particular, the consultation of experts for gathering complementary information was well guided by the framework's detailed *Fact Sheets* and accompanying definitions. Nevertheless, a detailed introduction to the framework and its underlying objectives proved to be necessary when consulting experts, guaranteeing that the information provided is in line with the definitions proposed by the assessment framework (e.g. which information is required and what are the data supposed to cover and express concretely, which delineations to other cost categories have to be considered, etc.).

As was to be expected, the data basis for assessing environmental and resource costs was lacking information at this stage of the implementation process and posed the biggest difficulties. The foreseen role of expert judgement could however act as the expected mediating element, which could substitute existing information deficits and deliver important estimates for the assessment. On this basis, the cost based approach, which draws on the information provided by the Programme of Measures, could be applied. The methodology's application to the assessment of environmental and resource costs will become easier to apply as implementation progresses and more detailed information on the costs of measures and their associated effects becomes publicly available. Nevertheless, at this stage the necessary judgements and related ramifications could be provided through expert judgement at case study level without greater difficulties. However, the certainty and accuracy of the assessment is likely to improve over subsequent implementation cycles.

The framework's application in illustrative cases aimed to test different steps of the methodology in order to gain insights on how the system will work out in practice. While the illustrative examples proved useful to this end, a number of aspects could not be addressed in this abbreviated testing and may warrant further substantiations through in-depth cases. In particular, a full application of the entire assessment framework to an international river basin is recommended, in order to study the framework's adaptability to national particularities more closely and investigate problems related to the transboundary context (cf. Chapter 5.3).

5.2.3 Potential influence of the applied analytical procedure on results

The methodological framework was derived from a theoretical analysis of the objectives of Article 9 and the underlying prerequisites for their implementation. Subsequently, an assessment framework was constructed which allows for examining the implementation path chosen in a particular investigation area and for mapping out its divergence from the ideal Article 9 implementation path as defined by the Article's objectives. The assessment framework was constructed with the goal of accommodating national particularities. Illustrative cases were used for validating the proposed procedures and testing their practicality for an area-wide application.

Instead, the methodological framework could also have been derived from a case study centred approach. By comparing present financing schemes for the provision of water services in different EU Member States and present related cost accounting procedures, an accommodating framework could be deduced and cross-referenced for the requirements of Article 9. The disadvantage of this procedure is that one cannot assume that present approaches to cost-recovery assessments in European Member States address all aspects relevant to the implementation of Article 9, as the analysis of present interpretation of the cost-recovery concept alone already documented. Hence, important factors could have been overlooked. Furthermore, the results obtained would have critically hinged upon a relevant selection of cases. Given the variety of institutional arrangements in the European water sector, the decision on relevant selection criteria is by no means straightforward, but decisive for guaranteeing an encompassing variety of cases that allows for deducing all relevant aspects for developing the methodological framework in line with the objectives of Article 9.

The methodological approach developed in this thesis relies at times to expert judgement. It can provide valuable insight by complementing publicly available information sources, judging the quality of information provided as well as by evaluating outcomes on the basis of local in-depth knowledge. Foremost, expert judgement ensures pragmatic implementation procedure in light of data deficiencies within early implementation phases. Despite these advantages, expert judgement also opens the decision-making process to subjectivity, rendering the selection of experts a decisive task. The influence of expert judgement may vary between different application cases in accordance with the available information base. To limit the influence of subjectivity introduced by expert judgement, the methodology developed in this thesis provides the necessary guidelines for establishing transparency on decision-making processes and for documenting all relevant information. Overall, the influence of expert judgement will diminish as implementation progresses and the information and data base expands.

5.3 Outlook on further research needs

While the analytical framework proved to be applicable in its present set-up, further research needs can be identified for extending the empirical knowledge base and improving the robustness of the proposed procedures. In this regard, future research should focus on the following five aspects.

First, as outlined in Section 5.2.2, testing so far has concentrated on validating the overall framework, checking its applicability and its practicality. The focus of this thesis was clearly on the development of the methodological framework itself and not on the practical/empirical assessment of cost-recovery for selected river basins. While the illustrative cases proved useful for verifying the assessment framework, a **large-scale application of the methodology** in an international river basin may provide important additional insights, especially regarding the specificities of the transboundary context. While the selected illustration cases had very different underlying legal, social, environmental and economic framework conditions, extending the information base to other countries would make the analysis more robust.

Second, the implementation of the illustration cases demonstrated that the proposed assessment framework allows for accommodating data of varying quality and accuracy, and permits to document drawbacks and shortcomings of the present **data collection** system for the analysis of WFD Article 9 implementation. These will differ on a regional (or even local) basis. Hence, one result of the assessment framework is the proposition for (regionally) adapted changes in data collection. However, the framework can be applied in a meaningful manner without changes in data collection.[363] One of the aims in the framework's development was to accommodate all national data collection and

363 Unless data collection is severely lacking – but in that case, the framework delivers a "to-do list" of required improvements in data and information collection for implementing WFD Article 9.

aggregation approaches and to offer a universal translation key for Article 9 implementation. Nevertheless, an **exchange on key indicators** may be recommended, e.g. within the existing pan-European statistical activities. Furthermore, exchange with activities pursued by international organisations such as the OECD may offer additional insights (OECD 2009), facilitate future implementation and reduce overlap of activities.

Third, **exchange at the European level** and a further concretisation within the CIS process may be beneficial regarding the exemption regulations of Article 9 as well as regarding environmental and resource costs. On the **exemption regulation**, a harmonised EU approach would favour comparable implementation. It will be of particular importance for cross-country comparisons of Article 9 implementation to see how the issue of exemptions has been treated in the analysis. Therefore, exchange and agreement on key aspects of the analysis would be valuable. This thesis proposes a first framework and set of guiding analytical questions and indicators, which may be further concretised as implementation progresses. **Environmental and resource costs** continue to pose major difficulties for implementation across European Member States. While pan-European research projects are under way, which will deliver insights on the applicability of economic monetisation approaches for the valuation of benefits, these require extensive economic expertise and are time consuming (and often expensive) in their execution. These analyses will be indispensible in critical and well-selected cases to aid sound political decision-making; however, their area-wide application will not prove feasible for the next implementation cycle (at least). Hence, pragmatic intermediate approaches need to be relied upon in the meantime. The pragmatic assessment proposed in this thesis draws in part on expert judgement and information on measures and instruments compiled within the overall WFD implementation process. An exchange on possible intermediate procedures for the assessment as well as on definitions of environmental and resource costs at the European level would aid comparable implementation.

Fourth, with regard to the **evaluation at EU level** on the implementation of Article 9 in the different Member States, the analysis should focus less on the numeric results, i.e. the reported level of cost-recovery or the average water service price reported, but more on how the assessment has been approached. The overarching fourth objective of Article 9 – transparency of financial flows and sustainability of service provision – should be at the centre of attention. Hence, it is important how the reported information was derived and how remaining difficulties for the assessment will be tackled in the upcoming implementation process. The analysis in Chapter 2 revealed the strong pertaining differences in terms of definitions on cost-recovery, on incentive pricing and in particular on cost categories. As long as the applied definitions and procedures and their compliance with WFD Article 9 objectives cannot be made transparent, pan-European comparisons will not be feasible. The proposed accounting framework may serve as an initial solution, which can be further expanded on and adapted to ongoing implementation needs as implementation progresses. For an improved implementation by European Member States in upcoming implementation cycles, the European Commission should therefore make its evaluation criteria explicit and ensure that all objectives are addressed in an integrated way. Narrowing the assessment down to individual indicators (e.g. average water price within a river basin district) will not aid the Article's overall implementation. Instead, work within the CIS process at the European level on the economic aspects of the WFD should focus on the harmonisation of assessment and evaluation procedures. This is also important because of the emphasis given to water pricing and cost-recovery as supporting elements of other aspects of EU environmental policy, i.e. the adaptation to climate change.

Fifth, this study showed the immense influence of underlying definitions and procedures on the assessment of cost-recovery. Furthermore, it outlined different approaches to water pricing and their differences with respect to the functions they may fulfil and e.g. the degree to which they allow for incorporating the polluter-pays principle. By offering an assessment framework for costs and revenue generating mechanisms, which accommodates national and regional specificities, the proposed methodology offers one stepping stone for a further development of a framework for **cross-country water price comparisons**. Yet, as outlined in Chapter 2.5, water price comparisons are additionally hampered even at the local level, by differences in the local framework conditions for water service provision (e.g. the natural and socio-structural conditions), as well as by differences in water service quality. The present assessment framework addresses these aspects only partially, in particular through objective 4 of Article 9. To enable meaningful water price comparisons, they would have to be incorporated more strongly through an extended set of indicators in the assessment framework.

6 Literature

Abernethy, C. L. (2005). Financing river basin organisations. In M. Svendsen (Ed.), *Irrigation and River Basin Management: Options for government and institutions.* (pp. 75–92). Wallingford: CABI Publishing.

Abgeordnetenhaus Berlin (Berlin House of Representatives) (2005). *Antrag der Fraktion Bündnis 90/Die Grünen: Wasserbewirtschaftungsplan vorlegen – Wasserwerk Jungfernheide nicht schließen* (Drucksache 15/3551).

Ammermüller, B. (2009). Die Finanzierung von Maßnahmen zur Reduktion diffuser Gewässerbelastungen aus der Landwirtschaft. *Zeitschrift für Umweltrecht, 20*(5), 250–253.

Ammermüller, B., Weiß, N., Seifert, A., et al. (2009): *Kartellrechtliche Wasserpreiskontrolle, VKU-Leitfaden*, VKU: Berlin.

Ammermüller, B., Fälsch, M., Kochmann, L., Holländer, R., Brosowski, A., & Klauer, B., et al. (2008). *Handbuch: Kosten-Nutzen-Abwägung zur Feststellung von Ausnahmen aufgrund unverhältnismäßiger Kosten im Kontext der EG-Wasserrahmenrichtlinie.* Leipzig.

Arbués, F., García-Valinas, M. A., & Martínez-Espineira, R. (2003). Estimation of residential water demand: a state-of-the-art review. *Journal of Socio-Economics, 32,* 81–102.

Arrojo, P. (2001). The impact of irrigation water pricing policy in Spain. In European Commission (Ed.). *Pricing water – economics, environment and society. Conference Proceedings, Sintra, 6 and 7 September 1999* (pp. 177–193). Luxembourg: Office for Official Publications of the European Commission.

Arrow, K., Solow, R., Portney, P. R., Leamer, E. E., Radner, R., & Schuman, H. (1993). *Report of the NOAA panel on contingent valuation: Federal Register 58, 4601–4614.*

ATT – Association of Drinking Water from Reservoirs, BGW – Federal Association of the German Gas and Water Industries, DBVW – German Alliance of Water Management Associations, DVGW – The German Technical and Scientific Association for Gas and Water, DWA-German Association for Water, W. and Waste, & VKU – Association of Local Utilities (2005). *Profile of the German water industry.* Bonn.

ATT – Association of Drinking Water from Reservoirs, BGW – Federal Association of the German Gas and Water Industries, DBVW – German Alliance of

Water Management Associations, DVGW – The German Technical and Scientific Association for Gas and Water, DWA-German Association for Water, W. and Waste, & VKU – Association of Local Utilities (2008). *Profile of the German water industry.* Bonn.
Avis, C., Tydeman, C., & Gelabert, E. R. (2000). *What role for water pricing? Ten actions for internalising sustainability.* Brussels.
Bakker, M., van Schagen, K., & Timmer, J. (2003). Flow control by prediction of water demand. *Journal of Water Supply Research and Technology, 52*(6), 417–424.
Balance, T., & Taylor, A. (2005). *Competition and economic regulation in water: The future of the European water industry.* London: IWA-Publishing.
Bals, H. (1973). Benutzungsgebühren und Beiträge. Eine Erwiderung. *Kommunale Steuer-Zeitschrift*, 11, 211–213.
Bals, H., & Nölke, A. (1990). Volkswirtschaftliche Kosten und kommunale Gebühren: Ansätze für eine ökologische Neuorientierung des Kommunalabgabenrechts. *Kommunale Steuer-Zeitschrift, 39*(11/12), 201–225.
Barraqué, B. (1999). Environmental, economic and ethical sustainability of water service industry. In . *Water 21: towards a sustainable European water policy. Final Report to the European Commission* . Brussels.
Barraqué, B. (2003). Past and future sustainability of water policies in Europe. *Natural Resources Forum*, 27, 200–211.
Bärthel, H. (1997). *Wasser für Berlin: Die Geschichte der Wasserversorgung.* Berlin: Verlag für Bauwesen.
Bartolini, F., Bazzani, G. M., Gallerani, V., Raggi, M., & Viaggi, D. (2007). The impact of water and agriculture policy scenarios on irrigated farming systems in Italy: an analysis based on farm level multi-attribute linear programming models. *Agricultural Systems, 93*, 90–114.
Bartsch, V. (2007). *Technische, natürliche und rechtliche Einflussfaktoren auf betriebliche Kennzahlen von Wasserversorgungsunternehmen.* Bonn: wvgw Wirtschafts- und Verl.-Ges. Gas und Wasser.
Bate, R. (1994). *Pick a number: a critique of contingent valuation methodology and its application in public policy.* (Competitive Enterprise Institute, Ed.). Washington D.C.
Bateman, I. J., & Turner, R. (1992). *Evaluation of the environment: the contingent valuation method* (CSERGE/GEC Working Paper No. 92–18). Norwich.
Bateman, I. J., & Willis, K. G. (1999). *Valuing environmental preferences: theory and practice of the contingent valuation method in the US, EU and developing countries.* Oxford: Oxford Univ. Press.
Baumann, D. D., Boland, J. J., & Hanemann, W. M. (1998). *Urban water demand management and planning.* New York: McGraw-Hill.

Baumol, W. J., & Bradford, D. F. (1970). Optimal departures from marginal cost pricing. *American Economic Review, 60*(3), 265–283.

Baumol, W. J., & Oates, W. E. (1988). *The theory of environmental policy.* Cambridge: Cambridge Univ. Press.

Baumol, W. J., Panzar, J. C., & Willig, R. D. (1982). *Contestable markets and the theory of industry structure.* San Diego: Harcourt Brace Jovanovich.

Bazzani, G. M. (2005). A decision support for an integrated multi-scale analysis of irrigation: DSIRR. *Journal of Environmental Management*, 77, 301–314.

Bazzani, G. M., Di Pasquale, S., Gallerani, V., Morganti, S., Raggi, M., & Viaggi, D. (2005). The sustainability of irrigated agricultural systems under the Water Framework Directive: first results. *Environmental Modelling and Software, 20*, 165–175.

Bazzani, G. M., Di Pasquale, S., Gallerani, V., & Viaggi, D. (2002). *Water regulation and irrigated agriculture under the EU Water Framework Directive: Paper prepared for presentation at the Xth EAAE Congress 'Exploring diversity in the European agri-food system', 28–31 August 2002.* Zaragoza (Spain).

Beecher, J. A., Mann, P. C., Hegazy, Y., & Stanford, J. D. (1994). *Revenue effects of water conservation and conservation pricing: issues and practices.* Ohio State University, Ohio, USA.

Berbel, J., & Gómez-Limón, J. A. (2000). The impact of water-pricing policy in Spain: an analysis of three irrigated areas. *Agricultural Water Management, 43*, 219–238.

Berbel, J., & Gutiérrez, C. (2004). *Sustainability of European irrigated agriculture under Water Framework Directive and Agenda 2000.* (European Commission, & Directorate General for Research, Eds.). Brussels.

Bhattacharyya, A., & Harries, T.R., Narayanen, R., Raffiee, K. (1995). Technical efficiency of rural water utilities. *Journal of Agriculture and Resource Economics, 20*, 373–391.

Bishop, R. C., & Heberlein, T. A. (1984). *Contingent valuation method and ecosystem damages from acid rain.* Staff Paper no 217, University of Wisconsin, Madison, from University of Wisconsin.

Blöch, H. (1999). The European Union Water Framework Directive: taking European water policy into the next millennium. *Water Science and Technology, 40*(10), 67–71.

Blöch, H. (2004). European water policy and the Water Framework Directive: An overview. *JEEPL*, 3, 170–178.

BMU- Federal Ministry for the Environment, Nature Protection and Nuclear Safety (Ed.) (2002). *Privatisierung in der Wasserwirtschaft* (Umweltpolitik). Bonn.

BMU-Federal Ministry for the Environment, Nature Protection and Nuclear Safety (2005). *Water Framework Directive – Summary of river basin district analysis 2004 in Germany.* Berlin.

BMU- Federal Ministry for the Environment, Nature Protection and Nuclear Safety (2006). *Water resource management in Germany, Part 1 – Fundamentals* (Environmental Policy). Berlin.

Bockstael, N. E., McConnell, K. E., & Strand, I. E. (1991). Recreation in measuring the demand for environmental quality. *Measuring the demand for environmental quality.* Amsterdam: North Holland.

Bodisco, C. de (2007). *The regional value of water in agriculture.* Dissertation, Vanderbilt University, Nashville, Tennessee.

Boiteux, M. (1949). La tarification des demandes en point: (reprinted as "Peak Load Pricing" Journal of Business. 30(2), 157–179). *Revue Generale de l'Electricite.*

Boiteux, M. (1956). Sur la gestion des monopoles publics astreints à l'equilibre budgétaire. *Econometrica, 24*(1), 22–40.

Boiteux, M. (1971). On the management of public monopolies subject to budgetary constraints. *Journal of Economic Theory, 3*, 219–240.

Boland, J. J., & Whittington, D. (2000). The political economy of water tariff design in developing countries: increasing block tariffs versus uniform price with rebate. *The political economy of water pricing reforms.* 1. print., pp. 215–235. New York NY u.a.: Oxford Univ. Press.

Bontemps, C., & Couture, S. (2002). Irrigation water demand for the decision maker. *Environment and Development Economics, 7*, 643–657.

Bosenius, U. (2001). Die Wasserrahmenrichtlinie. *Wasser & Boden*, (1+2), 27–32.

Bosenius, U., & Holzwarth, F. (2006). Grundlagen für eine gemeinsame Strategie zur Umsetzung der WRRL in Europa. *Handbuch der EU-Wasserrahmenrichtlinie. Inhalte, Neuerungen und Anregungen für die nationale Umsetzung,* pp. 11–25. Berlin: E. Schmidt.

Bosold, M. (1994). *Ökologisch orientierte Preisbildung in öffentlichen Unternehmen: Eine exemplarische Untersuchung für öffentliche Wasserversorgungsunternehmen unter Berücksichtigung des Allokationsproblems der Common-Property-Ressource Wasser* (Vol. 25). *Schriftenreihe Wirtschafts- und Sozialwissenschaften*, 25. Ludwigsburg: Verlag Wissenschaft und Praxis.

Böttcher, G. (1998). *Kalkulatorische Kosten in der Gebührenberechnung kommunaler Einrichtungen: Ein Beitrag zur Abstimmung abgabenrechtlicher, betriebswirtschaftlicher und kommunalrechtlicher Grundsätze.* (Mutius, A. von, Ipsen, J., & Knemeyer, F.-L., Eds.). Baden-Baden: Nomos-Verl.-Ges.

Boulding, K. E. (1980). *The implications of improved water application.* (Federal Reserve Bank of Kansas City, W. Water Resources, Ed.). Boulder: Westview Press.

Boyce, R. R., Brown, T. C., McClelland, G. H., Peterson, G. L., & Schulze, W. D. (1992). An experimental examination of intrinsic values as a source of the WTA-WTP disparity. *American Economic Review, 82*(6), 1366–1373.

Boyle, K. J. (2003). Introduction to revealed preference methods. *A primer on nonmarket valuation,* pp. 259–268. Dordrecht: Kluwer.

Brackemann, H., Epperlein, K., Grohmann, A., Höring, H., Kühleis, C., & Lell, O., et al. (2000). *Liberalisierung der deutschen Wasserversorgung: Auswirkungen auf den Gesundheits- und Umweltschutz, Skizzierung eines Ordnungsrahmens für eine wettbewerbliche Wasserwirtschaft.* (Umweltbundesamt, Ed.).

Breuil, L. et al. (2004). *Chapter 6: Country Report Italy.*

Briscoe, J. (1995). *Financing water and sanitation services: the old and new challenges.* Durban: The World Bank (Keynote address to The World Congress of the International Water Supply Association). Retrieved April 22, 2007.

Briscoe, J. (1996). *Water as an economic good: the idea and what it means in practice.* Washington D.C. (A paper presented at the World Congress of the International Commission on Irrigation and Drainage, Cairo, September 1996). Retrieved April 22, 2007.

Briscoe, J. (1997). Managing water as an economic good: Rules for reformers. *Water Supply, 15*(4), 153–177.

Brockmann, C. (2003). *Die Handlungsfähigkeit der Europäischen Union – untersucht am Beispiel der EU Wasserrahmenrichtlinie.* Dissertation, Universität Heidelberg, Heidelberg, from Universität Heidelberg.

Brod, E. (1988). Die Kalkulation von Entgelten nach betriebswirtschaftlichen Grundsätzen auf der Grundlage von § 6 KAG für Nordrhein-Westfalen. *Gemeindehaushalt, 89*(4), 81–83.

Brookshire D. & Coursey, D. (1987). Measuring the value of a public good: an empirical comparison of elicitation procedures. *American Economic Review, 77*, 554–566.

Brouwer, R. (2000). Environmental value transfer: State of the art and future prospects. *Ecological Economics, 32*(1), 137–152.

Brouwer, R. (2006). *Practical working definition: environmental and resource costs and benefits: Aquamoney Deliverable D 12.* Amsterdam.

Brouwer, R., & Spaninks, F. A. (1999). The validity of environmental benefits transfer: Further empirical testing. *Environmental and Resource Economics, 14*(1), 95–117.

Brown, S., & Sibley, D. S. (1986). *The theory of public utility pricing.* Cambridge: Cambridge Univ. Press.
Budäus, D. (1978). Finanzierungspotenzial einer kostendeckenden Gebührenpolitik öffentlicher Einrichtungen aufgrund von Spielräumen bei der Kostenerfassung und -bewertung. *Verwaltungsarchiv, 69*(4), 361–376.
BWB (n.y.). *Corporate Governance Kodex.* Retrieved November 29, 2009, from http://www.bwb.de/content/language1/downloads/CGK.pdf.
BWB (2004). *Geschäftsbericht 2004 (annual report),* from http://www.bwb.de.
BWB (2007). *Grundlagen der Tarifkalkulation – Dokumentation,* from www.bwb.de.
BWB (2008a). *Wasserabsatz: Die Daten seit 1992,* from www.bwb.de.
BWB (2008b). *Tarifkalkulation 2008.* Berlin. Retrieved May 04, 2008, www.bwb.de/content/language1/downloads/kalkulation_2008.pdf
BWB (2009a). *Geschäftsbericht 2008 (annual report).* Berlin.
BWB (2009b). Trinkwasserversorgung – Hausanschluss. Gültig ab 28. Januar 2009. Berlin.
Cardone, R., & Fonseca, C. (2003). *Financing and cost-recovery* (Thematic Overview Paper).
Carson, R. T. (1991). Constructed markets. *Measuring the demand for environmental quality,* pp. 121–163. Amsterdam: North Holland.
Carson, R. T., Wright, J., Alberini, A., & Flores, N. E. (1995). *A bibliography of contingent valuation studies and papers.* La Jolla, California.
Carson, R. T., Flores, N. E., & Mitchell, R. C. (1999). The theory and measurement of passive-use values. *Valuing environmental preferences: theory and practice of the contingent valuation method in the US, EU and developing countries.,* pp. 97–130. Oxford: Oxford Univ. Press.
Carson, R. T., Flores, N. E., & Meade N.F. (2001). Contingent valuation: controversies and evidence. *Environmental and Resource Economics, 19,* 173–210.
Carson, R. T., Mitchell, R. C., Hanemann, W. M., Kopp, R. J., Presser, S., & Ruud, P. A. (2003). Contingent valuation and lost passive use: Damages from the Exxon Valdez oil spill. *Environmental and Resource Economics, 25,* 257–286.
Carver, P. H., & Boland J. J. (1980). Short-run and long-run effects of price on municipal water use. *Water Resources Research, 16*(4), 609–616.
Castro, F., Da-Rocha, J. M., & Delicado, P. (2002). Desperately seeking theta's: Estimating the distribution of consumers under increasing block rates. *Journal of Regulatory Economics, 22*(1), 29–58.
CBRO (2009b). *I contributi irrigui.* Lugo.
CBRO (2009). *Consorzio di Bonifica della Romagna Occidentale, Website,* from http://www.bonificalugo.it/.

Chambouleyron, A. (2003). *Optimal water metering and pricing* (Economics Working Paper Archive at WUSTL).

Champ, P. A., Boyle, K. J., & Brown, T. C. (2003). *A primer on nonmarket valuation.* Dordrecht: Kluwer.

Chavez, C. A., & Quiroga, M. A. (2002). *Regulatory schemes for water provision in theory and practice: Background Document for the "Second Meeting on Tariff Reform in Urban Water Sector Reform", Russian Federation, June 2–4, 2002.* Moscow.

Chettri, R., & Venkatesan, R. (1983). Water resource pricing: Policy options. *Resources Policy*, 288–295.

Chiesura, A., & Groot, R. de (2003). Critical natural capital: a socio-cultural perspective. *Ecological Economics, 44*(2–3), 219–231.

Chohin-Kuper, A., Rieu, T., & Montginoul, M. (2003). *Water policy reforms: pricing water, cost-recovery, water demand and impact on agriculture. Lessons from the Mediterranean experience: Water Pricing Seminar (June 30 – July 2, 2003), Agencia Catalana del Agua & World bank Institute. Parallel session C: The impact of cost-recovery on Agriculture.* Retrieved November 23, 2009, from http://www.semide.net/documents/database/water-policy-reforms-pricing-water-cost-recovery.

Cicchetti, C.J. & Wilde, L.L. (1992). Uniqueness, irreversibility and the theory of non use values. *American Journal of Agricultural Economics*, 73, 1121–1125

Ciriacy-Wantrup, S. V. (1947). Capital returns from soil conservation practices. *Journal of Farm Economics, 29*, 1181–1196.

CIS – Common Implementation Process (2001). *Common Implementation Strategy for the Water Framework Directive (2000/60/EC): Strategic document as agreed by the Water Directors under Swedish presidency on 2 May 2001.*

CIS – Common Implementation Strategy (2003a). *WATECO Guidance Document: Economics and the environment – The implementation challenge of the Water Framework Directive: CIS Guidance Document No 1.* Luxembourg.

CIS – Common Implementation Strategy (2003b). *Overall Approach to the classification of ecological status and ecological potential: Guidance Document No. 13.* Luxembourg.

CIS – Common Implementation Strategy (2004a). *Information Sheet: Assessment of the recovery of costs for water services for the 2004 river basin characterisation report (Article 9).*

CIS – Common Implementation Strategy (2004b). *Information Sheet: Assessment of environmental and resource costs in the Water Framework Directive.*

CIS – Common Implementation Strategy (2004c). *Information Sheet: Methodology to prepare a baseline scenario.*

CIS – Common Implementation Strategy (2004d). *Information Sheet: River basin characterization: Economic analysis of water uses (Art 5 Annex III).*

CIS – Common Implementation Strategy (2005). *Pilot River Basin outcome report: Testing of the WFD Guidance Documents.* Luxembourg.

CIS – Common Implementation Strategy (2008). *Pilot River Basin Activity Report Phase II: 2005–2006: Water Framework Directive implementation pilot activities: Key challenges and recommendations from the Pilot River Basins.* Luxembourg, from http://ec.europa.eu/environment/water/water-framework/ prbs.htm.

CIS – Common Implementation Strategy (2009). *Guidance on exemptions to the environmental objectives* (Technical Report No. 27). Luxembourg.

Coase, R. H. (1946). The marginal cost controversy. *Economica, 13*, 169–182.

Coase, R. H. (1970). The theory of public utility pricing and its application. *The Bell Journal of Economics and Management Science, 1*(1), 113–128.

Cornish, G., Bosworth, B., Perry, C., & Burke, J. (2004a). *Water charging in irrigated agriculture: An analysis of international experience.* (Food and Agriculture Organisation of the United Nations, Ed.). Rome.

Cornish, G., Perry, C., & van Steenbergen, F. (2004b). *Charging for irrigation services: Guidelines for practitioners.*

Correia F.N. (1998). *Institutions for water resources management in Europe. Eurowater 1.* Rotterdam: Balkema.

Correia F.N. (1999). *Water 21: towards a sustainable European water policy: Final Report to the European Commission.* Brussels.

Courtecuisse, A. (2005). *Water prices and households' available income: Key indicators for the assessment of potential disproportionate costs: Illustration from the Artois-Picardie Basin (France).* Vienna (IWG-Env, International Work Session on Water Statistics, June 20–22 2005).

Crew, N., & Kleindorfer, P. (1979). *Public utility economics.* New York: St. Martins Press.

Cronauge, U. (2001). *Kommunale Unternehmen: Eigenbetriebe – Kapitalgesellschaften – Zweckverbände* (3., überarb. und erw. Aufl., Vol. 3). *Finanzwesen der Gemeinden, 3.* Berlin: Schmidt.

Cummings, R. G., Brookshire, D. S., & Schulze, W. D. (1986). *Valuing environmental goods: a state of the arts assessment of the contingent valuation method.* Totowa, N.J.: Rowman and Allanheld.

Dalhuisen, J. M., Groot, H. de, & Nijkamp, P. (2000). The economics of water: a survey of issues. *International Journal of Development Planning Literature.15*, 3–20.

Dalhuisen, J. M., Florax, R. j. G. M., Groot, H. L. F. de, & Nijkamp, P. (2003). Price and income elasticities of residential water demand: a meta-analysis. *Land Economics, 79*(2), 292–308.

Daly, H. E., Farley, J. (2004). *Ecological economics: Principles and applications.* Washington D.C.: Island Press.

Dandy, G., McBean, C., & Hutchinson, B. (1984). A model for constrained optimum water pricing and capacity expansion. *Water Resources Research, 20*(5), 511–520.

Dandy, G., Nguyen, T., & Davies, C. (1997). Estimating residential water demand in the presence of free allowances. *Land Economics, 73*(1), 125–139.

Deacon, R., & Shapiro, P. (1975). Private preference for collective goods revealed through voting referenda. *American Economic Review, 65*, 943–955.

Desvousges, W. H., Johnson, F. R., Dunford, R. W., Boyle K.J., Hudson, S. P., & Wilson, N. (1993a). Measuring natural resource damages with contingent valuation: tests of validity and reliability. *Contingent valuation: a critical assessment,* pp. 91–159. New York: North Holland.

Desvousges, W. H., Gable A. R., Dunford, R. W., & Hundson, S. P. (1993b). Contingent valuation: The wrong tool to measure passive-use lossess. *Choices*, 9–11.

Diakité, D., Semenov, A., & Thomas, A. (2009). A proposal for social pricing of water supply in Côte d'Ivoire. *Journal of Development Economics, 88*(2), 258–268.

Diamond, P. A., & Hausman, J. A. (1993). On contingent valuation measurement of nonuse values. *Contingent valuation: a critical assessment,* pp. 41–89. New York: North Holland.

Dinar, A. (1998). Irrigated agriculture and the environment: Problems and issues in water policy. In OECD (Ed.). *Sustainable management of water in agriculture. Issues and policies. The Athens Workshop.* (pp. 41–56). Paris: OECD Publications.

Dinar, A. (2000). *The political economy of water pricing reforms.* New York: Oxford Univ. Press.

Dinar, A., & Mody, J. (2004). Irrigation water management policies: allocation and pricing principles and implementation experience. *Natural Resources Forum*, (28), 112–122.

Domene, E., & Saurí, D. (2006). Urbanisation and water consumption: influencing factors in the metropolitan. *Urban Studies, 43*(9), 1605–1623.

Dudu, H., & Chumi, S. (2008). *Economics of irrigation water management: a literature survey with focus on partial and general equilibrium models* (Policy Research Working Papers – Sustainable Rural and Urban Development Team No. 4556). Washington D.C.

Dupuit, J. (1854). *Traité théorique et pratique de la conduite et de la distribution des eaux.* Paris.
Easter, K. W., & Liu, Y. (2005). *Cost-recovery and water pricing for irrigation and drainage projects.* Washington D.C.: The World Bank.
Edwards, S. F. (1992). Rethinking existence values. *Land Economics, 68,* 120–122.
Eisenreich, S. J. (2005). *Climate change and the European water dimension.* Ispra, Italy.
Ekelund, R. B. (1968). Jules Dupuit and the early theory of marginal cost pricing. *The Journal of Political Economy, 76*(3), 462–471.
Ekins P. (2000). *Economic growth and environmental sustainability.* London: Routledge.
Emschergenossenschaft (Ed.) (2003). *Veranlagungsgrundsätze: (Am 19.11.2003 durch die Genossenschaftsversammlung beschlossen).* Essen.
Emschergenossenschaft (Ed.) (2004). *Daten und Fakten.* Essen.
Emschergenossenschaft (Ed.) (2006). *Emscher:Zukunft – Masterplan 2006.*
Emschergenossenschaft (Ed.) (2008). *Geschäftsbericht 2007/2008.* Essen.
Emschergenossenschaft (Ed.) (2009a). *Flussgebietsplan Emscher.* Essen.
Emschergenossenschaft (2009b). *Fördermittel bei EG/LV. Internal document.*
Environmental Law Institute (1999). *Innovation, cost and regulation: Perspectives on business, policy and legal factors affecting the cost of environmental compliance.*
Espey, M., Espey J., & Shaw W. D. (1997). Price elasticity of residential demand for water: a meta-analysis. *Water Resources Research, 33*(6), 1369–1374.
European Commission (2000). *Communication from the Commission to the Council, the European Parliament and the Economic and Social Committee: Pricing policies for enhancing the sustainability of water resources.* Brussels (COM (2000) 477 final).
European Commission (Ed.). 2001. *Pricing water – economics, environment and society. Conference Proceedings, Sintra, 6 and 7 September 1999.* Luxembourg: Office for Official Publications of the European Commission.
European Commission (2002). *Decision No 1600/2002/EC of the European Parliament and of the Council of 22 July 2002: Laying down the Sixth Community Environment Action Programme.* Retrieved January 04, 2007, from http://europa.eu/eur-lex/pri/en/oj/dat/2002/l_242/l_24220020910en00010015.pdf.
European Commission (2007). *Communication from the Commission to the European Parliament and the Council: Towards sustainable water management in the European Union: First stage in the implementation of the Water Framework Directive 2000/60/EC, COM(2007) 128 final.*

European Commission (2009). *Commission Staff Working Document accompanying the White Paper: Adapting to climate change: Towards a European framework for action. Climate change and water, coasts and marine issues* (No. SEC(2009) 386). Brussels.

European Council (1999). *Gemeinsamer Standpunkt (EG) Nr. 41/1999 vom Rat festgelegt am 22. Oktober 1999 im Hinblick auf den Erlaß der Richtlinie 1999/./EG des Europäischen Parlaments und des Rates zur Schaffung eines Ordnungsrahmens für Maßnahmen der Gemeinschaft vom ... im Bereich der Wasserpolitik.* ABl. EG C 343/1 of 30 November 1999.

European Parliament (1999). *Vorschlag und geänderte Vorschläge für eine Richtlinie des Rates zur Schaffung eines Ordnungsrahmens für Maßnahmen der Gemeinschaft im Bereich der Wasserpolitik (KOM(97)0049 - C4–0192/97, KOM(97)0614 - C4–0120/98 und KOM(98)0076 - C4–0121/98 - 97/0067 (SYN)).* ABl. EG C 150 of 28 May1999, 388–419.

European Parliament (2000a). *Empfehlung für die zweite Lesung betreffend den Gemeinsamen Standpunkt des Rates im Hinblick auf den Erlass der Richtlinie des Europäischen Parlaments und des Rates zur Schaffung eines Ordnungsrahmens für Maßnahmen der Gemeinschaft im Bereich der Wasserpolitik (9085/3/1999, C5–0209/1999, 1997/0067(COD)), A5–0027/2000 endgültig, PE 231.746, 3 February 2000.*

European Parliament (2000b). *Legislative Entschließung des Europäischen Parlaments zum Gemeinsamen Standpunkt des Rates im Hinblick auf den Erlass der Richtlinie des Europäischen Parlaments und des Rates zur Schaffung eines Ordnungsrahmens für Maßnahmen der Gemeinschaft im Bereich der Wasserpolitik (9085/3/1999 – C5–0209/1999 – 1997/0067(COD)), A5–0027/2000, 16. Februar 2000; ABl. EG C 339 vom 29 November 2000, 136–154.*

European Parliament (2000c). *Bericht über den vom Vermittlungsausschuss gebilligten gemeinsamen Entwurf einer Richtlinie des Europäischen Parlaments und des Rates zur Schaffung eines Ordnungsrahmens für Maßnahmen der Gemeinschaft im Bereich der Wasserpolitik (C5–0347/2000 – 1997/0067 (COD)), A5–0214/2000, 29 August 2000.*

Ewers, H.-J., Botzenhart, K., Jekel, M., Salzwedel, J., & Kraemer, R. A. (2001). *Optionen, Chancen und Rahmenbedingungen einer Marktöffnung für eine nachhaltige Wasserversorgung. Endbericht Juli 2001.* (BMWi-Forschungsvorgaben No. 11/00).

Ewringmann, D. (2006). Der kostendeckende Wasserpreis. In forum a.s.b.l. (Ed.). *forum für Politik, Gesellschaft und Kultur. Wasser* (pp. 24–28). Luxembourg: Forum.

Faucheaux, S., & O'Connor, M. (1998). *Valuation for sustainable development: methods and policy indicators.* London: Elgar.

Feldman, S. L., Breese, J., & Obeiter, R. (1981). The search for equity and efficiency in the pricing of a public service: urban water. *Economic Geography, 57*(1), 78–93.
Feldstein, M. S. (1972). Equity and efficiency in public sector pricing: the optimal two-part tariff. *Quarterly Journal of Economics, 86*(2), 175–187.
Finger, M., Allouche, J., & Luís-Manso, P. (2007). *Water and liberalisation: European water scenarios.* London: IWA-Publishing.
Forst, J. (2001). Die Erhebung von Grundgebühren für die Wasserversorgung und die Abwasserbeseitigung. *Kommunale Steuer-Zeitschrift, 8*, 141–158.
Fraiture, C. de, & Perry, C. (2002). *Why is irrigation water demand inelastic at low price ranges?: Paper Presented at the Conference on "Irrigation Water Policies: Micro and Macro Considerations", 15–17 June 2002.* Agadir, Morocco.
Fredman, P. (1995). The existence of existence value — a study of the economic benefits of an endangered species. *Journal of Forest Economics, 1*(3), 307–327.
Freeman, A. M. (1993). *The measurement of environmental and resource values: Theory and methods.* Washington D.C.: Resources for the Future.
Friedl, U. (1996). Zielsetzungen und Wirkungen kommunaler Benutzungsgebühren. *Kommunale Steuer-Zeitschrift, 45*(10/11), 181–210.
Gabor, A. (1956). A note on block tariffs. *The Review of Economic Studies, 23*(1), 32–41.
Gallerani, V., Bazzani G.M., Viaggi D., Bartolini F., & Raggi M. The case of Italy European Commission & Directorate General for Research (Eds.). *Sustainability of European irrigated agriculture under Water Framework Directive and Agenda 2000,* pp. 141–164. Brussels.
Garcia, S., & Reynaud, A. (2004). Estimating the benefits of efficient water pricing in France. *Resource and Energy Economics, 26*, 1–25.
García-Valinas, M. A. (2005). Efficiency and equity in natural resources pricing: a proposal for urban water distribution. *Environmental and Resource Economics, 32*, 183–204.
Garrido, A. (2001). Pricing for water use in the agricultural sector. In European Commission (Ed.). *Pricing water – economics, environment and society. Conference Proceedings, Sintra, 6 and 7 September 1999* (pp. 105–117). Luxembourg: Office for Official Publications of the European Commission.
Gatersleben, B., Steg, L., & Vlek, C. (2002). The measurement and determinant of environmentally significant consumer behaviour. *Environment and Behaviour, 34*(3), 335–362.

Gawel, E. (1994). Zur Rechtfertigung von Wiederbeschaffungszeitwerten in der Kalkulation kommunaler Benutzungsgebühren. *Kommunale Steuer-Zeitschrift, 43*(5), 81–89.
Gawel, E. (1995). *Die kommunalen Gebühren: Ökonomische, ökologische und rechtliche Ansätze einer gesamtwirtschaftlichen Neuorientierung* (N.F., Bd. 64). *Finanzwissenschaftliche Forschungsarbeiten, N.F., Bd. 64.* Berlin: Duncker & Humblot.
Gawel, E. (1996). Die Gebühren vor der ökologischen Herausforderung: Möglichkeiten und Grenzen einer gesamtwirtschaftlichen Ausrichtung kommunaler Entgelte /. *Kommunale Steuer-Zeitschrift, 45*(2), 21–41.
Gibbons, D. (1986). *The economic value of water.* Washington D.C.: Resources For the Future.
Gleick, P. H., Singh, H., & Shi, H. (2001). *Emerging Threats to the World's Freshwater Resources.* Oakland, California: Pacific Institute for Studies in Development, Environment and Security.
Gómez-Limón, J. A., & Berbel, J. (2000). Multicriteria analysis of derived water demand functions: a Spanish case study. *Agricultural Systems, 63*(1), 49–72.
Gómez-Limón, J. A., & Riesgo, L. (2004). Irrigation water pricing: differential impacts on irrigated farms. *Agricultural Economics, 31*, 47–66.
Gordon-Walker, S., & Marr, S. (2002). *Study on the application of the competition rules to the water sector in the European Community: Prepared for the European Commission – Competition Directorate General* (WRc Report No. N°: UC6064). Swindon.
Goria, A. and Nicola Lugaresi (2002). *The Evolution of the National Water Regime in Italy.*
Görlach, B., & Interwies, E. (2005). Was ist Wasser wert? Die Ermittlung von Umwelt- und Ressourcenkosten nach der Wasserrahmenrichtlinie. *Ökologisches Wirtschaften*, (2), 33–34.
Görlach, B., & Pielen, B. (2007). *Disproportionate costs in the EC Water Framework Directive – the concept and its practical implementation: Paper presented at the envecon 2007 Applied Environmental Economics Conference, London, 23 March 2007.*
Görlach, B., Roo, C. de, Ehlers, M. H., & Bräuer, I. (2007). *Policy maker demand for economic information assessment in the implementation of the EC Water Framework Directive: Project Report, AquaMoney, Development and Testing of Practical Guidelines for the Assessment of Environmental and Resource Costs and Benefits in the WFD* ((unpublished manuscript)).
Green, C. (2000). If only life were that simple: Optimism and pessimism in economics. *Physics and Chemistry of the Earth, 25*(3), 205–212.

Green, C. (2003). *Handbook of water economics. Principles and practice.* Chichester: Wiley.
Grimble, R. J. (1999). Economic instruments for improving water use efficiency: theory and practice. *Agricultural Water Management, 40,* 77–82.
Grimeaud, D. (2001). Reforming EU water law: towards sustainability? *European Environmental Law Review, 10*(4), 125–135.
Grombach, P., Haberer, K., Merkl, G. & Trüeb, E. U. (2000). *Handbuch der Wasserversorgungstechnik* (3., völlig überarb. Aufl). München: Oldenbourg-Industrieverlag.
Groot, R. S. de (1992). *Functions of nature: Evaluation of nature in environmental planning, management and decision making.* Groningen: Wolters-Noordhoff.
Hall, D. C., & Hanemann, W. M. (1996). Urban water design based on marginal cost. *Marginal cost rate and wholesale water markets. Advances in the economics of the environment.* London: JAI Press.
Hanemann, W. M. (1991). Willingness to pay and willingness to accept: how much can they differ? *American Economic Review, 81,* 635–647.
Hanemann, W. M. (1994). Valuing the environment through contingent valuation. *Journal of Economic Perspectives, 8,* 19–43.
Hansjürgens, B., & Messner, F. (2002). Kostendeckende und verursachergerechte Wasserpreise in der EU-Wasserrahmenrichtlinie. *Wasser & Boden, 54*(7+8), 66–69.
Hansjürgens, B., & Messner, F. (2006). Erhebung kostendeckender Preise in der WRRL. *Handbuch der EU-Wasserrahmenrichtlinie. Inhalte, Neuerungen und Anregungen für die nationale Umsetzung,* pp. 399–424. Berlin: E. Schmidt.
Harris, J., Tate, D., & Renzetti, S. (2002). *Economic principles and concepts as applied to municipal water utilities.* Final Report. GeoEconomics Associates Incorporated.
Hattermann, F. F., Krysanova, V., Post, J., Dworak, T., Wrobel, M., & Kadner, S., et al. (2008). Understanding consequences of climate change for water resources and water-related sectors in Europe. *European Commission community research: The adaptiveness of IWRM. Analysing European IWRM research,* pp. 89–112. London: IWA Publishing.
Hausman, J. A. (1993). *Contingent valuation: a critical assessment.* New York: North Holland.
Hax, K. (1957). *Die Substanzerhaltung der Betriebe* (1. Aufl.). Köln: Westdt. Verl.
Hecht, D., & Werbeck, N. (2006). *Bewirtschaftung von Flusseinzugsgebieten: Institutionenökonomische Betrachtungen ausgewählter Regeln der Wasserrahmenrichtlinie.* Marburg: Metropolis-Verlag.

Heinz, I. (2005). *How can the WFD cost categories made more feasible?: 2nd international workshop on implementing economic analysis in the Water Framework Directive, Paris 17–18 February 2005.* Paris.

Hellegers, P., & Perry, C. (2004). *Water as an economic good in irrigated agriculture: Theory and practice* (Report 3.04.12). The Hague.

Hellegers, P. (2002). *Treating water in irrigated agriculture as an economic good.* Agricultural Economics Research Institute (LEI).

Herbke, N., Pielen, B., Ward, J. und Kraemer, R. A. (2006): *Urban Water Management – Case Studies: Berlin and Emscher Region ("Evaluation of Sustainability of Sewerage Systems in Metropolitan Areas – Case Studies Berlin and Emscher"), Background Paper for the Conference: Economic Analysis of the Milan Wastewater Treatment System: Sustainability Issues*, Organisational Models and Future Challenges, Mailand, 4. Mai 2006. Berlin: Ecologic.

Hirshleifer, J., Haven, J. de, & Milliman, J. (1960). *Water supply: economics, technology and policy,* University of Chicago, Chicago.

Hödl, E. B. (2005). *Wasserrahmenrichtlinie und Wasserrecht* (Vol. 32). *Recht-Monografien, 32.* Wien: NWV Neuer Wiss. Verl.

Hoekstra, A. (1998). Appreciation of water: four perspectives. *Water Policy, 1*(6), 605–622.

Hoffmeister, J., Tettinger, S., & Staben, N. (2008). Demografie und wirtschaftliche Entwicklung in Deutschland: Konsequenzen für die wasserwirtschaftliche Infrastruktur. *Wasser und Abfall,* (6), 10–16.

Höhndorf, M. (2009). *Expert interviews on 11 August 2009 and 28 November 2009. Berliner Wasserbetriebe.*

Holländer, R., Fälsch, M., Geyler, S., & Lautenschläger, S. (2009). *Leitfaden: Trinkwasserpreise in Deutschland – Wie lassen sich verschiedene Rahmenbedingungen für die Wasserversorgung anhand von Indikatoren abbilden? Im Auftrag von: Verband kommunaler Unternehmen (VKU), Berlin.* Leipzig.

Holländer, R., Zenker, C., Ammermüller, B., Geyler, S., & Lautenschläger, S. (2008). *Trinkwasserpreise in Deutschland – Welche Faktoren begründen regionale Unterschiede?: Im Auftrag von: Verband kommunaler Unternehmen (VKU), Berlin.* Leipzig.

Holtmeier, E. L. (1997). *Development of European water legislation: Environmental protection in Europe as well as on the federal and state level – Conference Proceedings.*

Holzwarth, F., Stratenwerth, T., Schwarz, F., Hornemann, C., Dworak, T., & Leipprand, A., et al. (2007). *Symposium report: Time to adapt: Climate change and the European water dimension – vulnerabilities, impacts, adaptation. 12 – 14 February 2007.* Berlin.

Horlacher, R. (2009). *Expert interview on 20 November 2009. Berliner Wasserbetriebe.*
Horowitz, J. K., & McConnell, K. E. (2002). A review of WTA / WTP Studies. *Journal of Environmental Economics and Management, 44*, 426–447.
Hotelling, H. (1938). The general welfare in relation to problems of taxation and of railway and utility rates. *Econometrica, 6*, 242–269.
Howe, C. W. (1982). The impact of price on residential water demand: some new insights. *Water Resources Research, 18*(4), 713–716.
Howe, C. W., & Linaweaver, F. P. (1967). The impact of price on residential water demand and its relationship to system design and price structure. *Water Resources Research, 3*(1), 13–32.
Hoyt, P. G. (1984). *Crop water production functions and economic implication for Colorado.* (US Department of Agriculture, Ed.).
IEEP (2000). *The environmental impact of irrigation in the EU: Report to the European Commission, DG Environment.*
Institute for Environmental Studies (IVM) (1998). *External economic benefits and costs in water and solid waste investments. Methodology, guidelines and case studies* (Report number R98/11). Amsterdam: Institute for Environmental Studies (IVM) and Economics For The Environment Consultancy Ltd (EFTEC).
Interwies, E., Kraemer, R. A., Kranz, N., Görlach, B., Dworak, T., & Borchardt, D., et al. (2004). *Grundlagenf ür die Auswahl der kosteneffizientesten Maßnahmenkombination zur Augnahme in das Maßnahmenprogramm nach Artikel 11 der Wasserrahmenrichtlinie: Handbuch* (Texte / Umweltbundesamt No. 02). Berlin.
Interwies, E., Pielen, B., & Strosser, P. (2003). *The economic analysis according to the Water Framework Directive in the Danube river basin: A cross-country assessment of implementation capacities and priority gaps: Prepared within the project output 1.1 – Activity 1.1–1.3: Applying EU economic guidelines for the economic analysis to the Danube river basin, UNDP/GEF Danube Regional Project. Berlin.*
ISTAT – National Institute of Statistics (2006). *Water resources assessment and water use in agriculture* (n. 18/2006). Rome.
James, D. E., Jansen H. M. A., & Opschoor J. B. (1978). *Economic approaches to environmental problems: Techniques and results of empirical analysis.* Amsterdam: Elsevier.
Johansson, R. C. (2000). *Pricing irrigation water – a literature survey* (Policy Research Working Papers – Rural Development Department No. 2449). Washington D.C.

Johansson, R. C., Tsur, Y., Terry L. Roe, T. L., Doukkali, R., & Dinar, A. (2002). Pricing irrigation water: a review of theory and practice. *Water Policy, 4,* 173–199.

Jones, T. (2003). *Pricing water.* Retrieved 19.112009, from http://www.oecdobserver.org/news/fullstory.php/aid/939/Pricing_water.html.

Kahlenborn, W., Buck, M., & Kraemer, R. A. (1999). *Kostendeckung bei Wasserpreisen und Abwassergebühren vor dem Hintergrund der künftigen Wasserrahmenrichtlinie der Europäischen Gemeinschaft: im Auftrag des Bundesministeriums für Umwelt, Naturschutz und Reaktorsicherheit. Berlin: Ecologic.*

Kahneman D., & Knetsch, J. L. (1992). Valuing public goods: the purchase of moral satisfaction. *Journal of Environmental Economics and Management, 22,* 57–70.

Kaika, M. (2003). The Water Framework Directive: A new directive for a changing social, political and economic European framework. *European Planning Studies, 11*(3), 299–316.

Kaika, M., & Page, B. (2003). The EU Water Framework Directive: Part 1. European policy-making and the changing topography of lobbying. *European Environment, 13*(6), 314–327.

Kessler, P. (2006). Neuausrichtung der Verwaltung auf die WRRL. *Handbuch der EU-Wasserrahmenrichtlinie. Inhalte, Neuerungen und Anregungen für die nationale Umsetzung,* pp. 45–57. Berlin: E. Schmidt.

Kim, E., & Lee, H. (1998). Spatial integration of urban water services and economies of scale. *RURDS, 10*(1), 3–18.

Kim, H. Y. (1995). Marginal cost and second-best pricing for water services. *Review of Industrial Organization,* (10), 323–338.

Klauer, B., Mewes, M., Sigel, K., Unnerstall, H., Görlach, B., & Bräuer, I., et al. (2007). *Verhältnismäßigkeit von Maßnahmekosten im Sinne der EG-Wasserrahmenrichtlinie – Komplementäre Kriterien zur Kosten-Nutzen Analyse* (UFZ-Bericht No. 02/2007). Leipzig.

Knetsch, J. L., & Sinden, J. A. (1984). Willingness to pay and compensation demanded: experimental evidence of an unexpected disparity in measures of value. *Quarterly Journal of Economics,* 507–521.

Kopp, R. J. (1992). Why existence value should be used in Cost-Benefit Analysis. *Journal of Policy Analysis and Management, 11,* 123–130.

Koziol, M. (2004). The consequences of demographic change for municipal infrastructure. *Deutsche Zeitschrift für Kommunalwissenschaften, 44*(1).

Kraemer, R. A. (2003). *Environmentally harmful subsidies in the water sector: Presented at the OECD Technical Expert Meeting on Environmentally Harmful Subsidies, Paris, 3–4 November 2003* (No. SG/SD(2003)13). Paris.

Kraemer, R. A., & Buck, M. (1997). *Water subsidies and the environment* (OCDE/GD(97)220). Paris.
Kraemer, R. A., Castro, Z. G., da Motta, R. S., & Clifford, R. (2003). *Environment Network: Economic Instruments for Water Management: Experiences from Europe and Implications for Latin America and the Caribbean.* (Inter-American Development Bank, Ed.). Washington D.C.: Inter-American Development Bank.
Kraemer, R. A., & Jäger, F. (1998). Germany. *Institutions for water resources management in Europe. Eurowater 1,* pp. 83–325. Rotterdam: Balkema.
Kraemer, R. A., Pielen, B., & Leipprand, A. (2003). Economic instruments for water management: Extra-regional experiences and their applicability in Latin America and the Caribbean Inter-American Development Bank (Ed.), *REGIONAL POLICY DIALOGUE STUDY SERIES: Environment Network. Economic Instruments for Water Management: Experiences from Europe and Implications for Latin America and the Caribbean,* pp. 3–55. Washington D.C.: Inter-American Development Bank.
Krutilla, J. (1967). Conservation reconsidered. *American Economic Journal*, 57, 777–786.
Krysanova, V., Hattermann, F. F., & Habeck, A. (2005). Expected changes in water resources availability and water quality with respect to climate change in the Elbe river basin (Germany). *Nordic Hydrology, 36*(4–5), 321–333.
Landespressedienst von Berlin (1999). *BWB-Teilprivatisierung: Berliner Senat gibt Zuschlag an RWE/Vivendi – Zukunftsfonds kommt.* (18 June 1999). Berlin.
Larson, D. M. (1992). Can nonuse values be measured from observable behavior? *American Journal of Agricultural Economics, 74*, 1114–1120.
Lazo, J. K., Schulze, W. D., McClelland, G. H., & Doyle, J. K. (1992). Can contingent valuation measure nonuse values? *American Journal of Agricultural Economics, 74*, 1126–1132.
Leist, H. -J (2007). *Wasserversorgung in Deutschland: Kritik und Lösungsansätze* (Vol. 35). *Hochschulschriften zur Nachhaltigkeit, 35.* München: oekom-Verl.
Libbe, J. (2007). *Demographic change in Europe – Impacts on the provision of sustainable water management: Final Conference of the INTERREG III C project MunServNet, 29 October 2007, Berlin.*
Liu, J., Savenije, H., & Xu, J. (2003). Water as an economic good and water tariff design: comparison between IBT-con and IRT-cap. *Physics and Chemistry of the Earth, 28*, 209–217.
Loehman, E. T. (2004). Cost-recovery, efficiency and economic organisation for water utilities. *Contributions to Economic Analysis and Policy, 3*, 1–44.

Loh, M., & Coghlan, P. (2003). *Domestic water use study: Perth, Western Australia 1998–2021.* Perth, Western Australia. Retrieved November 24, 2009, from http://www.watercorporation.com.au/_files/PublicationsRegister/12/Domestic_water_use_study.pdf.

Loomis, J. (1996). Measuring the economic benefit of removing dams and restoring the Elwha River: Results of a contingent valuation survey. *Water Resources Research, 32*(2), 441–447.

Louviere, J. J., Hensher, D. A., & Swait, J. D. (2000). *Stated choice methods: Analysis and applications.* New York: Cambridge University Press.

Lyman, R. A. (1992). Peak and off-peak residential water demand. *Water Resources Research, 28*(9), 2159–2167.

Madariaga, B. & McConnell, K. (1987). Some issues in measuring existence values. *Water Resources Research*, 23, 936–942.

Maestu, J. (2005). *Cost-recovery – the way ahead in Implementing Article 9: International Experience II, Spain.* 7–8 July 2005, Leipzig.

Markandya, A., Harou, P., Bellù, L. G., & Cistulli, V. (2002). *Environmental economics for sustainable growth: A handbook for practitioners.* (International Bank for Reconstruction and Development, Ed.). Cheltenham: Elgar.

Marr, S., & Gordon-Walker, S. (2002). *Study on the Application of the Competition Rules to the Water Sector in the European Community*, Report for the Competition Directorate-General of the European Commission.

Martínez, Y., & Albiac, J. (2004). Agricultural pollution control under Spanish and European environmental policies. *Water Resources Research, 40,* W10501.

Martinez-Espiñeira, R., & Nauges, C. (2004). Is all domestic water consumption sensitive to price control? *Applied Economics, 36*, 1697–1703.

Massarutto, A. (1999). *Agriculture, water resources and water policies in Italy* (FEEM Working Paper No. No. 33.99). Milan.

Massarutto, A. (2003). Water pricing and irrigation water demand: economic efficiency versus environmental sustainability. *European Environment,* 13, 100–119.

Massarutto, A. (2002a). *Irrigation water demand in Europe: the impact of Agenda 2000 and the Water Framework Directive* (Working Paper Series in Economics No. 01–02–eco). Udine.

Massarutto, A. (2002b). *The full-cost-recovery of irrigation: rationale, methodology, European experience: Paper prepared for the international conference "Irrigation Water Policies: Micro and Macro Considerations", Agadir, 15–17 June 2002.*

Massarutto, A. (2004a). Water pricing: a basic tool for a sustainable water policy? In E. Cabrera & R. Cobacho (Eds.), *Challenges for the new water policies in XXI century.* Amsterdam: Balkema.
Massarutto, A. (2004b). Full cost-recovery of water services: The European experience. *Zeitschrift für angewandte Umweltforschung, 15/16*(3–5), 511–527.
Massarutto, A. (2005). An ecological-economic approach for assessing carrying capacity of river basins. In Agence de l'Eau Seine-Normandie (Ed.). *Opening the black box? Second International Workshop on Implementing Economic Analysis in the Water Framework Directive*. Paris, 17 – 18 February . Paris: Agence de l'Eau Seine-Normandie.
Massarutto, A. (2006). Urban water conflicts in Italy: an ecological-economic analysis International Hydrological Programme of UNESCO (Ed.). *An analysis of the origins and nature of water-related unrest and conflicts in the urban context,* pp. 139–168. Paris: UNESCO.
Massarutto, A., & Paccagnan, V. (2007). *Deliverable No 1: WFD harmonisation procedures review: The WFD implementation in arid and semi-arid countries – institutional and economic issues: INECO – institutional and economic instruments for sustainable water management in the mediterranean region.*
McNeill, D. (1998). Water as an economic good. *Natural Resources Forum, 22*(4), 253–261.
Merkel, W., Leuchs, W., & Odenkirchen, G. (2008). Herausforderungen des globalen Klimawandels für die Wasserwirtschaft in Deutschland: Praxisberichte, Handlungsfelder und Forschungsbedarf. *Wasser – Abwasser, 149*(4), 332–337.
Merrett, S. (2007). *The price of water: Studies in water resource economics and management* (2. ed.). London: IWA Publishing.
Metropolitan Consulting Group (2006). *VEWA – Vergleich Europäischer Wasser- und Abwasserpreise: Commissioned by: Bundesverband der deutschen Gas- und Wasserwirtschaft (BGW).* Bonn: wvgw Wirtschafts- und Verl.-Ges. Gas und Wasser.
Milgrom, P. (1993). Is sympathy an economic value? Philosophy, economics and the contingent valuation method. *Contingent valuation: a critical assessment,* pp. 417–41. New York: North Holland.
Ministerium für Umwelt und Forsten des Landes Rheinland Pfalz (2004). *Fragebogen zur Umsetzung der Wasserrahmenrichtlinie, Wasserversorgung: Statistik von Jahresabschluss, Kostenrechnung und Entgelte öffentlich bestimmter Einrichtungen, Betriebe und Unternehmen der Wasserversorgung für das Geschäftsjahr 2002.* Mainz.

Mitchell, D., & Hanemann, W. M. (1994). *Water pricing and cost data: getting the right numbers*. University of Hawaii, Water Resources Centre.
Mitchell, R. C., & Carson, R. T. (1989). *Using surveys to value public goods: The contingent valuation method.* Washington D.C.: Resources for the Future.
Mohamed, A. S., & Savenije, H. (2000). Water demand management: positive incentives, negative incentives or quota regulation? *Physics and Chemistry of the Earth, 25*(3), 251–258.
Molle, F., & Berkoff, J. (2007). *Irrigation water pricing policy in context: the gap between theory and practice*. Wallingford: CABI Publishing.
Möller, K., & Burgschweiger, J. (2008). *Wasserversorgungskonzept für Berlin und das von den BWB versorgte Umland (Entwicklung bis 2040): Sponsored by Berliner Wasserbetriebe*.
Monteiro, H. (2005). *Water pricing models: a survey*. Lisboa.
Moor, A. de, & Calami, P. (1997). *Subsidizing unsustainable development: undermining the earth with public funds*. San Jose, Coast Rica: Earth Council.
Moss, T. (2004). The governance of land use in river basins: prospects for overcoming problems of institutional interplay with the EU Water Framework Directive. *Land Use Policy, 21*, 85–94.
Mostert, E. (2003). The European Water Framework Directive and water management research. *Physics and Chemistry of the Earth*, (28), 523–527.
MUNLV-Ministerium für Umwelt und Naturschutz, L. u. V. N. (Ed.) (2005). *Ergebnisbericht Emscher: Wasserrahmenrichtlinie in NRW – Bestandsaufnahme.*
MUNLV-Ministerium für Umwelt und Naturschutz, L. und Verbraucherschutz NRW (2008a). *Bewirtschaftungsplan Nordrhein-Westfalen – Entwurf.* Düsseldorf. Retrieved December 01, 2009, from http://www.flussgebiete.nrw.de/Bewirtschaftungsplanung/index.jsp.
MUNLV-Ministerium für Umwelt und Naturschutz, L. und Verbraucherschutz NRW (2008b). *Maßnahmenprogramm Nordrhein-Westfalen – Entwurf.* Düsseldorf. Retrieved December 01, 2009, from http://www.flussgebiete.nrw.de/.
MUNLV-Ministerium für Umwelt und Naturschutz, L. und Verbraucherschutz NRW (2008c). *Planungseinheiten-Steckbriefe: Emscher.* Düsseldorf. Retrieved December 01, 2009, from http://www.emscher.nrw.de/.
MUNLV-Ministerium für Umwelt und Naturschutz, Landwirtschaft und Verbraucherschutz NRW (2009). *Uhlenberg übergibt Förderbescheid über 25 Millionen Euro zur Umgestaltung des Emschersystems.* (29 May 2009). Düsseldorf. Retrieved October 04, 2009, from MUNLV-Ministerium für Umwelt und Naturschutz, Landwirtschaft und Verbraucherschutz NRW: http://www.umwelt.nrw.de/ministerium/presse/presse_aktuell/presse090529.php.

Mutschmann, J., & Stimmelmayr, F. (1999). *Taschenbuch der Wasserversorgung* (12., überarb. Aufl.). Braunschweig: Vieweg, F /VVA; Vieweg.

Nafo, I. (2009). *Expert interview on 7 August 2009 and 7 October 2009. Emschergenossenschaft.*

Navrud, S. (2001). Economic valuation of inland recreational fisheries: empirical studies and their policy use in Norway. *Fisheries Management and Ecology, 8*, 369–382.

Newberry, D. M. (2000). *Privatisation, restructuring and regulation of network industries:* MIT Press.

Nguyen, D. (1976). The problem of peak loads and inventories. *The Bell Journal of Economics, 7*(1), 232–241.

Nobis-Wicherding, H. (2009). *Expert interviews on 27 July 2009 and 20 November 2009. Berliner Wasserbetriebe.*

OECD (1972). *Recommendation of the council on guiding principles concerning international economic aspects of environmental policies: Council Document no. C(72)128.* Paris.

OECD (1987). *Pricing of water services.* Paris: OECD.

OECD (1989). *Recommendation of the council concerning the application of the polluter-pays principle to accidental pollution: Council Document no. C(89)88.* Paris.

OECD (1996). *Subsidies, tax disincentives and the environment: concepts.* Paris: OECD.

OECD (Ed.). 1998. *Sustainable management of water in agriculture: Issues and policies. The Athens Workshop.* Paris: OECD Publications.

OECD (1999a). *The price of water: Trends in OECD countries.* Paris: OECD.

OECD (1999b). *Household water pricing in OECD countries.* Paris: Working Party on Economic and Environmental Policy Integration, OECD (ENV/EPOC/GEEI(98)12/FINAL).

OECD (1999c). *Agricultural Water Pricing in OECD Countries: Working Party on Economic and Environmental Policy Integration* (ENV/EPOC/GEEI(98)11/FINAL). Paris: OECD.

OECD (Ed.). 2000. *OECD Seminar Social and Environment Interface: Proceedings.* 22–24 September 1999. Paris.

OECD (2001). *Transition to full-cost pricing of irrigation water for agriculture in OECD countries* (COM/ENV/EPOC/AGR/CA(2001)62/FINAL). Paris.

OECD (2002). *Transition to full-cost pricing of irrigation water for agriculture in OECD countries* (COM/ENV/EEPOC/AGR/CA(2001)62/FINAL). Paris.

OECD (2003). *Social issues in the provision and pricing of water services.* Paris: OECD Publishing.

OECD 2005 (Ed.). *Workshop on Agriculture and Water: Sustainability, Markets and Policies.* Paris: OECD.

OECD (2009). *Managing water for all: An OECD perspective on pricing and financing.* Paris: OECD Publishing.

Oelmann, M. (2005). *Zur Neuausrichtung der Preis- und Qualitätsregulierung in der deutschen Wasserwirtschaft.* Köln: Kölner Wissenschaftsverl.

Oelmann, M., Böschen, I., Kschonz, C., & Müller, G. (2009). *10 Jahre Wasserpartner Berlin: Eine Bilanz der öffentlich-rechtlichen Partnerschaft zwischen dem Land Berlin, RWE Aqua und Veolia Wasser.* Bad Honnef (Study conducted on behalf of RWE Aqua and Veolia Water), from www.wasserpartner-berlin.de.

OFWAT (2006). *Water and sewerage service unit costs and relative efficiency 2005-06 report.* Birmingham.

OFWAT (2007). *Regulatory accounting guidelines 4.03. Guidelines for the analysis of operating costs and assets: Operative: financial year 2006–07.* Birmingham.

Olmstead, S. M., Hanemann, W. M., & Stavins, R. N. (2007). Water demand under alternative price structures. *Journal of Environmental Economics and Management, 54,* 181–198.

Olmstead, S. M., & Stavins, R. N. (2008). *Comparing price and non-price approaches to urban water conservation.* RPP–2008–05, Harvard University, Cambridge, MA, from Harvard University.

Opschoor, J. (2006). Water and merit goods: In response to the paper by Marco Schouten and Klaas Schwartz: "Water as a political good: implications for investments". *International Environmental Agreements,* 6, W423–428.

Overath, H., & Merkel, W. (2004). *Kennzahlen für die Wasserversorgung: Feld-Test des Kennzahlensystems der IWA (International Water Association): Nationales Teilprojekt Deutschland.* Mülheim an der Ruhr.

Palmquist, R. B. (1991). Hedonic methods. *Measuring the demand for environmental quality.* Amsterdam: North Holland.

Pearce, D. (1992). *Economic valuation and the natural world: Background paper for the World Development Report 1992* (Policy Research Working Papers – World Development Report).

Pearce, D., Atkinson, G., & Mourato, S. (2006). *Cost-benefit analysis and the environment: Recent developments.* (OECD, Ed.). Paris.

Pearce, D., & Turner, R. (1990). *Economics of natural resources and the environment.* New York: Harvester Wheatsheaf.

Perry, C. J., Rock, M., & Seckler, D. (1997). *Water as an economic good: a solution or a problem?* (Research Report No. 14). Colombo: International Irrigation Management Institute. Retrieved April 19, 2007.

Pielen, B., Herbke, N., & Interwies, E. (2004). *Analysis of the legislation and emerging regulation at the EU country level: Country report: Germany*. Berlin (Work Package 4 (Phase 2), European Research Project Euromarket (Water Liberalisation Scenarios)).

Pielen, B., & Holländer, R. (2006*). Existence values and contingent valuation – a literature review*, Working Paper. Leipzig.

Pieters, J. (2002). *What Makes a Subsidy Environmentally Harmful: Developing a Checklist Based on the Conditionality of Subsidies. OECD Workshop on Environmentally Harmful Subsidies, Paris, 7–8 November 2002.* (No. SG/SD (2002)18). Paris.

Portney, & P. R. (1994). The contingent valuation debate: why economists should care. *Journal of Economic Perspectives, 8*, 3–17.

Pressman, I. (1970). A mathematical formulation of the peak load pricing. *The Bell Journal of Economics, 1*(2), 304–322.

Quiggin, J. (1993). Existence value and benefit-cost analysis: a third view. *Journal of Policy Analysis and Management, 12*, 195–199.

Ramsey, F.P. (1927). A contribution to the theory of taxation. *Economic Journal*, 37, 47–61.

Randall, A., & Stoll, J. R. (1980). Consumer surplus in a commodity space. *American Economic Review, 70*, 449–455.

Randall, A., & Stoll, J. R. (1983). Existence values in a total valuation framework. In R. Rowe & L. Chestnut (Eds.), *Managing air quality and scenic resources at national parks and wilderness areas*. Boulder: Westview Press.

Randall, A. (1993). Passive-use and contingent valuation—valid for damage assessment. *Choices*, 12–15.

Reif, T. (2002). *Preiskalkulation privater Wasserversorgungsunternehmen: Betriebswirtschaftliche Erfordernisse und rechtliche Rahmenbedingungen unter dem Gesichtspunkt der Unternehmenserhaltung.* Bonn: wvgw Wirtschafts- und Verl.-Ges. Gas und Wasser.

Reinhardt, M. (2006). Kostendeckungs- und Verursacherprinzip nach Art. 9 der EG-Wasserrahmenrichtlinie. *Natur und Recht, 12*, 737–744.

Renwick, M. E., & Archibald, S. O. (1998). Demand side management policies for residential water use: Who bears the conservation burden? *Land Economics, 74*(3), 343–359.

Renwick, M. E., & Green, R. (2000). Do residential water demand side management policies measure up? An analysis of eight California water agencies. *Journal of Environmental Economics and Management, 40*, 37–55.

Renzetti, S. (1992). Evaluating the welfare effects of reforming municipal water prices. *Journal of Environmental Economics and Management, 22*(1), 147–163.

Renzetti, S. (2000). An empirical perspective on water pricing reforms. *The political economy of water pricing reforms.* 1. print., pp. 123–140. New York: Oxford Univ. Press.

Riordan, C. (1971). Multistage marginal cost model of investment-pricing decisions: Application to urban water supply treatment facilities. *Water Resources Research, 7*(3), 463–478.

Rogers, P., Bhatia, R., & Huber, A. (1998). *Water as a social and economic good: How to put the principle into practice* (TAC Background Papers No. 2). Stockholm. Retrieved April 21, 2007.

Rogers, P., Silva, R., & Bhatia R. (2002). Water is an economic good: How to use prices to promote equity, efficiency, and sustainability. *Water Policy, 4*(1), 1–17.

Rosenberger, R. S., & Loomis, J. B. (2003). Benefits Transfer. *A primer on non-market valuation,* pp. 445–482. Dordrecht: Kluwer.

Rosenthal, D. H., & Nelson, R. H. (1992). Why existence values should not be used in Cost-Benefit Analysis. *Journal of Policy Analysis and Management, 11,* 116–122.

Roth, E. (2001). *Water pricing in the EU: a review.* Brussels (Publication Number 2001/002).

Rowe, R. D., d'Arge, R. C., & Brookshire, D. S. (1980). An experiment on the economic value of visibility. *Journal of Environmental Economics and Management, 7*(1), 1–19.

RPA – Risk & Policy Analysts Limited (2000). *Economic instruments in relation to water abstraction: Research report prepared for DETR.*

RPA – Risk & Policy Analysts Limited (2004). *CEA and developing a methodology for assessing disproportionatecCosts. Final report and annexes. 13 July 2004,* from http://www.defra.gov.uk/environment/water/wfd/economics/.

Rumm, P., Keitz, S. von, & Schmalholz, M. (2006). *Handbuch der EU-Wasserrahmenrichtlinie: Inhalte, Neuerungen und Anregungen für die nationale Umsetzung.* Berlin: E. Schmidt.

Russell, C., & Shin, B. S. (1996). Public utility pricing: Theory and practical limitations. *Marginal cost rate and wholesale water markets. Advances in the economics of the environment.* London: JAI Press.

Russell, C. S. (2001). *Applying economics to the environment.* Oxford: Oxford Univ. Press.

Sagoff, M. (1988). On teaching a course on ethics, agriculture, and the environment. *Journal of Agricultural and Environmental Ethics, 1*(1).

Saliba, B. C., & Bush, D. B. (1987). *Water markets in theory and practice: market transfers, water values and public policy.* Boulder: Westview Press.

Sauer, J. (2005). *The economics and efficiency of water supply infrastructure.* Berlin: Logos Verlag.

Savenije, H. (2000). Water scarcity indicators: the deception of the numbers. *Physics and Chemistry of the Earth, 25*(3), 199–204.

Savenije, H. (2002). Why water is not an ordinary economic good, or why the girl is special. *Physics and Chemistry of the Earth, 27*(11), 741–744. Retrieved April 22, 2007.

Savenije, H., & van der Zaag, P. (2002). Water as an economic good and demand management: Paradigms with pitfalls. *Water International, 27*(1), 98–104. Retrieved April 22, 2007.

Scheele, U. (2000). *Auf dem Weg zu neuen Ufern? Wasserversorgung im Wettbewerb.* Discussion Paper Series, No. V–214–2000, Universität Oldenburg, Oldenburg.

Schmitz, A. (2009). *Expert interview on 11 August 2009. Berliner Wasserbetriebe.*

Schmutzer, S. (2006). Umsetzung der EG-Wasserrahmenrichtlinie in die Kommunalabgabengesetze: Handlungsbedarf für die Länder. *DVBL, 121*(4), 228–235.

Schön, J. (2009). *Expert interview on 7 August 2009. Emschergenossenschaft.*

Schouten, M., & Schwartz, K. (2006). Water as a political good: implications for investments. *International Environmental Agreements, 6*, 407–421.

Schulz, A. (2009). *Expert interview on 7 August 2009. Emschergenossenschaft.*

Schulze, D. (1994). Der Einfluss der Kostenrechnung in der kommunalen Abwasserentsorgung auf die finanzielle Belastung der Bürger. *Finanzwirtschaft*, (9), 205–208.

Semrau, M., Brinkmann, S., & Wegner, R. (2009). Das ökologische Konzept für den Umbau der Emscher – eine WRRL-konforme Planung einer anthropogen überformten Flusslandschaft. *WasserWirtschaft*, (4), 17–21.

Landespressedienst von Berlin (1999). *BWB-Teilprivatisierung: Berliner Senat gibt Zuschlag an RWE/Vivendi – Zukunftsfonds kommt.* (18.06.1999), Berlin.

Shaw, W. D. (2005). *Water resource economics and policy: An introduction.* Cheltenham: Elgar.

Shogren, J. F., Seung Y. S., Dermot, J. H., & Kliebenstein, J. B. (1994). Resolving differences in willingness to pay and willingness to accept. *American Economic Review, 84*, 255–270.

Simon, J. (2009, September 23). *Herausforderungen der deutschen Wasserwirtschaft.* Wasserfachliche Aussprachetagung (wat). Leipzig.

Smets, H. (2000). Implementing the right to drinking water In OECD countries. In OECD (Ed.). *OECD Seminar Social and Envirnonment Interface. Proceedings.* 22–24 September 1999 (pp. 107–148). Paris.

Smith, K. V. (1993). Nonmarket valuation of environmental resources: an interpretative appraisal. *Land Economics, 69*, 1–26.
Somma, F. (2002). *Aqualibrium country report Italy.*
Speck, S., & Ekins, P. (2000). *Recent trends in the application of economic instruments in the EU Member States plus Norway and Switzerland and an overview of economic instruments in Central and Eastern Europe – Update of Database of Environmental Taxes and Charges.* London.
Stadtwerke Essen (2003). *Abwasserbeseitigungskonzept (ABK) für die Stadt Essen 2003.* Essen. Retrieved December 01, 2009, from http://www.stadtwerke-essen.de/abwasser/abwasserbeseitigungskonzept-2/.
Steenbock, R. (1984). *Grundlagen der Betriebsabrechnung. Abschreibungen und Zinsen.* Köln.
Steiner, P. (1957). Peak loads and efficient pricing: *Quarterly Journal of Economics, 71*(4), 585–610.
Strosser, P. (2007). *Integrating economics into water policy and management: Implementing the EU Water Framework Directive.* Retrieved October 31, 2009, from http://aca-web.gencat.cat/aca/documents/ca/agenda/rp1_02.pdf.
StUA Herten (Staatliches Umweltamt Herten) (Ed.) (2004). *Dokumentation der wasserwirtschaftlichen Grundlage – Bestandsaufnahme Flussgebietseinheit Rhein: Bearbeitungsgebiet Niederrhein, Arbeitsgebiet Emscher,* from www.emscher.nrw.de/offen/start.htm.
Syme, G., Shao, Q., Po, M., & Campbell, E. (2004). Predicting and understanding home garden water use. *Landscape and Urban Planning, 68*(1), 121–128.
Tardieu, H., & Préfol, B. (2002). Full cost or "sustainability cost" pricing in irrigated agriculture: Charging for water can be effective, but is it sufficient? *Irrigation and Drainage*, (51), 97–107.
Tietenberg, T. (1996). *Environmental and natural resource economics.* Boston: Scott, Foresman and Co.
Timmins, C. (2002). Measuring the dynamic efficiency costs of regulators preferences: Municipal water utilities in the arid West. *Econometrica, 70*(2), 603–629.
Treseler, U., & Gütling, K. (2006). Der Umbau des Emscher-Systems und die Rolle der Behörden. *Wasser und Abfall*, (3), 19–23.
Tsur, Y., & Dinar, A. (1997). The relative efficiency and implementation costs of alternative methods for pricing irrigation water. *World Bank Economic Review*, (11), 243–262.
Tsur, Y., Dinar, A., Doukkali, R. M., & Roe, T. L. (2002). *Efficiency and equity Implications of irrigation water pricing,* from http://www.worldbank.org/agadirconference.

Turner, R. (1992). *Speculations on weak and strong sustainability* (CSERGE Working Paper No. GEC 92–26).
Turner, R. (1999). Economic values in environmental valuation. *Valuing environmental preferences: theory and practice of the contingent valuation method in the US, EU and developing countries.* Oxford: Oxford Univ. Press.
Turner, R., & Dubourg, W. R. (1993). *Water resources scarcity: an economic perspective.* Cserge Working Paper, PA 93–06, University of East Anglia, Norwich.
Turvey, R. (1976). Analyzing the marginal cost of water supply. *Land Economics, 52*(5), 158–168.
UBA-Umweltbundesamt (1998). *Vergleich der Trinkwasserpreise im europäischen Rahmen: Forschungsbericht 296 21 427. Texte / Umweltbundesamt.* Berlin: Umweltbundesamt.
UBA-Umweltbundesamt (Ed.) (2001). *Nachhaltige Wasserversorgung in Deutschland: Analyse und Vorschläge für eine zukunftsfähige Entwicklung. Beiträge zur nachhaltigen Entwicklung.* Berlin: Erich Schmidt.
UNECE (1992). Dublin International Conference on Water and the Environment.
UNECE (2001). Convention on Access to Information, Public Participation in Decision-Making and Access to Justice in Environmental Matters. Adopted on 25 June 1998 (entered into force on 30 October 2001).
UNEP (2005). *Ecosystems and human well-being: Wetlands and water synthesis – a report of the Millennium Ecosystem Assessment.* (Millennium Ecosystem Assessment, Ed.). Washington D.C.: World Resources Institute.
United Nations (1992). *Agenda 21: the United Nations programme of action from Rio.* New York.
Unnerstall, H. (2005). *Verursachergerechte Kostendeckung für Wasserdienstleistungen – die Anforderungen des Art. 9 WRRL und ihre Umsetzung* (UFZ Diskussionspapiere – Department Umwelt- und Planungsrecht). Leipzig.
Unnerstall, H. (2007). Anforderungen an die Kostendeckung im Abwassersektor nach der EU-Wasserrahmenrichtlinie. *Zeitschrift für Wasserrecht, 46*(3), 129–140.
van Bömmel, B. (2009). *Expert interview on 8 October 2009. Emschergenossenschaft.*
van der Veeren, R. (2005). *In pursuit of optimal measure packages: Dutch handbook on costeffectiveness analyses for the EU Water Framework Directive.*
van der Zaag, P., & Savenije, H. (2000). Towards improved management of shared river basins: lessons from the Maseru Conference. *Water Policy, 2*(2), 47–63.

Varela-Ortega, C., Sumpsi, J. M., Garrido, A., Blanco, M., & Iglesias, E. (1998). Water pricing policies, public decision making and farmers' response: implications for water policy. *Agricultural Economics, 19*, 193–202.

Venkatachalam, L. (2004a). The contingent valuation method: a review. *Environmental Impact Assessment Review, 24*, 89–124.

Venkatachalam, L. (2004b). *Economic valuation of watershed services of Commons: Marginal opportunity cost approach within the environmental accounting framework.* Nagarbhavi, Bangalore: Institute for Social and Economic Change.

Viaggi, D. (2003). *Experiences with economic incentives in irrigated agriculture in Italy,* from www.worldband.org/rural.

Vohra, R. (1990). On the inefficiency of two-part tariffs. *Review of Economic Studies, 57*(3), 415–438, from http://www.jstor.org/pss/2298022.

Wackerbauer, J. (2007). Privatisation and regulation of water supply. *CESifo DICE Report*, (2), 3.

Walsh, R. G., Loomis, J., & Gillman, R. A. (1984). Valuing option, existence and bequest demands for wilderness. *Land Economics, 60*, 14–29.

Ward, F. A. (2007). Decision support for water policy: a review of economic concepts and tools. *Water Policy, 9*, 1–31.

Ward, F. A., & Michelsen, A. (2002). The economic value of water in agriculture: concepts and policy applications. *Water Policy*, (4), 423–446.

Ward, W. A., & Loomis, J. (1986). The travel cost demand model as an environmental policy assessment tool: a review of the literature. *Western Journal of Agricultural Economics, 11*(2), 164–178.

Weisbrod, B.A. (1964). Collective-consumption services of individualized-consumption goods. *Quarterly Journal of Economics*, 18, 265–276.

Wetzel, U. (1996). Die Kosten der Wasserversorgung als Faktor im Wasserpreis: Stellungnahme der EUREAU-Kommission III. *Wasser – Abwasser*, (5), 252–255.

Williamson, O. (1966). Peak load pricing and optimal capacity under indivisibility constraints. *American Economic Review, 56*(9), 810–827.

Willig, R. D. (1978). Pareto-superior nonlinear outlay schedule. *The Bell Journal of Economics, 9*(2), 56–69.

WWF – World Wide Fund for Nature & EEB – European Environmental Bureau (2006). *Survey of the economic elements of the Article 5 report of the EU Water Framework Directive.* Brussels.

Yard, S. (2004). Costing fixed assets in Swedish municipalities: effects of changing calculation methods. *International Journal of Production Economics, 87*, 1–15.

Young, R., & Haveman, R. (1985). Economics of water resources: a survey. *Handbook of natural resource and energy economics,* pp. 465–529.
Zarnikau, J. (1994). Spot market pricing of water resources and efficient means of rationing water resources during scarcity. *Resource and Energy Economics, 16*(3), 189–210.
Zwehl, W. von (1988). Die Kalkulation kommunaler Benutzungsgebühren in betriebswirtschaftlicher Sicht. *ZfB, 58*(1), 155–170.

7 Annex

Box A2.1: Categories of the 'Total Economic Value'

Direct and indirect use values

Use values are defined as values derived by individuals from the actual use of the environment (BATEMAN AND TURNER 1993). They can again be divided into two categories namely, direct use values (e.g. water used for irrigation purposes) and indirect use values (e.g. erosion control). Another delineation made in the literature divides use values into primary values or marketed goods and services or consumptive values and secondary values or non-marketed goods and services or non-consumptive values (cf. MARKANDYA ET AL. 2002, BATEMAN 1993, PEARCE AND TURNER 1990).

Non-use values

It is generally recognized that only the contingent valuation method is applicable to the estimation of non-market economic values of natural resources (CARSON ET AL. 2003, CARSON ET AL. 2001, LOUVIERE ET AL. 2000, BATEMAN AND TURNER 1993, SMITH 1993, MITCHELL AND CARSON 1989).[364]

(1) Option value (and quasi-option value)

WEISBROD (1964) first introduced the concept of option value. It refers to the willingness to pay of individuals for ensuring that a particular resource base is available for future use (VENKATACHALAM 2004A) over and above any expected consumer benefit that the person would receive from future use of the resource. The concept is based on two basic arguments, namely uncertainty and irreversibility linked to environmental issues (CARSON ET AL. 2003). In the sense that someone would be willing to pay in order to guarantee that the resource exists should they decide to visit it in the future, option value can be considered as corresponding to an insurance premium (RANDALL 1991, WALSH ET AL. 1984). It has been argued by some authors that option value should be classified in the category of use instead of non-use value, due to its close linkage to

364 For a detailed description and assessment of the contingent valuation method cf. MITCHELL AND CARSON (1989) or VENKATACHALAM (2004).

the expected value of future use of the resource (see CARSON ET AL. 1997). But although option value is related to use, *'in a formal sense it is a passive-use value since the benefit of keeping one's option open can occur in complete isolation from the resource, and occurs simply because of the fact that the resource exists'* (VINCENT ET AL. 1995:3).[365]

(2) Bequest value

Bequest value constitutes another category of non-use value. As the name suggests, it is an individual's willingness to pay for preserving a natural resource for the use of future generations (RANDALL AND STOLL 1983, FREEMAN 1993). Bequest value is generally thought of as originating from altruistic motives (see e.g. MADARIAGA AND MCCONNELL 1987). Related to these altruistic motives, some authors have raised concerns of double-counting in valuation studies, if the value elicited from each individual in the economy would reflect not only her own benefit but also the benefit derived by others (DIAMOND AND HAUSMAN 1994, CICCHETTI AND WILDE 1992, MILGROM 1993). Other authors rebut this argumentation (see e.g. HANEMANN 1994).

(3) Existence value

The concept of existence value dates back to KRUTILLA (1967). He expanded upon WEISBROD'S option value concept and argued that people care not only about the environment because they, or their heirs, can get some part of benefit by preserving environmental resources, but may value the mere knowledge that rare species or unique environments exist, even if they never intent to benefit from these resources in a direct or indirect way (MARKANDYA ET AL. 2002, PORTNEY 1994, FREEMAN 1993). Since the concept has first been introduced by KRUTILLA, it has been subject to controversial debate in the literature on natural resource valuation (KOPP 1992, ROSENTHAL AND NELSON 1992, DESVOUSGES ET AL. 1993B, DIAMOND AND HAUSMAN 1993, RANDALL 1993, QUIGGIN 1993, BATE 1994,

365 Some authors add another category to non-use values which is called the 'quasi-option value'. It arises out of uncertainty related to the future value of natural resources. Faced with information uncertainty, individuals may be willing to pay for delaying a decision on the use of a particular natural resource until further information is available on its (future) value (MITCHELL AND CARSON 1989). To illustrate the concept of quasi-option value, VINCENT ET AL. (1995) provide the example of environmental resources which have medical properties of which we are at the moment still unaware. NAVRUD (2001) points out that the quasi-option value can be considered as the economists' parallel to the precautionary principle.

EDWARDS 1995).[366] BROOKSHIRE ET AL. (1986), for example, argue that existence value consists of two distinct components, an economic one, which is consistent with utility maximizing behaviour, and an ethical one, which is inconsistent with normative benefit-cost assumptions (FREDMAN 1995). This ethical dimension is also emphasized by several authors dealing with different aspects of environmental ethics (e.g. SAGOFF 1988). However, MITCHELL & CARSON (1989) refute the argument put forward by BROOKSHIRE ET AL. (1986) and claim that the contingent valuation (CV) method is also capable of measuring, what they call, 'inherent components' of existence value (FREDMAN 1995). Other authors raise concerns regarding the technical measurement of existence value (e.g. DIAMOND AND HAUSMAN (1994), DESVOUSGES ET AL. (1993A).[367] However, at this time, most economists agree that non-use values should be included in the calculation of benefits (CARSON ET AL. 1993) and are consistent with economic theory.

Source: Pielen and Holländer (2006).

---

366 While the environmental economics literature on option value developed rapidly in the early 1970s, the discussion on existence values did not gather pace before the 1980s (CICCHETTI AND WILDE, 1992).

367 See HANEMANN (1994) for a detailed discussion of CVM critique. He addresses the four most common objections which he summarises as (1) surveys are vulnerable to response effects; (2) the survey process creates the value; (3) ordinary people are ill-trained for valuing the environment; (4) survey responses cannot be verified. He also discusses (and rejects) the claim of critics (e.g. as expressed in HAUSMAN 1993) that the results of contingent valuation studies are inconsistent with economic theory.

Figure A2.1: Classifying goods and services: supply and demand conditions

Supply	Demand	
	Private	**Public**
Individual	Cup op coffee, well, cesspit, flood proofing	Lighthouse, breakwater, wastewater treatment works
Collective	Piped water supply and sanitation, flood warnings, canal irrigation	Conserving blue whales, national defence, flood embankment

Source: Green (1997) (in Green 2003).

Figure A2.2: Water service classification applied in Article 5 reports

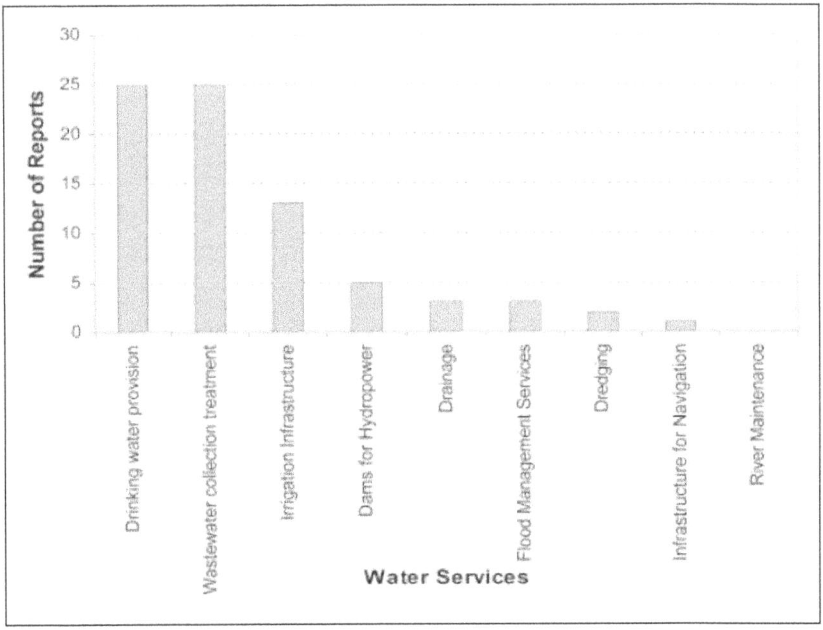

Source: EEB and WWF (2006).

Figure A4.1: Main actors in the German water sector

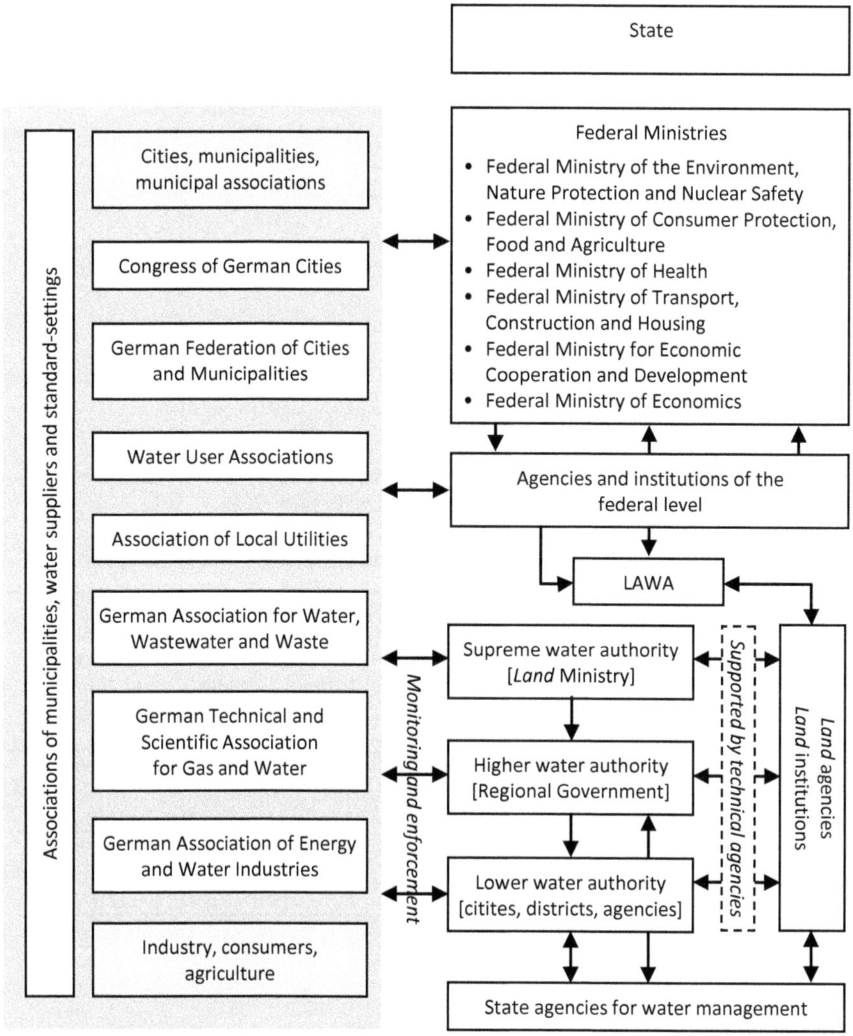

Source: PIELEN ET AL. *(2004)*

Figure A4.1: Development of annual wastewater service costs for a 4-person household in the Emscher region and North Rhine-Westfalia

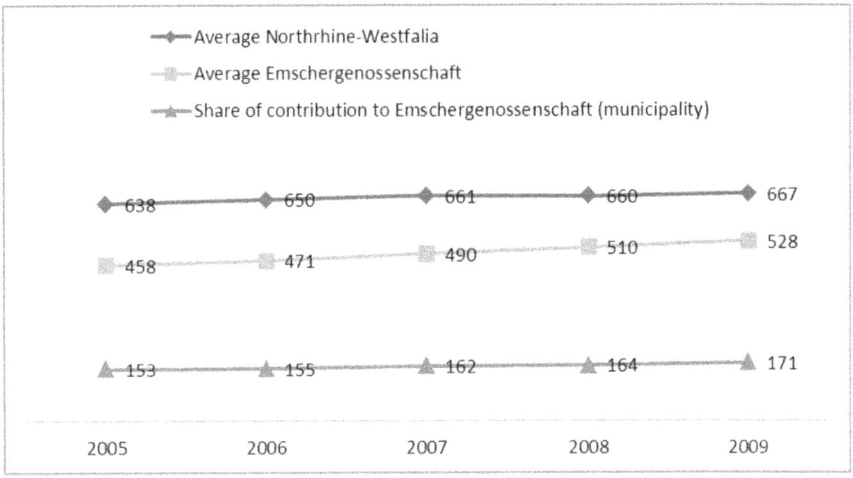

Source: Emschergenossenschaft (2009b), Bund der Steuerzahler NRW (2009).

Table A4.2: Size and capacity of sewerage treatment plants in the Emscher river basin (2008)

Wastewater treatment plant	Size of the service area	Capacity	Daily run-off [m³/d]			Annual quanti-ty of wastewater
	[km²]	[EW]	Min	Mittel	Max	[m³/a]
Emscher-Estuary	429	2.400.000	757.756	1.262.380	2.158.516	462.031.232
Duisburg-Old Emscher	47	500.000	117.000	155.165	374.000	56.790.300
Bottrop	214	1.340.000	273.838	381.604	619.840	139.667.217
Dortmund-Deusen	152	625.000	45.615	180.161	263.463	65.938.917
Total	842	4.865.000				313.215.398

Note: The reduced total value is explained by the river treatment plants.

Source: Emschergenossenschaft (2009a).

Kommunalwirtschaftliche Forschung und Praxis

Herausgegeben von Prof. Dr. Wolf Gottschalk

Band 1 Ulf Klostermann: Controlling zur Unterstützung von Least-Cost Planning im Elektrizitätsunternehmen. 1998.

Band 2 Heinz Bolsenkötter / Michael Poullie: Rechnerisches Unbundling in der Strom- und Gasversorgung. Rechnungslegung nach dem neuen Energiewirtschaftsgesetz. 3., neubearbeitete Auflage. Die 1. und 2. Auflage sind im Jahre 2000 unter dem Titel *Rechnungslegung nach der Elektrizitätsbinnenmarkt-Richtlinie und nach § 9 EnWG* erschienen. 2003.

Band 3 Klaus Traube / Wolfgang Schulz: Aktuelle Bewertung der Kraft-Wärme-Kopplung. Ökologische und ökonomische Wirkung eines mittelfristigen Ausbaus der Kraft-Wärme-Kopplung zur Nah- / Fernwärmeversorgung in Deutschland. 2001.

Band 4 Axel Thomas: Unternehmensinternes Kostenmanagement in Energieversorgungsunternehmen. Grundlagen effektiver Kostenrechnungsverfahren auf der Basis der SAP-Software. 2001.

Band 5 Georg Raffetseder: Erfolgs- und Mißerfolgsfaktoren der Verwaltungsmodernisierung. Zur Steuerung von Reformprozessen. 2001.

Band 6 Andreas Schmidt: Stadtwerke auf neuen Märkten. Gemeinderechtliche Chancen umweltschonender Energiedienstleistungen. 2002.

Band 7 Hannes Rehm / Sigrid Matern-Rehm: Kommunale Finanzwirtschaft. 2003.

Band 8 Alexander Meckies: Die persönliche Haftung von Geschäftsleitern in Kapitalgesellschaften der öffentlichen Hand. 2003.

Band 9 Nicole Weiß: Liberalisierung der Wasserversorgung. Gestaltungsspielräume und Grenzen einer Reform der Wasserversorgung unter besonderer Berücksichtigung der Rechtslage in Bayern. 2004.

Band 10 Matthias Ganske: Corporate Governance im öffentlichen Unternehmen. 2005.

Band 11 Bohdan Kalwarowskyj: Kommunale Dauerverlustbetriebe und verdeckte Gewinnausschüttung. 2006.

Band 12 Michael Schöneich (Hrsg.): Stadt-Werke. Festschrift für Gerhard Widder. 2007.

Band 13 Stefan Thomas: Aktivitäten der Energiewirtschaft zur Förderung der Energieeffizienz auf der Nachfrageseite in liberalisierten Strom- und Gasmärkten europäischer Staaten: Kriteriengestützter Vergleich der politischen Rahmenbedingungen. 2007.

Band 14 Ulrich Büdenbender / Peter Rosin / Carola Reichold: Berücksichtigung des für ein Netz vor der Energierechtsreform 2005 gezahlten Kaufpreises bei der Kalkulation und Regulierung der Netzentgelte. 2008.

Band 15 Arnt Baer: Abnahmepflichten und Vergütungspflichten in der Energiewirtschaft. Gemeinschaftsrechtliche und verfassungsrechtliche Problemstellungen. 2008.

Band 16 Nikolaus Richter / Stefan Thomas: Perspektiven dezentraler Infrastrukturen im Spannungsfeld von Wettbewerb, Klimaschutz und Qualität. Endbericht der Forschungspartnerschaft INFRAFUTUR. 2009.

Band 17 Joachim Wieland: Rechtsprobleme der Gasnetzbewertung nach der Gasnetzentgeltverordnung. 2009.

Band 18 Jürgen Weigt: Die Förderung der Stromerzeugung aus erneuerbaren Energien in der Europäischen Union. Probleme und Herausforderungen im Spannungsfeld zwischen Binnenmarkt und Umweltschutz. 2009.

Band 19 Sebastian Bolay: Einführung von Energiemanagement und erneuerbaren Energien. Eine Untersuchung von Erfolgsfaktoren in deutschen Kommunen. 2009.

Band 20 Oliver Rottmann: Herausforderungen für die Innensteuerung von Stadtwerken aus der Interdependenz der Außensteuerungspostulate. 2011.

Band 21 Britta Ammermüller: Assessing Cost Recovery. A New Comparative Framework in Line with WFD Article 9. 2011.

www.peterlang.de

www.ingramcontent.com/pod-product-compliance
Ingram Content Group UK Ltd.
Pitfield, Milton Keynes, MK11 3LW, UK
UKHW021831100926
8556IPUK00004B/267

9 783631 613719